Advances in
COMPUTERS
VOLUME 78

Advances in COMPUTERS

Improving the Web

EDITED BY

MARVIN V. ZELKOWITZ

Department of Computer Science
University of Maryland
College Park, Maryland
USA

VOLUME 78

ELSEVIER

AMSTERDAM • BOSTON • HEIDELBERG • LONDON • NEW YORK • OXFORD
PARIS • SAN DIEGO • SAN FRANCISCO • SINGAPORE • SYDNEY • TOKYO
Academic Press is an imprint of Elsevier

ACADEMIC
PRESS

Academic Press is an imprint of Elsevier

32 Jamestown Road, London, NW1 7BY, UK
Radarweg 29, PO Box 211, 1000 AE Amsterdam, The Netherlands
30 Corporate Drive, Suite 400, Burlington, MA 01803, USA
525 B Street, Suite 1900, San Diego, CA 92101-4495, USA

First edition 2010

Copyright © 2010 Elsevier Inc. All rights reserved

No part of this publication may be reproduced, stored in a retrieval system or transmitted in any form or by any means electronic, mechanical, photocopying, recording or otherwise without the prior written permission of the publisher Permissions may be sought directly from Elsevier's Science & Technology Rights Department in Oxford, UK: phone (+44) (0) 1865 843830; fax (+44) (0) 1865 853333; email: permissions@elsevier.com. Alternatively you can submit your request online by visiting the Elsevier web site at http://elsevier.com/locate/permissions, and selecting *Obtaining permission to use Elsevier material*

Notice
No responsibility is assumed by the publisher for any injury and/or damage to persons or property as a matter of products liability, negligence or otherwise, or from any use or operation of any methods, products, instructions or ideas contained in the material herein

Library of Congress Cataloging-in-Publication Data
A catalog record for this book is available from the Library of Congress

British Library Cataloguing-in-Publication Data
A catalogue record for this book is available from the British Library

ISBN: 978-0-12-381019-9

ISSN: 0065-2458

For information on all Academic Press publications
visit our web site at elsevierdirect.com

Printed and bound in USA

10 11 12   10 9 8 7 6 5 4 3 2 1

Working together to grow
libraries in developing countries

www.elsevier.com | www.bookaid.org | www.sabre.org

ELSEVIER   BOOK AID International   Sabre Foundation

Contents

CONTRIBUTORS ix
PREFACE xiii

Search Engine Optimization—Black and White Hat Approaches

Ross A. Malaga

1. Introduction 2
2. Background 4
3. The SEO Process 5
4. Black Hat SEO 18
5. Legal and Ethical Considerations 31
6. Conclusions 34
   References 37

Web Searching and Browsing: A Multilingual Perspective

Wingyan Chung

1. Introduction 42
2. Literature Review 43
3. A Multilingual Perspective 53
4. Case Studies 57

5. Summary and Future Directions 63
   Acknowledgments 66
   References 66

Features for Content-Based Audio Retrieval

Dalibor Mitrović, Matthias Zeppelzauer, and Christian Breiteneder

1. Introduction 72
2. Background 74
3. Audio Feature Design 84
4. A Novel Taxonomy for Audio Features 94
5. Audio Features 99
6. Related Literature 133
7. Summary and Conclusions 139
   Acknowledgments 139
   References 139

Multimedia Services over Wireless Metropolitan Area Networks

Kostas Pentikousis, Jarno Pinola, Esa Piri, Pedro Neves, and Susana Sargento

1. Introduction 154
2. WiMAX Overview 157
3. Multimedia over WiMAX Reference Scenarios 170
4. Advances in Telephony and the Emergence of Voice over IP 178
5. VoIP over WiMAX 191
6. Remote Surveillance and IPTV over WiMAX 209
7. Summary and Outlook 216
   Acknowledgments 219
   References 219

An Overview of Web Effort Estimation

Emilia Mendes

1. Introduction 224
2. How to Measure a Technique's Prediction Accuracy? 248
3. Which Effort Estimation Technique to Use? 253
4. Web Effort Estimation Literature Survey 254
5. Conclusions 266
   References 267

Communication Media Selection for Remote Interaction of *Ad Hoc* Groups

Fabio Calefato and Filippo Lanubile

1. Introduction 272
2. Task-Classification Frameworks 277
3. Group Research 281
4. CMC Theories 285
5. Development of a Comprehensive Theoretical Framework 300
6. Conclusions 308
   Acknowledgments 309
   References 309

AUTHOR INDEX 315
SUBJECT INDEX 327
CONTENTS OF VOLUMES IN THIS SERIES 339

Contributors

Prof. Christian Breiteneder is a Full Professor for Interactive Systems with the Institute of Software Technology and Interactive Systems at the Vienna University of Technology. Dr. Breiteneder received the Diploma Engineer degree in Computer Science from the Johannes Kepler University in Linz in 1978 and a Ph.D. in Computer Science from the University of Technology in Vienna in 1991. Before joining the institute he was Associate Professor at the University of Vienna and had postdoc positions at the University of Geneva, Switzerland, and GMD (now Frauenhofer) in Birlinghoven, Germany. His current research interests include interactive media systems, media processing systems, content-based multimodal information retrieval, 3D user interaction, and augmented and mixed reality systems. His email address is breiteneder@ims.tuwien.ac.at.

Dr. Fabio Calefato received the M.Sc. and Ph.D. degrees in Computer Science from the University of Bari in 2002 and 2007, respectively. He is currently associated with the Collaborative Development Group at the University of Bari as postdoctoral research assistant. His main research interests focus on collaborative software engineering and computer-mediated communication theories and tools. He is also a member of the Eclipse Italian Community Group, since 2006, and AI*IA (Italian Association for Artificial Intelligence), since 2008. He can be contacted at calefato@di.uniba.it.

Prof. Wingyan Chung is an Assistant Professor in the Department of Operations and Management Information Systems in the Leavey School of Business at Santa Clara University. He received his Ph.D. in Management Information Systems from the University of Arizona and an M.S. and B.B.A. from the Chinese University of Hong Kong. His research interests include knowledge management, Web analysis and mining, data and text mining, business intelligence, information visualization, and human–computer interaction. He has over 50 refereed publications in *Journal of Management Information Systems*, *Communications of the ACM*, *IEEE Computer*, *International Journal of Human–Computer Studies*, *Decision Support Systems*, and

Journal of the American Society for Information Science and Technology, among others. A certified teacher since 1998, Dr. Chung has more than 10 years of teaching and curriculum development experiences in the United States and Hong Kong. He teaches information systems courses at both undergraduate and graduate levels. Contact him at wchung@scu.edu.

Prof. Filippo Lanubile is an Associate Professor of Computer Science at the University of Bari. From 1995 to 1997, he was a Research Associate in the Experimental Software Engineering Group at the University of Maryland. His research interests lie in the areas of software engineering and CSCW, focusing on social software and distributed software development. He is a recipient of a NASA Group Achievement Award (1996), an IBM Eclipse Innovation Award (2006), and an IBM Faculty Award (2008). He has been the Program Chair of the 3rd IEEE International Conference on Global Software Engineering (2008). He can be contacted at lanubile@di.uniba.it.

Prof. Ross A. Malaga is an Associate Professor of Management and Information Systems at the School of Business, Montclair State University. He received his Ph.D. from George Mason University. Dr. Malaga has published extensively in the area of electronic commerce in *Communications of the ACM*, *Electronic Commerce Research*, and the *Journal of Organization Computing and Electronic Commerce*. He serves on the editorial review boards of *Information Resources Management Journal* and the *Journal of Electronic Commerce in Organizations*. He can be reached at malagar@mail.montclair.edu.

Prof. Emilia Mendes is an Associate Professor in Computer Science at the University of Auckland in New Zealand. She has active research interests in the areas of empirical web and software engineering, evidence-based research, hypermedia, computer science and software engineering education, in which areas she has published widely and over 120 refereed publications, which include two books. Prof. Mendes is on the editorial board of the *International Journal of Web Engineering and Technology*, the *Journal of Web Engineering*, the *Journal of Software Measurement*, the *International Journal of Software Engineering and Its Applications*, the *Empirical Software Engineering Journal*, the *Advances in Software Engineering Journal*, the *Software Quality Journal*, and *Springer Journal of Internet Services and Applications*. She worked in the software industry for 10 years before obtaining a Ph.D. in 1999 in Computer Science from the University of Southampton (UK) and moving to Auckland (NZ). Her email address is emilia@cs.auckland.ac.nz.

CONTRIBUTORS

Dalibor Mitrović is a Teaching and Research Associate with the Institute of Software Technology and Interactive Systems at the Vienna University of Technology. He received an M.Sc. degree in Computer Science in 2005 from the Vienna University of Technology, Austria. He pursues a Ph.D. in Computer Science focusing on multimodal information retrieval. His research interests include audio retrieval, real-time feature extraction, and computer vision. He can be reached at dalibor.mitrovic@computer.org.

Pedro Neves received his B.S. and M.S. degrees in Electronics and Telecommunications Engineering from the University of Aveiro, Portugal, in 2003 and 2006 respectively. From 2003 to 2006 he joined the Telecommunications Institute (IT), Aveiro, Portugal, and participated in the DAIDALOS-I and DAIDALOS-II (*Designing Advanced network Interfaces for the Delivery and Administration of Location independent, Optimized personal Services*) European-funded projects. Since 2006 he joined Portugal Telecom Inovação, Aveiro, Portugal, and he was involved in the WEIRD and, more recently, HURRICANE European-funded projects. Furthermore, since 2007 he is also pursuing a Ph.D. in Telecommunications and Informatics Engineering at the University of Aveiro. He has been involved in six book chapters, as well as more than 25 scientific papers in major journals and international conferences. His research interests are focused on broadband wireless access technologies, mobility and QoS management in *all-IP* heterogeneous environments, multicast and broadcast services, as well as mesh networks.

Dr. Kostas Pentikousis is a Senior Research Scientist at VTT Technical Research Centre of Finland, the foremost multidisciplinary applied research organization in Northern Europe. He holds a Ph.D. in Computer Science from the State University of New York at Stony Brook (2004). He received his B.Sc. and M.Sc. degrees in Computer Science from Aristotle University of Thessaloniki (1996; summa cum laude; ranked first) and State University of New York at Stony Brook (2000), respectively. Since 1996, he has been working in R&D positions in both industry and academia. He has been involved in several contract and joint research projects, including the EU-funded Ambient Networks (Phase 2), PHOENIX, WEIRD, and 4WARD; and the Future Internet program of the Finnish Strategic Centre for Science, Technology and Innovation in the field of ICT (TIVIT). Dr. Pentikousis has published more than 70 papers and book chapters in areas such as network architecture and design, mobile computing, applications and services, local and wide-area networks, and simulation and modeling, and has presented several tutorials on these topics. Dr. Pentikousis was an ERCIM Fellow in 2005 and is a member of IEEE, ACM, ICST, and TEK, the Finnish Association of Graduate Engineers.

Jarno Pinola received his M.Sc. degree in Telecommunications from the University of Oulu, Finland, in June 2008. He has worked at VTT Technical Research Centre of Finland since 2007 concentrating on wireless access technologies and mobility management. He has worked in both European and national projects covering these topics on different layers of the communication protocol stack. In the area of wireless communications, his current research interests include WiMAX system performance evaluations as well as LTE protocol studies. In addition, the energy-efficiency at the upper protocol levels in the next-generation cellular networks and wireless networking is part of his research agenda. He can be reached at jarno.pinola@vtt.fi.

Esa Piri received his M.Sc. from the University of Oulu, Finland, in Spring 2008. During his studies, he specialized in information networks and wrote his Master's Thesis on mobility management issues in heterogeneous networks. Currently, he is working as a Research Scientist in the field of seamless networking at VTT Technical Research Centre of Finland in Oulu, Finland. He can be contacted via e-mail at esa.piri@vtt.fi.

Dr. Susana Sargento joined the Department of Computer Science of the University of Porto in September 2002, and has been in the University of Aveiro and the Institute of Telecommunications since February 2004. She earned a Ph.D. in 2003 in Electrical Engineering. Dr. Sargento has been involved in several national and European projects, taking leaderships of several activities in the projects, such as the QoS and *ad hoc* networks integration activity in the FP6 IST-Daidalos Project. Her main research interests are in the areas of next generation and heterogeneous networks, infrastructure, mesh, and *ad hoc* networks, where she has published more than 150 scientific papers.

Matthias Zeppelzauer was born in Vienna, Austria, on October 15, 1980. He received the M.Sc. degree in Computer Science in 2005 from the Vienna University of Technology. Since 2006 Matthias Zeppelzauer works as a Ph.D. student at the Interactive Media Systems (IMS) group at the Vienna University of Technology. He has been employed in different research projects focusing on content-based audio and video retrieval. His research interests include content-based retrieval and recognition of sound, time series analysis, data mining, and multimodal media understanding. He can be reached at zeppelzauer@ims.tuwien.ac.at.

Preface

This is volume 78 in the series of *Advances in Computers* books. Currently, we publish three volumes, each containing 5 to 7 chapters, annually on new technology affecting the information technology industry. This series began in 1960 and is the oldest continuously published book series chronicling the ever changing computer landscape.

In this volume we again look at the World Wide Web. No longer an interesting research curiosity, the Web has become a dominant force in managing the flow of information internationally. It has become the primary source of information—both good and bad—for many researchers, it has become the backbone of multitrillion dollar (US) worldwide commerce, it has become the primary source of information between companies and governments to customers and the general public, and it is often the critical resource that manages the information flow within a company. In this volume we present six chapters that show how the Web is changing to address more complex issues in helping organizations manage their information technology resources.

In Chapter 1, "Search Engine Optimization—Black and White Hat Approaches," Ross A. Malaga explores the role of search engines in managing the information flow through the Web. With the number of accessible web pages now being in the billions, one needs an index of the Web in order to access any desired page. Companies, such as Google, Yahoo, and Microsoft, are competing to be the search engine of choice. Each of these search engines try to present the user with the most relevant pages for any given query. On the other hand, companies want their own pages to be ranked high so that the search engines will display their pages to a given user's query. Various techniques are employed by the search engines to rank pages and by companies to try and boost their own rankings. In this chapter, Dr. Malaga discusses this competing tension and describes the various approaches used to develop these rankings.

Chapter 2's "Web Searching and Browsing: A Multilingual Perspective" by Wingyan Chung presents a different challenge presented by the emerging Web. As an outgrowth of the 1960s US Defense Advanced Research Projects Agency's (DARPA) ARPANET, the Web had its roots in English text and the Roman character set. But there has been significant growth in areas where other languages predominate, such as Japanese, Chinese, Arabic, among others. In this chapter, Dr. Chung discusses the issues in opening up the Web to other character sets and the issue of developing non-English search engines.

Dalibor Mitrović, Matthias Zeppelzauer, and Christian Breiteneder in Chapter 3's "Features for Content-Based Audio Retrieval" discuss still another attribute of the Web that has changed since its introduction. Initially, information was stored as textual documents, and search engines were needed to find appropriate documents containing appropriate textual information. But as the Web has grown, audio, video, graphics, and pictures are quickly overtaking the growth of text files. Tools such as iTunes for downloading music and video and sites such as YouTube.com store thousand of new audio and video files each day. Searching for the specific item is now based upon short textual tags placed with each such file. But there is increasing interest in searching such files by the features in the files themselves—how to locate features in a picture or specific sounds in an audio file. In this chapter, the authors discuss the searching of audio music files by the contents of the music itself.

Chapter 4 is related to the audio search issues in Chapter 3 but from a still different point of view. With the growth of the Web, more information is being processed and transmitted—not only text files. When the Web started to grow in the 1990s, you could say that the Web was a digital network that was being transmitted over analog telephone lines. However, somewhere around 10 years ago, the balance changed. Most telecommunications are now over digital lines, with telephony now being one of the technologies carried over this digital medium. The term "digital lines" is also becoming somewhat anachronistic with more and more communications over wireless radio links. Kostas Pentikousis, Jarno Pinola, Esa Piri, Pedro Neves, and Susana Sargento in "Multimedia Services over Wireless Metropolitan Area Networks" discuss these issues, including the current trend of using the Internet as a telephony network using "voice over IP" (VoIP) technologies.

With all of this development of new Web technologies, some of the criteria and assumptions (such as cost, schedules, reliability) concerning development of Web applications does not carry over from earlier "mainframe" development. Emilia Mendes looks at the software development problems of Web development in Chapter 5's "An Overview of Web Effort Estimation." How can we estimate the effort required to build new Web applications and how does this differ from older models? In this chapter, Dr. Mendes looks at various approaches to perform effort estimation for new Web applications.

In the last chapter, Fabio Calefato and Filippo Lanubile in "Communication Media Selection for Remote Interaction of *Ad Hoc* Groups" look at some of the implications of the technology described in the earlier chapters. With the increased speeds of the digital lines of the Web, with the increased capacity of the emerging search engines, and with an increasing set of objects (e.g., text, pictures, video, telephony) being transmitted, one of the obvious impacts of these changes is that it is no longer necessary for groups to be working together in order to use and share such objects. Thus, we have an increasing dispersion of the workforce with social networks such as Facebook or Myspace providing mechanisms for communication over a distance. How can such groups communicate effectively? In this chapter, the authors discuss the various mechanisms that the Web provides for managing such interactions.

I hope that you find these chapters of interest. For me, volume 78 marks a significant milestone in my association with the *Advances in Computers* book series. The first volume I put together was volume 40 in 1995, which makes this the 39th volume produced under my direction. Or in other words, I have produced as many volumes as all previous series editors combined. I have found the process both interesting and rewarding, although at times a bit stressful in maintaining production schedules without sacrificing quality. I am always looking for new and interesting topics to appear in these pages. If you have any suggestions of topics for future chapters, or if you wish to contribute such a chapter yourself, I can be reached at mvz@cs.umd.edu.

Marvin Zelkowitz
College Park, Maryland

Search Engine Optimization—Black and White Hat Approaches

ROSS A. MALAGA

Management and Information Systems, School of Business, Montclair State University, Montclair, New Jersey, USA

Abstract

Today the first stop for many people looking for information or to make a purchase online is one of the major search engines. So appearing toward the top of the search results has become increasingly important. Search engine optimization (SEO) is a process that manipulates Web site characteristics and incoming links to improve a site's ranking in the search engines for particular search terms. This chapter provides a detailed discussion of the SEO process. SEO methods that stay within the guidelines laid out by the major search engines are generally termed ''white hat,'' while those that violate the guidelines are called ''black hat.'' Black hat sites may be penalized or banned by the search engines. However, many of the tools and techniques used by ''black hat'' optimizers may also be helpful in ''white hat'' SEO campaigns. Black hat SEO approaches are examined and compared with white hat methods.

1. Introduction 2
2. Background 4
   2.1. Search Engines History and Current Statistics 4
   2.2. SEO Concepts 5
3. The SEO Process 5
   3.1. Keyword Research 5
   3.2. Indexing 9
   3.3. On-Site Optimization 11
   3.4. Link Building 15

- 4. Black Hat SEO 18
  - 4.1. Black Hat Indexing Methods 19
  - 4.2. On-Page Black Hat Techniques 19
  - 4.3. Cloaking 22
  - 4.4. Doorway Pages 23
  - 4.5. Content Generation 25
  - 4.6. Link Building Black Hat Techniques 25
  - 4.7. Negative SEO 30
- 5. Legal and Ethical Considerations 31
  - 5.1. Copyright Issues 31
  - 5.2. SEO Ethics 33
  - 5.3. Search Engine Legal and Ethical Considerations 33
- 6. Conclusions 34
  - 6.1. Conclusions for Site Owners and SEO Practitioners 35
  - 6.2. Future Research Directions 35
  - References 37

1. Introduction

The past few years have seen a tremendous growth in the area of search engine marketing (SEM). SEM includes paid search engine advertising and search engine optimization (SEO). According to the Search Engine Marketing Professional Organization (SEMPO), search engine marketers spent over $13.4 billion in 2008. In addition, this figure is expected to grow to over $26 billion by 2013. Of the $13.4 billion spent on SEM, about 10% ($1.4 billion) was spent on SEO [1].

Paid advertising are the small, usually text-based, ads that appear alongside the query results on search engine sites (see Fig. 1). Paid search engine advertising usually works on a pay-per-click (PPC) basis. SEO is a process that seeks to achieve a high ranking in the search engine results for certain search words or phrases. The main difference between SEO and PPC is that with PPC, the merchant pays for every click. With SEO each click is free (but the Web site owner may pay a considerable amount to achieve the high ranking). In addition, recent research has shown that users trust the SEO (called organic) results and are more likely to purchase from them [2].

Industry research indicates that most search engine users only clicked on sites that appeared on the first page of the search results—basically the top 10 results. Very few users clicked beyond the third page of search results [3]. These results confirm the research conducted by Granka et al. [4], in which they found that almost 80% of

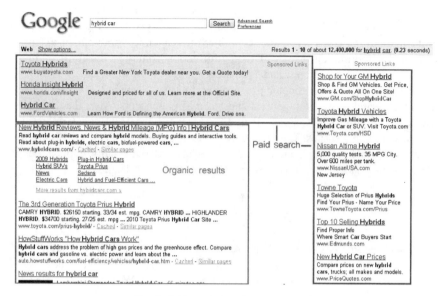

FIG. 1. Paid versus organic search results.

the clicks on a search engine results page came went to those sites listed in the first three spots.

SEO has become a very big business. Some of the top optimizers and SEO firms regularly charge $20,000 or more per month for ongoing optimization. It is not uncommon for firms with large clients to charge them $150,000 or more on a monthly basis [5].

Because of the importance of high search engine rankings and the profits involved, search engine optimizers look for tools, methods, and techniques that will help them achieve their goals. Some focus their efforts on methods aimed at fooling the search engines. These optimizers are considered "black hat," while those that closely follow the search engine guidelines would be considered "white hat." There are two main reasons why it is important to understand the methods employed by black hat optimizers. First, some black hats have proven successful in achieving high rankings. When these rankings are achieved, it means that white hat sites are pushed lower in the search results. However, in some cases these rankings might prove fleeting and there are mechanisms in place to report such sites to the search engines. Second, some of the tools and methods used by black hat optimizers can actually be used by white hat optimizers. In many cases, it is just a matter of scope and scale that separates black and white hat.

While there are some studies dealing with SEO, notably Refs. [6–9], academic research in the area of SEO has been relatively scant given its importance in the online marketing field. This chapter combines the academic work with the extensive practitioner information. Much of that information comes in the form of blogs, forum discussions, anecdotes, and Web sites.

The remainder of this chapter proceeds as follows. Section 2 provides a background on search engines in general and basic SEO concepts. After that a detailed discussion on the SEO process including keyword research, indexing, on-site factors, and linking ensues. The section that follows focuses on black hat SEO techniques. Legal and ethical implications of SEO are then discussed. Finally, implications for management, conclusions, and future research directions are detailed.

2. Background

A search engine is simply a database of Web pages, a method for finding Web pages and indexing them, and a way to search the database. Search engines rely on spiders—software that followed hyperlinks—to find new Web pages to index and insure that pages that have already been indexed are kept up to date.

Although more complex searches are possible, most Web users conduct simple searches on a keyword or key phrase. Search engines return the results of a search based on a number of factors. All of the major search engines consider the relevance of the search term to sites in its database when returning search results. So, a search for the word "car" would return Web pages that had something to do with automobiles. The exact algorithms used to determine relevance are constantly changing and are a trade secret.

2.1 Search Engines History and Current Statistics

The concept of optimizing a Web site so that it appears toward the top of the results when somebody searches on a particular word or term has existed since the mid-1990s. Back then the search engine landscape was dominated by about 6–10 companies, including Alta Vista, Excite, Lycos, and Northern Lights. At that time, SEO largely consisted of keyword stuffing. That is adding the search term numerous times to the Web site. A typical trick employed was repeating the search term hundreds of times using white letters on a white background. Thus, the search engines would "see" the text, but a human user would not.

The search engine market and SEO have changed dramatically over the past few years. The major shift had been the rise and dominance of Google. Google currently handles more than half of all Web searches [10]. The other major search engines used in the United States are Yahoo and MSN. Combined, these three search engines are responsible for over 91% of all searches [10]. It should be noted that at the time this chapter was written, Microsoft had just released Bing.com as its main search engine.

The dominance of the three major search engines (and Google in particular) combined with the research on user habits meant that for any particular search term, a site must appear in the top 30 spots on at least one of the search engines or it was effectively invisible. So, for a given term, for example "toyota corolla," there were only 90 spots available overall. In addition, 30 of those spots (the top 10 in each search engine) are highly coveted and the top 10 spots in Google are extremely important.

2.2 SEO Concepts

Curran [11] states, "search engine optimization is the process of improving a website's position so that the webpage comes up higher in the search results [search engine results page (SERP)] of major search engines" (p. 202). This process includes manipulation of dozens or even hundreds of Web site elements. For example, some of the elements used by the major search engines to determine relevance include, but are not limited to: age of the site, how often new content is added, the ratio of keywords or terms to the total amount of content on the site, and the quality and number of external sites linking to the site [12].

3. The SEO Process

In general, the process of SEO can be broken into four main steps: (1) keyword research, (2) indexing, (3) on-site optimization, and (4) off-site optimization.

3.1 Keyword Research

A search engine query is basically just a word or phrase. It is the result of the query to a specific word or phrase that is of interest to search engine optimizers. The problem is that there are usually many words or phrases that can be used for a particular search. For example, if a user was looking to purchase a car—say a Toyota Prius, she might use any of the following words or phrases in her search:

- Car
- Automobile
- New car
- Toyota

- Prius
- Toyota Prius
- New Toyota Prius
- Toyota Prius New York City
- Toyota Prius NYC
- NYC Toyota Prius

It is easy to see that this list can keep going. In terms of SEO, which term or terms should we try to optimize our site for?

Keyword research consists of building a large list of relevant search words and phrases and then comparing them along three main dimensions. First, we need to consider the number of people who are using the term in the search engines. After all, why optimize for a term that nobody (or very few people) use? Fortunately, Google now makes search volume data available via its external keyword tool (available at https://adwords.google.com/select/KeywordToolExternal). Simply type the main keywords and terms and click on Get Keyword Ideas. Google will generate a large list of relevant terms and provide the approximate average search volume (see Fig. 2).

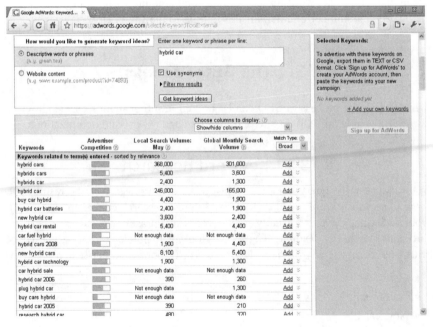

FIG. 2. Google external keyword tool search volume results for hybrid car.

Clearly, we are looking for terms with a comparatively high search volume. So, for example, we can start building a keyword list with:

- Toyota Prius—388,000
- Hybrid car—165,000
- Hybrid vehicle—33,100
- Hybrid vehicles—60,500
- Hybrid autos—4400

Many search engine optimizers also consider simple misspellings. For example, we can add the following to our list:

- Hybird—27,100
- Hybird cars—2900
- Pruis—9900

Once we have generated a large list of keywords and phrases (most optimizers generate lists with thousands of terms), the second phase is to determine the level of competition for each term. To determine the level of competition, simply type the term into Google and see how many results are reported in the top right part of the page (see Fig. 3).

To compare keyword competition, optimizers determine the results to search (R/S) ratio. The R/S ratio is calculated by simply dividing the number results (competitors) by the number of searches over a given period of time. On this scale lower numbers are better. So, we might end up with a list like that in Table I.

Comparing R/S ratios is more effective than just looking at how many people are searching for a particular word or phrase as it incorporates the level of competition. In general, optimizers want to target terms that are highly searched and have a low level of competition. However, the R/S ratio can reveal terms that have a relatively

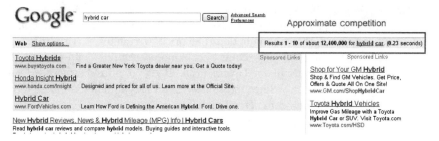

FIG. 3. Competition in Google results.

TABLE I
SEARCH RESULTS AND COMPETITION

Term	No. of searches per month	Competition	R/S
Toyota Prius	388,000	5,490,000	14.14
Hybrid car	165,000	12,900,000	78.18
Hybrid vehicle	33,100	17,900,000	540.78
Hybrid vehicles	60,500	7,880,000	130.24
Hybird	27,100	1,230,000	45.38

low level of searches, but also a very low level of competition. For instance, Table I shows that the misspelled word "hybrid" has a lower search volume than many of the other terms. However, when the competition is also considered via the R/S ratio, the misspelled word appears to be a good potential target for SEO.

The third factor to consider, at least in most cases, is the commercial viability of the term. To determine commercial viability we must understand a bit about consumer buying behavior. The traditional consumer purchase model consists of five phases: (1) need recognition, (2) information search, (3) option evaluation, (4) purchase decision, and (5) postpurchase behavior.

Once a potential consumer becomes aware of a need she begins to search for more information about how to fulfill that need. For example, a person who becomes aware of a need for a car would begin by gathering some general information. For example, she might research SUVs, trucks, sedans, etc. At this point the consumer does not even know what type of vehicle she wants. She might use search terms like "how to buy a car" or "types of cars." Since the consumer does not know what type of car she wants at this point, these terms would be considered to have low commercial viability.

In the next phase, the consumer begins to narrow down the choices and evaluate the various options. Some exemplar search terms in this phase might include "car review," "SUV recommendation," and "best cars." These terms have a bit more commercial viability, but would still not be considered high viable.

During the fourth phase, consumers have made a choice and are now just looking for where to purchase—comparing options like price, warranties, service, trust, etc. At this point the search terms become much more specific. For example, the consumer might use terms like "Toyota Prius 2009," "prius best price," and "new jersey Toyota dealer." Since the consumer is ready to purchase these terms are considered to have high commercial viability.

A good optimizer will actually target multiple terms—some for the site's homepage and some for the internal pages of the site. For instance, the site for a Toyota dealer in Montclair New Jersey might use "New Jersey Toyota Dealer" as the main

SEO target for the homepage. The same site might use "Toyota Prius 2009 Best Price" for an internal page that lists the features of that car and the details of the vehicles on the lot.

Clearly, determining commercial viability is a combination of art and science. It requires the optimizer to think like a consumer. Microsoft researchers have conducted research into this area [13]. They have broken search into three main categories: navigational, informational, and transactional. In addition, queries are also categorized as either commercial or noncommercial based on the nature of the search term used, resulting in the 3 × 2 grid shown in Table II. For example, terms that include words such as "buy," "purchase," or "price" would be considered commercial in nature. The researchers determined the categorization of commercial and noncommercial by asking human reviewers to rate various terms along those dimensions. Obviously, this approach has serious limitations. However, Microsoft has developed a Detecting Commercial Online Intent tool, which is available at http://adlab.microsoft.com/Online-Commercial-Intention/. Many optimizers use this site to gauge the commercial viability and search type of their keywords.

Finally, some optimizers have attempted to capture the consumer earlier in the process—during the option evaluation phases in particular. This is typically accomplished by developing review and recommendation type sites. There are, to date, no reliable data on how well these types of sites perform in terms of moving the visitor from the information phase to the transactional phase.

3.2 Indexing

Indexing is the process of attracting the search engine spiders to a site, with the goal of getting indexed (included in the search engine's database) and hopefully ranked well by the search engine quickly. All of the major search engines have a site submit form where a user could submit a site for consideration. However, most SEO experts advise against this approach. It appears that the major search engines prefer "discovering" a new site. The search engines "discover" a new site when the spiders find a link to that site from

TABLE II
COMMERCIAL ONLINE INTENTION (ADAPTED FROM REF. [13])

	Commercial	Noncommercial
Navigational	Toyota	Hotmail
Informational	Hybrid Car	San Francisco
Transactional	Buy Toyota Prius	Collide lyrics

other sites. So the main approach to indexing involves getting links to a site from other sites that were frequently visited by the spiders.

An increasingly popular approach to generating quick links to a new site is via Web 2.0 properties. The term Web 2.0 seems to have been coined by Tim O'Reilly whose O'Reilly Media sponsored a Web 2.0 conference in 2004 [14]. There does not appear to be any standard definition of Web 2.0 as the concept is continually evolving. However, Web 2.0 incorporates concepts such as user-developed content (blogs, wikis, etc.), social bookmarking, and the use of really simple syndication (RSS). For the purposes of indexing, user-developed content and social bookmarking sites are key.

User-developed content sites enable users to quickly and easily publish written, audio, and/or video content online. Many content sites employ a simple user interface for text input that in many ways mimics traditional word processing software. Most of these sites allow users to upload pictures and embed multimedia content (audio and video). Among the most widely used Web 2.0 content sites are blogs and wikis from multiple providers, and sites that allow users to create simple Web sites (e.g., Squidoo and Hubpages).

A blog (short for Weblog) is simply a Web site where a user can post comments and the comments are typically displayed in reverse chronological order. The comments can range broadly from political commentary to product reviews to simple online diaries. Modern blog software can also handle pictures, audio files, and video files. The blog tracking site Technorati had recorded over 112 million blogs in its system as of November 2007 [15].

A wiki is a type of software that allows users to create a Web site in a collaborative manner. Wikis use a simple markup language for page creation. The most well-known wiki site is Wikipedia (www.wikipedia.org)—an online encyclopedia to which anyone can contribute. In fact, the site has over 9 million articles and more than 75,000 contributors. Personal wikis are easy to create on a user's own domain or on free wiki sites, such as Wetpaint.com.

According to Wikipedia, "social book marking is a way for Internet users to store, organize, share, and search bookmarks of web pages" [16]. These bookmarks are typically public, meaning anyone can see them. Bookmarks are usually categorized and also tagged. Tagging allows the user to associate a bookmark with any chosen words or phrases. Tags are completely created by the user, not the social bookmarking site. Some of the more popular social bookmarking sites include Del.icio.us, Digg, and StumbleUpon.

An optimizer who wants to get a new site indexed quickly can build one or more Web 2.0 content sites and include links to the new site. The optimizer can also use social bookmarking to bookmark the new site. Since the search engine spiders visit these sites frequently, they will find and follow the links to the new site and index them. This approach has an added benefit, in that the content site or social bookmark itself might rank for the chosen term. If done correctly a good optimizer can dominate the top positions in the search engines using a combination of their own sites, Web 2.0 content sites, and social bookmarks.

Malaga [17] reports on using Web 2.0 SEO techniques to dominate the search engine results pages. In one experiment the researcher was able to get sites indexed in less than a day. In addition, after only a few days the researcher was able to obtain two top 10 rankings on Google and five top 30 rankings. The results were similar for results on Yahoo and MSN.

3.3 On-Site Optimization

On-site optimization is the process of developing or making changes to a Web site in order to improve its search engine rankings. There are dozens, perhaps hundreds, of on-site factors that are considered in determining a site's ranking for a particular term.

Some of the main on-site factors used by the search engines to determine rank include title tag, meta description tag, H1 tag, bold text, keyword density, and the constant additional of relevant unique content.

3.3.1 Meta Tags

Meta tags are HTML structures that are placed in the HEAD section of a Web page and provide metadata about that page. Research has shown that meta tags are an important component in SEO. Zhang and Dimitroff [18], for instance, examined a large number of SEO components in controlled experiments. Their results show that sites that make proper use of meta tags achieve better search engine results. In terms of SEO, the most frequently used (and perhaps important) meta tags include title, description, and keyword. Search engine optimizers are also usually interested in the robots meta tag.

Officially the title tag is not really a meta tag; however since it works like a meta tag, most optimizers think of it like one. The title tag is used by the search engines in a number of ways. First, it appears to be an important component in the ranking algorithms of the major search engines. Malaga [6] showed that aligning the title tag with the targeted SEO term can result in major improvement in search engine rankings. Second, the search engines typically use the title tag as a site's name in the search engine results page (see Fig. 4).

There are two common mistakes that many Web developers make when it comes to the title tag. First, they use the name of the company or the domain name as the title tag. In other words, they do not align the title tag with the chosen SEO target

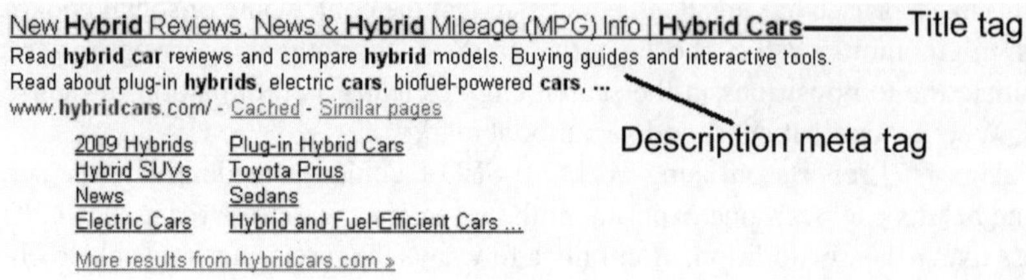

Fig. 4. Use of title and description meta tags in search results.

term. Second, they use the same title tag for every page on the site. If we consider the example of the Toyota dealer the homepage might have a title tag of "Montclair New Jersey Toyota Dealer." However, we would want the internal pages to have their own title tags. For example, the Prius page might use "Toyota Prius" for the title and the Corolla page would use "Toyota Corolla."

The description meta tag is used to explain what the Web page is about. It is usually a short sentence or two consisting of about 25–30 words. Again, the search engines use the description tag in two ways. First, it appears that at least some of the search engines use the description tag as part of their ranking algorithm. Second, some of the search engines will display the contents of the description tag below the site link in the search engine results page (see Fig. 4).

The keyword meta tag is simply a list of words or phrases that relate to the Web page. For example, terms that might be used on the Toyota Prius page are "prius," "Toyota prius," and "hybrid car." The list of keywords can be quite long. It appears that the major search engines do not give the keyword tag much or any weight in determining a site's rank. But some anecdotal evidence exists to support the notion that Yahoo still uses this tag in its ranking algorithm.

The major search engines will index each Web page they find and will follow all of the links on that page. There are some circumstances in which we might not want the search engines to index a page or follow the links on it. For example, we might not want a privacy policy or terms of use page to be indexed by the search engines. In this case the optimizer can use a robot tag set to "noindex." The major search engines will still crawl the site, but most will not include the page in their database. There are some cases where we might not want the search engines to follow the links on a particular page. Including sponsored links on a page would be a good example. In this case the optimizer would use a robot tag set to "nofollow." These terms can be combined so that a page is not indexed and the links on the page are also not followed. It should be noted that while the major search engines currently abide by "noindex" and "nofollow" statements, they might change their policies at any time.

Optimizers and Web developers often handle robots restrictions at the domain level. They can include a file named robots.txt in the root directory. That file lists directories and files that should and should not be followed or indexed.

If we pull all meta tag information together, we might wind up with the following example for a Prius review on a hybrid car blog that is targeting the term "prius car review":

```
<head>
<title>Prius Car Review</title>
<meta name="description" content="Unbiased Prius Car Review from the Hybrid Car Experts"/>
<meta name="keyword" content="prius, toyota prius, hybrid car review, hybrid truck review, hybrid SUV review, hybrid cars"/>
</head>
```

If the same blog also contained a page with sponsored links that the author did not want the search engines to index or follow the head might appear as follows:

```
<head>
<title>Hybrid Car Review—Sponsors</title>
<meta name="description" content="Links to interesting sites about hybrid cars"/>
<meta name="keyword" content="hybrid car review, hybrid truck review, hybrid SUV review, hybrid cars, hybrid car blog, hybrid car links"/>
<meta name="robots" content="noindex">
<meta name="robots" content="nofollow">
</head>
```

3.3.2 Other On-Site Elements

While meta tags are important the text content on the page may be just as, if not more, important. For instance, Zhang and Dimitroff [9] found that sites with keywords that appear in both the site title tag and throughout the site's text have better search engine results than sites that only optimize the title tag.

3.3.2.1 Latent Semantic Indexing.
All of the major search engines place a good deal of importance on relevant text content. The search engines determine relevance by analyzing the text on the Web page. For this purpose, the search engines tend to ignore HTML tags, Javascript, and cascading style sheet (CSS) tags. "Understanding" what a particular Web page is about is actually quite a

complex task. A simple approach is to just match up the text on a Web site with the query. The problem with this technique is that there are many ways to search for the same thing. To overcome this problem, the major search engines now use latent semantic indexing (LSI) to "understand" a Web page. Google's Matt Cutts has confirmed that Google uses LSI [19].

Latent semantic indexing (sometimes called latent semantic analysis) is a natural language processing method that analyzes the pattern and distribution of words on a page to develop a set of common concepts. Scott Deerwester, Susan Dumais, George Furnas, Richard Harshman, Thomas Landauer, Karen Lochbaum, and Lynn Streeter were granted a patent in 1988 (U.S. Patent 4,839,853) for latent semantic analysis. A basic understanding of LSI is important in terms of writing for the search engines.

The technical details of LSI are beyond the scope of this chapter—the interested reader is referred to Ref. [20]. However, a short example should aid in understanding the overall concept. If we go back to our Toyota dealer we might expect its site to include terms like car, truck, automobile, as well as names of cars (Corolla, Prius, Highlander, etc.). To keep the example simple, let us assume that one page on our Toyota dealer's site might read "Toyota trucks are the most fuel efficient." If a search engine just used a simple matching approach then a query for "Toyota Tacoma" would not return the page—since that exact term does not appear on the page. However, using LSI the search engine is able to "understand" that the page is about Toyota trucks and that a search for Toyota Tacoma should return the page. It does this by comparing the site with all of the others that include terms such as "Toyota," "truck," and "Tacoma" and judges that it is likely that these terms are related. The goal of LSI is to determine just how closely related the terms are. For example, if the user searched on the term "Chevy Silverado" it should not return the Toyota truck page. Although both pages are about "trucks," using LSI the search engine can determine that the relation among the terms is not very close.

It should be noted that Google probably does not actually compare pages with every other page on the Web, as this would be extremely computationally intensive. However, Google does attempt to determine the "theme" of individual pages and of sites. Therefore, optimizers attempt to organize their sites based on the "themes" they are attempting to optimize for. For example, if we were trying to optimize for "Toyota trucks," "Toyota sedans," and "Toyota hybrids" each of these would become a page on the dealer's site. In addition, each page would be written around the "theme."

3.3.2.2 Updated Content. There seems to be a general belief among optimizers that unique fresh text content is an essential aspect of SEO [21]. Many optimizers go to great lengths, such as hiring copy writers, to constantly add fresh content to a site. However, some optimizers (and this author) have found that some sites with no fresh content may still rank well. For example, Darren Rowse writing on ProBlogger [22] discusses the case of a blog that still ranked third on Google for the term "digital photography" 9 months after any new content was added. In addition, this author had a site that ranked first on Google for "discount ipod accessories" 2 years after any new content was added to the site. While the data are scant and anecdotal, it appears that adding fresh content may help get a site ranked well to begin, but does not appear to be necessary to maintain a high ranking.

3.3.2.3 Formatting.

The content on the page is important for SEO, but so is the formatting. Formatting gives the search engines clues as to what the Web developer thinks are the important aspects of the page. Some of the formatting components considered by the search engines include header tags (H1, H2, etc.), bold, and emphasis.

Header tags are used to identify various sections of a page. Header 1 (H1) tags deliminate major sections on a page. Higher header numbers (H2, H3, etc.) identify subsections. The major search engines consider text in an H1 HTML header to be of increased importance [23]. Therefore, many optimizers place the main key terms within an H1 tag structure. Lesser SEO terms might be placed in higher number header tags (H2, H3, etc.). The bold and emphasis (em) tags are also used by optimizers to indicate important text on a page.

Some early Web browsers, such as Lynx, were text based. That is, they could not display any graphics. The <alt> tag was originally used to display information about a Web-based image when displayed in a text-based browser. Today the overwhelming majority of Web traffic is from graphical browsers (such as Internet Explorer, FireFox, Opera, and Chrome). However, optimizers still use the <alt> tag as another way of serving text content to the search engines. At one time search engines appeared to make use of the text in an <alt> tag. According to Jerry West [24], today the search engines not only do not use the <alt> tag in determining rankings, but also may actually penalize sites that use, or overuse, the tag. The <alt> tag remains important for people with disabilities, so it is good Web development practice to include descriptive <alt> tags for all images.

3.4 Link Building

All of the major search engines consider backlinks in their ranking algorithms. A backlink is a hyperlink from a site to the target site. All the major search engines also consider the relevance of the text used in the backlink (called the anchor text). For example, a link to the Toyota Web site that said "car" would be considered relevant, but one that said "flowers" would be irrelevant.

Yahoo and MSN use the number of backlinks in their algorithms. However, Google places particular importance on backlinks. Google does not just consider the number of links, but also the "quality" of those links. Google assigns each page in its index a Page Rank (PR). PR is a logarithmic number scale from 0 to 10 (with 10 the best). Google places more weight on backlinks that came from higher PR sites.

Exactly how PR is currently calculated is a closely held secret. However, we can gain a basic understanding of the concept by examining the original PR formula (U.S. Patent 6,285,999):

$$PR(A) = (1 - d) + d(PR(t_1)/C(t_1) + \cdots + PR(t_n)/C(t_n)),$$

where PR is the Page Rank of a particular Web page, $t_1, \ldots, t_n$ are Web pages that link to page A, C is the number of links from a page to another page on the Web, and d is a damping factor.

Clearly, determining PR is an iterative process. Also, each page provides a portion of its PR to each page it links to. So, from a SEO perspective we want incoming links from sites with a high PR. In addition, it is beneficial to obtain links from pages that have fewer outgoing links. Consider the example in Fig. 5 (adapted from http://www.whitelines.nl/html/google-page-rank.html).

In this example, the entire Web consists of only three pages (A, B, and C). In addition, the only links are the ones indicated. Each page has its PR set initially to 1.0. An initial PR needs to be assumed for each page. However, after 15 iterations the actual PR emerges and can be seen in Fig. 6.

In addition to external sites, it must be noted that links from pages within a site (internal links) also count toward PR. Therefore, optimizers must consider the internal linking structure of the site. For example, many sites include links to privacy policies, terms of service, about us, and other pages that do not allow visitors to take

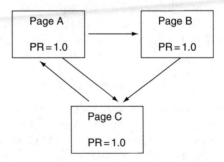

Fig. 5. Page Rank Example 1.

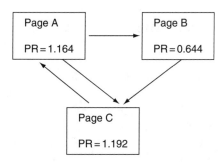

Fig. 6. Page Rank Example 2.

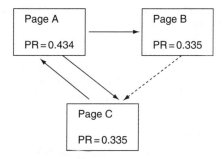

Fig. 7. Page Rank example with NoFollow.

commercial action. However, since links to these pages appear on every page in the site they may obtain a high PR. Optimizers can attempt to manipulate PR by using the Robots NoFollow meta tag.

For instance, assume that the above example represents pages on a site. If we use NoFollow to essentially eliminate the link between page B and page C (note that the link still exists, but does not count toward PR), we would wind up with the structure shown in Fig. 7 after 20 iterations. In this example, adding the NoFollow leads to a major change in PR for all of the pages on the site. In fact, it reduced the PR for all of the pages—a very nonoptimal outcome!

Clearly, the result in this example is highly unlikely due to the circular link structure on the site and lack of external links. But it does point out how an inexperienced optimizer can run into problems when attempting to manipulate PR on a site.

4. Black Hat SEO

Some of the basic tenants of black hat SEO include automated site creation by using existing content and automated link building. While there is nothing wrong with automation in general, black hat SEOs typically employ techniques which violate the search engines Webmaster guidelines.

For example, both Google and Yahoo provide some guidance for Webmasters. Since Google is the most widely used search engine and their guidance is the most detailed we will discuss their policies.

Google's Webmaster guidelines (available at http://www.google.com/support/webmasters/bin/answer.py?answer=35769) offer quality guidelines that are important for SEO. Google states:

> These quality guidelines cover the most common forms of deceptive or manipulative behavior, but Google may respond negatively to other misleading practices not listed here (e.g., tricking users by registering misspellings of well-known websites). It's not safe to assume that just because a specific deceptive technique isn't included on this page, Google approves of it. Webmasters who spend their energies upholding the spirit of the basic principles will provide a much better user experience and subsequently enjoy better ranking than those who spend their time looking for loopholes they can exploit.

In general, Google would like Webmasters to develop pages primarily for users, not search engines. Google suggests that when in doubt the Webmaster should ask "does this help my users?" and "would I do this if the search engines did not exist?"

The Google Quality Guidelines also outline a number of specific SEO tactics which Google finds offensive. These tactics are discussed in detail in the sections below.

The astute reader might now ask, what will Google (or any other search engine) do if I violate its quality guidelines. Google has two levels of penalties for sites that are in violation. Those sites that use the most egregious tactics are simply banned from Google. For example, on February 7, 2006 Google banned BMW's German language site (www.bmw.de) for using a "doorway page." This is a page that shows different content to search engines and human visitors. Sites that use borderline tactics may be penalized instead of banned. A penalty simply means that the site loses ranking position.

We can use the penalty system to provide a working definition of black and white hat SEO. In general, black hat SEO consists of methods that will most likely lead to Google penalizing or banning the site at some point. White hat SEOs are methods that Google approves of and will therefore not lead to any penalty. There are some techniques that are borderline and some that are generally white hat, but may be overused. We might define optimizers that fall into this category as gray hat. Gray hat techniques may lead to a penalty, but will not usually result in a ban.

If black hat strategies lead to a site ban, why do it? Black hats tend to fall into two categories. First, as it typically takes a bit of time for Google to ban a site, there are individuals who use this delay to temporarily achieve a top ranking and make a bit of money from their site. As they use software to automate the site creation and ranking process, they are able to churn out black hat sites.

The second category consists of SEO consulting firms that use black hat techniques. These companies achieve a temporary high ranking for their clients, collect their money, and move on. For example, according to Google employee Matt Cutts' blog [25], the SEO consulting company Traffic Power was banned from the Google index for using black hat strategies. In addition, Google also banned Traffic Powers' clients.

4.1 Black Hat Indexing Methods

While links from other sites might enable a site to get indexed quickly, it usually takes time for a site to begin ranking well. As mentioned above, Google appears to place new sites into the "sandbox" for an unspecified period of time in order to see how the site evolves.

One of the primary tricks black hat SEOs use to attract search engine spiders is called Blog Ping (BP). This technique consists of establishing hundreds or even thousands of blogs. The optimizer then posts a link to the new site on each blog. The final step is to continually ping the blogs. Pinging automatically sends a message to a number of blog servers that the blog has been updated. The number of blogs and continuous pinging attracts the search engine spiders that then follow the link.

It should be noted that many white hat SEOs use the BP technique in an ethical manner. That is, they post a link to the new site on one (or a few) blog and then ping it only after an update. This method has been shown to attract search engine spiders in a few days [6].

As many of the techniques for indexing involve getting links to the site, we will discuss them in more detail in Section 4.6.

4.2 On-Page Black Hat Techniques

Black hat optimizers use a variety of on-page methods. Most of these are aimed at providing certain content only to the spiders, while actual users see completely different content. The reason for this is that the content used to achieve high rankings may not be conducive to good site design or a high conversion rate (the rate at which site visitors perform a monetizing action, such as make a purchase). The three main methods that fall into this category are keyword stuffing with hidden content, cloaking, and doorway pages.

4.2.1 Keyword Stuffing

Keyword stuffing involves repeating the target keyword or term repeatedly on a page. Take, for example, our Toyota Prius site. It might repeat the term "Toyota Prius" over and over again on the page. Obviously, repeating this term on the page would not look good to human visitors, so early optimizers used HTML elements such as the same foreground and background colors, or very small fonts to hide the content from human visitors. The search engines quickly got wise to these tricks and will now ignore content that is obviously not visible to humans.

Black hat optimizers generally look for any way to stuff their keywords into the page code. As mentioned above, use of the <alt> tag on all images is good Web development practice. However, in the early days of SEO a typically keyword stuffing trick was to repeat the word or phrase within the <alt> tag text. This is the reason that the search engines no longer consider <alt> tag text in their ranking algorithms.

Today black hat SEOs have found new keyword stuffing methods. For instance, optimizers (both black hat and white hat) have experimented with more obscure meta tags such as abstract, author, subject, and copyright. These tags can be used responsibly by white hat SEOs and as a place to stuff keywords for black hats. While the major search engines seem to ignore these tags, they may influence ranking in less well-known search engines.

4.2.1.1 Cascading Style Sheet—Keyword Stuffing.

More recently, optimizers have taken to using CSS to hide elements. The elements the optimizer wants to hide are placed within hidden div tags, extremely small divisions, off-page divisions, and using z position to hide content behind a visible layer (see the sample code below for complete details).

Hidden Division:

```
<div style:visibility ="hidden">Toyota Prius</div>
```

By setting the division's visibility setting to "hidden," none of the text in the division is displayed to human visitors. However, the text can be found and indexed by search engine spiders.

Small Division:

```
#hidetext a
{
width:1px;
height:1px;
overflow:hidden;
}
<div id="hidetext">Toyota Prius</div>
```

In this example, the text in the division is theoretically visible to both humans and search engines. However, the layer is so small (only 1 pixel in size) that it will likely be completely overlooked by human visitors.

Positioning Content Off-Screen:

```
#hideleft {
position:absolute;
left:-1000px;
}
<div id="hideleft">Toyota Prius</div>
```

The content in this division is also theoretically visible to both humans and search engines. In this case, however, the layer is positioned so far to the left (1000 pixels) that it will not be seen by humans.

Hiding Layers Behind Other Layers:

```
<div style="position: absolute; width: 100px; height: 100px;
z-index: 1" id="hide">
This is the text we want to hide</div>
<div style="position: absolute; width: 100px; height: 100px;
z-index: 2; background-color: #FFFFFF" id="showthis">
This is the text we want to display
</div>
```

This code example uses the CSS layer's z index property to position one layer on top of the other. The z index provides a measure of three dimensionality to a Web page. A z index of 1 is content that site directly on the page. z index values greater than 1 refer to layers that are "coming out of the screen." z index values less than 1 are used for layers that are further behind the screen. In the example above, the hidden text is put in a layer with a z index value of 1. The visible text appears in a layer with a z index value of 2. Thus, due to the positioning, the second layer is aligned directly over the first. This causes the text in the first layer to appear only to the search engines.

Google, for one, has begun removing content contained within hidden div tags from its index. It has explicitly banned sites that use small divisions for keyword stuffing purposes [26]. However, these policies may cause a problem for legitimate Web site developers who use hidden divisions for design purposes. For example, some developers use hidden CSS layers and z index positioning to implement mouse over multilevel menus (menus that expand when the user places their mouse over a menu item) and other interactive effects on their sites.

4.2.1.2 Keyword Stuffing Using Code.

Javascript is widely used by Web developers to add interactivity to their sites. One of the most popular uses of Javascript is to enable cascading menus. However, some site visitors might turn off Javascript in their browser settings. Therefore, good Javascript coding specifies that the developer should include content in the NoScript tag. The NoScript tag provides alternative navigation or text for those who have Javascript turned off.

NoScript Example:

```
<script type="text/javascript">
[Javascript menu or other code goes here]
</script>
<noscript>Toyota Prius, Prius, Prius Car, Prius Hybrid
</noscript>
```

While NoScript has legitimate uses, it can also be abused by black hat optimizers. Many use the NoScript tag for keyword stuffing. There is some good anecdotal evidence to suggest that this tactic has helped some sites achieve a high ranking for competitive terms [27].

4.3 Cloaking

As we have already seen, some of the tricks used in black hat SEO are not conducive to a good visitor experience. Cloaking overcomes this problem. The main goal of cloaking is to provide different content to the search engines and to human visitors. Since users will not see a cloaked page, it can contain only optimized text—no design elements are needed. So the black hat optimizer will set up a normal Web site and individual, text only, pages for the search engines. The Internet protocol (IP) addresses of the search engine spiders are well known. This allows the optimizer to include simple code on the page that serves the appropriate content to either the spider or human (see Fig. 8).

Some black hat optimizers are taking the cloaking concept to the next level and using it to optimize for each individual search engine. Since each search engine uses a different algorithm, cloaking allows optimizers to serve specific content to each different spider.

Since some types of cloaking actually may provide benefits to users, the concept of cloaking and what is, and is not, acceptable by the search engines has evolved

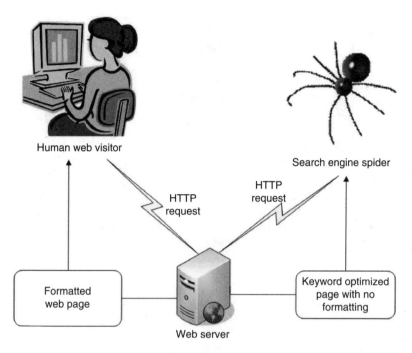

FIG. 8. Cloaking.

over the past few years. One topic of much debate is the concept of geolocation. Geolocation uses a visitor's IP address to determine their physical location and changes the site's content accordingly. For instance, a site that sells baseball memorabilia might use geolocation to direct people who live in the New York City area to a Yankees page and those who live in the Boston area to a Red Sox page.

Clearly, geolocation allows site developers to provide more highly targeted content. The main question is if the site serves different content to the search engines than to most users, is it still considered cloaking? Maile Ohye [28] posting on the Google Webmaster Central Blog chimed in on the controversy. According to Ohye as long as the site treats the Google spider the same way as a visitor, by serving content that is appropriate for the spider's IP location, the site will not incur a penalty for cloaking.

4.4 Doorway Pages

The goal of doorway pages is to achieve high rankings for multiple keywords or terms. The optimizer will create a separate page for each keyword or term. Some optimizers use hundreds of these pages. Doorway pages typically use a fast meta

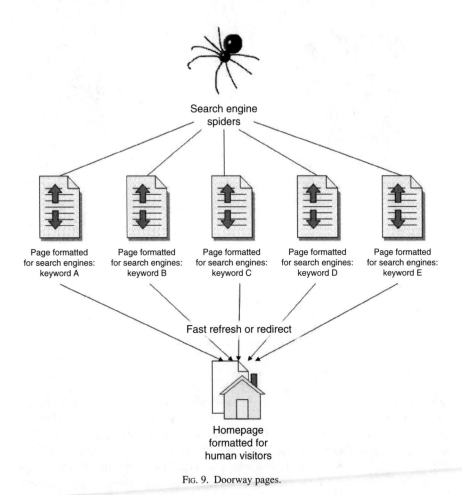

FIG. 9. Doorway pages.

refresh to redirect users to the main page (see Fig. 9). A meta refresh is an HTML command that automatically switches users to another page after a specified period of time. Meta refresh is typically used on out of date Web pages—you often see pages that state you will be taken to the new page after 5 s. A fast meta refresh occurs almost instantly, so the user is not aware of it. All of the major search engines now remove pages that contain meta refresh. Of course, the black hats have fought back with a variety of other techniques, including the use of Javascript for redirects. This is the specific technique that caused Google to ban bmw.de and ricoh.de.

4.5 Content Generation

Black hat optimizers make extensive use of tools that allow them to automatically generate thousands of Web pages very quickly. The ability to quickly create new sites is extremely important for black hats since their sites are usually targeted at low competition terms and are also likely to be banned by the search engines. Site generators pull relevant content from existing sites based on a keyword list. This content includes not only text, but also videos and images.

As mentioned previously, unique content appears to be an important element in many of the search engine algorithms. In fact, Google will penalize a site that includes only duplicate content. It should be noted that some duplicate content on a site is generally okay, especially now that many sites syndicate their content via RSS. As long as a site has some unique content, it will usually not receive a penalty from Google. The best site generators have the ability to randomize the content so that it appears unique.

To understand how site generators work, let us use the "Yet Another Content Generator" (YACG) site generation tool (this is an open source tool that can be found at getyacg.com). YACG is written in PHP and is highly flexible and customizable. After downloading YACG and installing it on a Web server, the optimizer simply uploads a list of keywords. The longer the list of keywords the larger the site. For example, this author used a list of 40 weight loss terms. Next, some basic setup information, such as the domain name, contact information, and some API keys (needed to pull in content from certain sites), was entered. Within a matter of minutes YACG has created a site with over 189 content pages, as seen in Fig. 10. The tool also includes a sitemap and RSS feed based on the pages generated.

YACG has separated the look and feel of the site (template system) from the content generation (the hook system). YACG templates use PHP and HTML and provide the optimizer with complete control over site design. In fact, a paid version of YACG called the YACG Mass Installer will actually grab the design of any site the optimizer chooses to use as a template for generated sites.

The YACG hook system consists of scripts that pull in content from various sites. For example, hooks for Wikipedia, Flickr, and YouTube are available in the basic open source version. However, since YACG hooks are simple to code in PHP optimizers can add hooks for virtually any site.

4.6 Link Building Black Hat Techniques

All of the major search engines consider the number and quality of incoming links to a site as part of their algorithm. Links are especially important for ranking well on Google. Therefore, black hat optimizers use a variety of methods to increase their site's backlinks (links from other sites).

Fig. 10. YACG-generated site.

4.6.1 Guestbook Spamming

One of the simplest black hat linking techniques is guest book spamming. Optimizers simply look for guest book programs running on authority (usually .edu or .gov) sites. Links from authority sites are given more weight than links from ordinary sites. Black hats then simply add a new entry with their link in the comments area. This is becoming more difficult for black hats as guest books have decreased in popularity and those that do exist may use a nofollow tag to deter this behavior.

4.6.2 Blog Spamming

While guest books may have decreased in popularity over the past few years, the popularity of blogs (Weblogs) have certainly increased. Black hat optimizers have obviously noticed this trend and have developed techniques to take advantage of them. Two ways that black hats use blogs to generate incoming links are blog comment spamming and trackback spamming.

Blog comment spamming is similar to guest book spamming, in that the optimizer leaves backlinks in the comments section of publicly available blogs. However, many blogging systems have added features to handle comment spam. At the simplest level, the blog owner can require that all comments receive his or her approval before they appear on the site. Another simple technique is to use nofollow tags on all comments.

Many blogging systems now have a simple plugin that will require the commenter to complete a "Completely Automated Public Turing test to tell Computers and Humans Apart" (CAPTCHA). A CAPTCHA system presents a visitor with an obscured word, words, or phrase. The obscuring is usually achieved by warping the words, distorting the background, or segmenting the word by adding lines. While humans can see through the obscuring technique, computers cannot. Since computers cannot solve a CAPTCHA, systems that use it are usually not vulnerable to automated spamming software. However, some CAPTCHA systems have been hacked, allowing black hat optimizers to successfully bypass this countermeasure on certain sites.

A trackback is a link between blogs that is used when one blog refers to comment on another blog. When this occurs, the original blog will generate a link to the blog that made the comment. For example, legitimate blog A makes a post on a blog that uses trackbacks. A black hat optimizer then "comments" on the post with his backlink. The optimizer then sends blog a trackback ping (just a signal that indicates that the optimizer had something to say about a post on blog A). Blog A will then automatically display a summary of the comments provided by the black hat and a link to the black hat's site. Trackbacks bypass many of the methods that can be used to handle comment spam. For this reason many blog systems no longer use trackbacks.

Most serious black hat optimizers use scripts or software to find blogs that have comments sections that do not require approval, do not have CAPTCHA, and do follow. Once the system builds a large list of the appropriate type of blogs it then submits the backlink to all of them. Similar systems are available that can find blogs that have trackbacks enabled and implement trackback spamming. Using blog spamming systems a black hat optimizer can easily generate thousands, even tens of thousands, of quick backlinks.

It should be noted that many white hat optimizers also use scripts or software to automate their own blog comment campaigns. However, instead of taking a scatter shot approach, as the black hats do, white hat optimizers submit useful comments to relevant blogs. They use the automated tools to find those blogs and manage their comment campaigns.

4.6.3 Forum Spamming

The underlying concepts involved with guest book and blog spamming has also been implemented in online forums. Many forums allow users to post links—either in the body of their forum post or in their signature lines. Black hats have developed automated tools that find open forums, develop somewhat relevant posts, and insert the posts with a link back to the target site.

4.6.4 Stats Page Spamming

Stats page spamming, also known as referrer spam, is a bit more complicated than the other types of link spam discussed above. Some Web server statistics packages, such as AWStats, publish their statistics, including their referrer lists, publicly on the domain on which they run. A black hat can repeatedly make a Web site request using the target site as the referrer (requesting site). As shown in Fig. 11, when the

#	Hits		Referrer
	Top 86 of 86 Total Referrers		
1	1196	37.27%	- (Direct Request)
2	1038	32.35%	http://www.easymusclegain.com/
3	234	7.29%	http://www.easymusclegain.com/index.php
4	128	3.99%	http://easymusclegain.com/
5	95	2.96%	http://search.yahoo.com/search
6	83	2.59%	http://hubpages.com/hub/How-To-Gain-Muscle-Mass
7	63	1.96%	http://www.google.com/search
8	48	1.50%	http://www.easymusclegain.com/musclegainarticles.htm
9	39	1.22%	http://www.easymusclegain.com/musclebuildingarticles/determine_one_rep_max.htm
10	24	0.75%	http://hubpages.com/hub/The-Best-Muscle-Building--Weight-Loss-Programs
11	24	0.75%	http://www.keywordspy.com/find-competitor-keywords/easymusclegain.com
12	20	0.62%	http://www.ask.com/web
13	18	0.56%	http://www.easymusclegain.com
14	16	0.50%	http://www.easymusclegain.com/rnr.php
15	12	0.37%	http://search.msn.com/results.aspx
16	11	0.34%	http://search.live.com/results.aspx
17	11	0.34%	http://www.easymusclegain.com/About_Us.html
18	10	0.31%	http://blog.360.yahoo.com/blog-PPfUhH4SbrUX_EqoV0_m4LSOLAeH
19	8	0.25%	http://www.easymusclegain.com/musclebuildingarticles/deadlift.htm
20	6	0.19%	http://cc.msnscache.com/cache.aspx
21	6	0.19%	http://www.google.co.uk/search
22	6	0.19%	http://www.google.com.au/search
23	4	0.12%	http://bwp.findarticles.com/search
24	4	0.12%	http://cache.search.yahoo-ht2.akadns.net/search/cache
25	4	0.12%	http://www.bigfinder.de/index.php
26	4	0.12%	http://www.easymusclegain.com/index.htm
27	4	0.12%	http://www.google.com/custom

FIG. 11. Referrer statistics example.

statistics package publishes the referrer list the target site will appear as a backlink. Like the other types of link spam discussed, there are scripts and software available that will help the black hat find appropriate sites (those that publish referrer links) and continuously request Web pages from them so that the referrer appears on the top referrers list.

Clearly, a simple way around this type of link spam is to set stats packages so they do not publish their results in a publicly accessible area. Site owners can also ban specific IP addresses (or address ranges) or ignore requests from certain referrers. However, these measures typically require at least some level of technical expertise, which may be beyond most site owners.

4.6.5 Link Farms

Unlike link spamming, a link farm is a set of sites setup in order to link to each other. Like most black hat sites many link farms are developed using automated tools. Link farms usually work by including simple code on a links page. This code generates a huge list of links from a database. All sites that include the code are listed in the database. So, a site owner who includes the code on his page will have a link to his site appear on every other site in the link farm.

As the search engines caught on to link farming, they began to ban sites that participated in these schemes. Of course, the black hat optimizers started using more subtle techniques. For example, instead of linking to every site in the database some link farms will only present relevant links, which makes the outgoing links page look more natural. Some will also randomize the links, so that every site does not have the same list of links—which is fairly easy for the search engines to spot.

There is a fine line between legitimate link exchange management systems and link farms. Link exchange management systems help optimizers (both white hat and black hat) find sites they might want to link with. The systems will automate the sending of a link request. These systems also check to see that the other site still contains the optimizer's link and will remove the reciprocal link if necessary.

4.6.6 Paid Links

Realizing the value of good links, many entrepreneurs started selling links from their sites. In fact, some set up companies to act as link brokers, while others became link aggregators. The aggregators would buy up links from various sites and resell them as a package. There are no reliable statistics on the size of the link selling industry; however, in its heyday (around 2004–2005) it was not uncommon for links from high PR sites to sell for hundreds or even thousands of dollars per month.

Since Google's algorithm gives links a great deal of weight, the company quickly caught on to these schemes and began to take action in 2005 [29]. Google appears to have two ways it deals with paid links. First, its algorithm looks for paid links and will penalize sites that sell them and those that purchase them. It attempts to find paid links in a number of ways. It can, for example, look for links that follow text like "sponsored links" or "paid links." It also looks for links that appear out of place from a contextual perspective. For example, a site about computers that contains links to casino sites might be penalized. Second, Webmasters can now report sites they believe are involved in buying or selling links. These sites then undergo a manual review.

The penalties for buying or selling links appear to vary. At the low end of the penalty range, Google might simply discount or remove paid links when determining a site's PR. Sites that sell links may lose their ability to flow PR to other sites. Finally, Google can ban repeat or egregious violators from its index.

Of course, the companies that sell links have attempted to develop techniques aimed at fooling Google. For example, some companies now sell links that appear within the text of a site. These links appear natural to Google. In addition, some companies specialize in placing full site or product reviews. Some bloggers, for example, will provide a positive site or product review, with associated links, for a fee.

4.6.7 HTML Injection

One popular off-site black hat method is HTML injection, which allows optimizers to insert a link in search programs that run on another site. This is achieved by sending the search program a query which contains special HTML characters. These characters cause the insertion of data specified by the user into the site. For example, WebGlimpse is Web site search program widely used on academia and government Web sites. The Stanford Encyclopedia of Philosophy Web site located at plato.stanford.edu, which is considered an authority site, uses the WebGlimpse package. So an optimizer that would like a link from this authority site could simply navigate to http://plato.stanford.edu/cgi-bin/webglimpse.cgi?nonascii=on&query=%22%3E%3Ca+href%3Dhttp%3A%2F%2F##site##%3E##word##%3C%2Fa%3E&rankby=DEFAULT&errors=0&maxfiles=50&maxlines=30&maxchars=10000&ID=1. The optimizer then replaces ##site## with the target site's URL and ##word## with the anchor text.

4.7 Negative SEO

One of the most insidious black hat methods is manipulating competitors' search engine results. The result is a search engine penalty or outright ban (a term black hatters call bowling). The incentive for this type of behavior is fairly obvious.

If a black hat site is ranked third for a key term, the optimizer who can get the top two sites banned will be ranked first.

There are a number of techniques that can be used for bowling. For instance, the HTML injection approach discussed above can be used to change the content that appears on a competitor's site. If a black hat optimizer is targeting a site that sells computers, for example, the HTML injected might be <H1>computer, computer, computer... The extensive use of keywords over and over again is almost guaranteed to lead to a penalty or outright ban in all the major search engines.

A recent article in Forbes [30] discussed the tactic of getting thousands of quick links to a site a black hat wants to bowl. Quickly piling up incoming links is viewed in a negative light by most search engines.

Since this type of SEO is ethically questionable, most optimizers who conduct such campaigns are required to sign nondisclosure agreements. Therefore, uncovering actual cases where this approach has worked (produced a ban) is extremely difficult. However, a competitor in a 2006 SEO competition believes he was inadvertently bowled [31]. The goal of the competition was to achieve the best ranking for a made-up term. The winner received $7000. One competitor offered to donate any winnings to Celiac Disease Research. So many people began linking to the site that it quickly started ranking well (typically in the top five on the major search engines). However, over a period of a few months, as the number of incoming links kept increasing, the site began to lose rank in Google (while maintaining rank on Yahoo and MSN). The site was never completed bowled and eventually regained much of it rank.

According to the Forbes article, Google's Matt Cutts states, "We try to be mindful of when a technique can be abused and make our algorithm robust against it. I won't go out on a limb and say it's impossible. But Google bowling is much more inviting as an idea than it is in practice."

5. Legal and Ethical Considerations

Obviously, the field of SEO raises important legal and ethical considerations. The main legal concern is copyright infringement. Ethical considerations are more complex as there are currently no standards or guidelines in the industry.

5.1 Copyright Issues

The main legal issues associated with SEO (particularly black hat) involve intellectual property—primarily copyright, but also trademark. Automated content generators are of particular concern in this area as they are designed to steal content from various sites and, in some cases, even the look and feel of a particular site.

The use of content from another site without attribution is clearly a violation of copyright law. However, on the Web copyright issues can become somewhat complex. For example, Wikipedia (www.wikipedia.org) use the GNU Free Documentation License (GFDL), which explicitly states, "Wikipedia content can be copied, modified, and redistributed if and only if the copied version is made available on the same terms to others and acknowledgment of the authors of the Wikipedia article used is included (a link back to the article is generally thought to satisfy the attribution requirement)." So simply including content from Wikipedia on a site would not constitute a copyright violation as long as a link back to the source article was included.

YouTube.com is another interesting example of the complexities of Web copyright. The site has very strict guidelines and enforcement mechanisms to prevent users from uploading copyrighted material without appropriate permissions. However, when a user submits their own videos to YouTube they, "grant YouTube a worldwide, nonexclusive, royalty-free, sublicenseable and transferable license to use, reproduce, distribute, prepare derivative works of, display, and perform the User Submissions in connection with the YouTube Web site... You also hereby grant each user of the YouTube Web site a nonexclusive license to access your User Submissions through the Web site, and to use, reproduce, distribute, display, and perform such User Submissions as permitted through the functionality of the Web site and under these Terms of Service (from http://www.youtube.com/t/terms?hl=en_US)." Since YouTube makes it very easy to include videos on a Web site (via the embed feature), it appears that doing so does not violate copyright.

So, at what point does the optimizer (either white or black hat) cross the line and become a copyright violator? If, for example, you read this chapter and then write a summary of it in your own words it is not a violation of copyright. Does this assessment change if instead of writing the summary yourself, you write a computer program that strips out the first two sentences of each paragraph in order to automatically generate the summary? What if you take those sentences and mix them with others extracted from a dozen other articles on SEO? This is essentially what some of the content generators do.

From a practical perspective, black hats can actually use copyright law as a weapon in their arsenal. All of the major search engines provide a mechanism for making a complaint under the Digital Millennium Copyright Act (DMCA). Some of the search engines will remove a site from its index as soon as a DMCA complaint is received. In addition, sites that contain backlinks, such as YouTube, will also remove the content when a complaint is received. All of these sites will attempt to contact the "infringing" site and allow it to provide a counter-notification—basically a defense against the complaint. There is at least one anecdotal report (see http://www.ibrian.co.uk/26-06-2005/dmca-the-new-blackhat-for-yahoo-search/) that this approach resulted in the temporary removal of a site from Yahoo.

It should be noted that submitting a false DMCA complaint is not without serious risks and potential consequences. As stated on the Google DMCA site (http://www.google.com/dmca.html), "Please note that you will be liable for damages (including costs and attorneys' fees) if you materially misrepresent that a product or activity is infringing your copyrights. Indeed, in a recent case (please see http://www.onlinepolicy.org/action/legpolicy/opg_v_diebold/ for more information), a company that sent an infringement notification seeking removal of online materials that were protected by the fair use doctrine was ordered to pay such costs and attorneys' fees. The company agreed to pay over $100,000."

5.2 SEO Ethics

According to Merriam–Webster's dictionary (http://www.merriam-webster.com/dictionary/ethics) ethics are, "the principles of conduct governing an individual or group." In some industries, these principles are determined by a governing body. However, there is no governing body in the field of SEO. The SEMPO is a voluntary industry association. However, even SEMPO has not put forth a set of ethical principles for SEO professionals.

As a practical matter, Google's Webmaster guidelines currently serve as a *de facto* set of standards in the SEO industry. Violating those guidelines generally puts an optimizer in the black hat camp.

One of the major problems in determining ethical behavior in the SEO industry is that some tools and techniques can be used for either white or black hat SEO. For example, if done in moderation and with good content the blog and ping indexing approach is generally considered white hat. However, when it is abused the technique becomes black hat.

5.3 Search Engine Legal and Ethical Considerations

As search engines, especially Google, have become the starting point for most commercial activity on the Web, it is appropriate to ask if they are acting in an ethical and legal manner. For instance, can a search engine selectively increase or decrease the organic results for certain sites based on commercial, political, or other interests. Clearly, the major search engines have censored their search results in various countries—most notably China. Could a search engine raise the search rank of its business partners or certain politicians? Since the search algorithms are a trade secret it would be very difficult to determine if such a move was made intentionally or as a natural result of the algorithm.

Currently most of the revenue generated by search engines comes from PPC ads which are clearly labeled as "sponsored links." However, could a search engine ethically and legally "sell" the top spots in its organic results without labeling them as sponsored or ads? In fact, in the early days of PPC advertising (prior to 2002) most search engines did not readily distinguish between paid and nonpaid listings in the search results. This led to a June 27, 2002 U.S. Federal Trade Commission (FTC) recommendation to the search engines that "any paid ranking search results are distinguished from nonpaid results with clear and conspicuous disclosures" [32]. The major search engines have complied with this recommendation and it is likely that any search engine that "sells" organic results without disclosing they are paid will incur a FTC investigation.

Finally, do the search engines have any legal or ethical obligations when it comes to its users? Search engines have the ability to gather a tremendous amount of data about a person based on his or her search patterns. If that user has an account on the search engine (usually for ancillary services like email) then these data can become personally identifiable. What the search engines and users must be aware of is that as global entities, the data collected by search engines may fall under various jurisdictions. For example, in 2004 Yahoo! Holdings (Hong Kong) in response to a subpoena from Chinese authorities provided the IP address and other information about dissident Shi Tao. Based on the information provided, Tao was sentenced to 10 years in prison for sending foreign Web sites the text of a message warning journalists about certain actions pertaining to the anniversary of the Tiananmen Square massacre [33]. While Yahoo! acted according to local laws, many would claim that they acted unethically. In fact, Yahoo! executives were called before a Congressional committee to testify about the ethics of its actions.

6. Conclusions

The growth in the number of Web searches, along with more online purchases, and the ability to precisely track what site visitors are doing (and where they have come from) has led to explosive growth in the SEM industry. This growth is expected to continue at around a 13% annual rate over the next few years, this is opposed to a 4% growth rate for offline advertising [34]. Given this growth and the profit incentives involved it is no wonder that some people and companies are looking for tools and techniques to provide them with an advantage.

SEO is a constantly changing field. Not only are the major search engines continually evolving their algorithms, but also others are entering (and frequently leaving) the Web search space. For example, while writing this chapter Microsoft launched its latest search engine—Bing. The press has also given much attention to Wolfram Alpha and Twitter's search capability. By the time you read this we should have a better idea if these new initiatives have been a success or a flop. In either case SEO practitioners and researchers need to keep abreast of the most recent developments in this field.

6.1 Conclusions for Site Owners and SEO Practitioners

The implications of black hat SEO are particularly important for anyone who owns a Web site or anyone who is involved in the online marketing industry. Web site owners need to be aware of black hat SEOs on two levels. First, they need to understand if the competition is using black hat approaches and if they are achieving success with them as this might impact the prospects of the owner's site. Second, the site owner who might want to hire a search engine optimizer needs to be aware of the techniques the person or firm might use. Some SEO companies use black hat techniques to get quick results, and payment, from their clients. This was the case with a company called Traffic Power. In 2005, Traffic Powers' sites and many of its clients' sites were banned by Google for using black hat SEO approaches [25].

Online marketers also need to understand and keep up with the most current SEO methods and tools, including the black hat approaches. As was mentioned above, many black hat methods work well for white hat SEO when done in moderation. In addition, many black hat tools may prove useful to white hat SEOs.

Everyone involved in SEO must realize that today's "white hat" SEO methods might become "black hat" in the future as the search engines change their algorithms and policies. We have already seen this occur with the case of purchased links.

Finally, white hat SEOs need to understand when the competition is using black hat techniques. The major search engines have processes in place to report black hat sites. For instance, Google's Webmaster Tools site allows Webmasters to "Report spam in our index" and "Report paid links."

6.2 Future Research Directions

While SEO has been around since the mid-1990s, academic researchers have only recently begun to take an interest in this area. Therefore, there are numerous directions for future research in both black hat and white hat SEO. In general, the area of SEO research can be broken into three main categories.

First, we need to have a better understanding of which techniques produce the best results. This is true for both white hat and black hat approaches. For instance, it is not clear exactly which SEO techniques and the scope of those techniques will result in a penalty or outright ban in the search engines.

As the search engine industry evolves researchers should try to keep pace. For example, Google has recently begun to provide integrated search results. As shown in Fig. 12, instead of just showing a list of Web sites, Google now provides products, video, and images that match the query. It is unclear exactly how these should be handled from a SEO perspective.

Second, researchers can play a very important role in helping the search engines improve their algorithms and develop other measures to deal with black hat SEO. Much of the original research and development of many of the major search engines comes directly from academia. For example, Google began as a Ph.D. research project at Stanford University and the university provided the original servers for the search engine. Ask.com is a search engine that started out as Teoma. The underlying algorithm for Teoma was developed by professors at Rutgers University in New Jersey.

While more work needs to be done, some researchers have already begun conducting research into preventing black hat SEO methods. Krishnan and Raj [35] use a seed set of black hat (spam) pages and then follow the links coming into those pages. The underlying concept is that good pages are unlikely to link to spam pages.

FIG. 12. Integrated Google search results.

Thus by following back from the spam pages, as process they term antitrust, they find pages that can be removed from or penalized by the search engines. Kimuara et al. [36] developed a technique based on network theory aimed at detecting black hat trackback blog links.

Third, academic researchers can help in understanding how visitors and potential customers use search engines. As mentioned above, Microsoft has taken a first step in this direction with its research into Online Commercial Intention (OCI). However, the current OCI research is based on human interpretation of search terms. A promising area for future research would be to analyze what search engine visitors actually do after performing certain queries.

Another interesting research area is understanding the demographic and behavioral characteristics of the people who visit each search engine and how these impact online purchases. There is some anecdotal evidence to suggest that the major search engines do perform differently in terms of conversion rate—at least for paid search [37]. Taking this concept a step further, an optimizer might decide to focus a SEO campaign on a certain search engine based on a match between demographics, the product or service offered, and the conversion rate.

REFERENCES

[1] SEMPO, The State of Search Engine Marketing 2008, 2008. Retrieved May 1, 2009, from http://www.sempo.org/lcarning_center/research/2008_execsummary.pdf.
[2] R. Sen, Optimal search engine marketing strategy, Int. J. Electron. Comm. 10 (1) (2005) 9–25.
[3] iProspect, iProspect Search Engine User Behavior Study, 2006. Retrieved June 15, 2009, from http://www.iprospect.com/premiumPDFs/WhitePaper_2006_SearchEngineUserBehavior.pdf.
[4] L.A. Granka, T. Joachims, G. Gay, Eye-tracking analysis of user behavior in WWW search, in: Proceedings of the 27th Annual International ACM SIGIR Conference on Research and Development in Information Retrieval, Sheffield, United Kingdom, July 25–29, 2004. Retrieved June 15, 2009, from http://www.cs.cornell.edu/People/tj/publications/granka_etal_04a.pdf.
[5] R. Bauer, SEO Services Comparison & Selection Guide, 2008. Retrieved June 10, 2009, from http://www.scribd.com/doc/2405746/SEO-Pricing-Comparison-Guide.
[6] R. Malaga, The value of search engine optimization: an action research project at a new e-commerce site, J. Electron. Comm. Organ. 5 (3) (2007) 68–82.
[7] R. Malaga, Worst practices in search engine optimization, Commun. ACM 51 (12) (2008) 147–150.
[8] M.S. Raisinghani, Future trends in search engines, J. Electron. Comm. Organ. 3 (3) (2005) i–vii.
[9] J. Zhang, A. Dimitroff, The impact of metadata implementation on webpage visibility in search engine results (Part II), Inform. Process. Manage. 41 (2005) 691–715.
[10] E. Burns, U.S. Core Search Rankings, February 2008, 2008. Retrieved May 1, 2009, from http://searchenginewatch.com/showPage.html?page=3628837.
[11] K. Curran, Tips for achieving high positioning in the results pages of the major search engines, Inform. Technol. J. 3 (2) (2004) 202–205.
[12] D. Sullivan, How Search Engines Rank Web Pages, 2003. Retrieved June 15, 2009, from http://searchenginewatch.com/webmasters/article.php/2167961.

[13] H.K. Dai, L. Zhao, Z. Nie, J. Wen, L. Wang, Y. Li, Detecting Online Commercial Intention (OCI), in: Proceedings of the 15th International Conference on the World Wide Web, Edinburgh, Scotland, 2006, pp. 829–837.
[14] T. O'Reilly, What Is Web 2.0—Design Patterns and Business Models for the Next Generation of Software, 2005. Retrieved May 1, 2009, from http://www.oreillynet.com/pub/a/oreilly/tim/news/2005/09/30/what-is-web-20.html.
[15] Technorati, Welcome to Technorati, 2008. Retrieved May 1, 2009, from http://technorati.com/about/.
[16] Wikipedia, Social Bookmarking, 2008. Retrieved May 1, 2009, from http://en.wikipedia.org/wiki/Social_bookmarking.
[17] R. Malaga, Web 2.0 techniques for search engine optimization—two case studies, Rev. Bus. Res. 9 (1) 2009.
[18] J. Zhang, A. Dimitroff, The impact of webpage content characteristics on webpage visibility in search engine results (Part I), Inform. Process. Manage. 41 (2005) 665–690.
[19] A. Beal, SMX: Cutts on Themes and Latent Semantic Indexing, 2007. Retrieved June 10, 2009, from http://www.webpronews.com/blogtalk/2007/06/11/smx-cutts-on-themes-and-latent-semantic-indexing.
[20] S. Deerwester, S. Dumais, G. Furnas, T. Landauer, R. Harshman, Indexing by latent semantic analysis, J. Am. Soc. Inform. Sci. 1 (6) (1990) 291–407.
[21] Practical Ecommerce, Importance of New, Fresh Content for SEO, 2009. Retrieved June 10, 2009, from http://www.practicalecommerce.com/podcasts/episode/803-Importance-Of-New-Fresh-Content-For-SEO.
[22] D. Rowse, How Much Does Fresh Content Matter in SEO? 2007. Retrieved June 10, 2009, from http://www.problogger.net/archives/2007/05/19/how-much-does-fresh-content-matter-in-seo/.
[23] A. K'necht, SEO and Your Web Site—Digital Web Magazine, 2004. Retrieved June 10, 2009, from http://www.digital-web.com/articles/seo_and_your_web_site/.
[24] R. Nobles, How Important Is ALT Text in Search Engine Optimization? 2005. Retrieved June 10, 2009, from http://www.webpronews.com/topnews/2005/08/15/how-important-is-alt-text-in-search-engine-optimization.
[25] M. Cutts, Confirming a Penalty, 2006. Retrieved June 10, 2009, from http://www.mattcutts.com/blog/confirming-a-penalty/.
[26] M. Cutts, SEO Tip: Avoid Keyword Stuffing, 2007. Retrieved June 10, 2009, from http://www.mattcutts.com/blog/avoid-keyword-stuffing/.
[27] S. Spencer, Bidvertiser SO Does Not Belong in Google's Top 10 for "marketing", 2007. Retrieved June 10, 2009, from http://www.stephanspencer.com/tag/noscript.
[28] M. Ohye, How Google Defines IP Delivery, Geolocation, and Cloaking, 2008. Retrieved June 10, 2009, from http://googlewebmastercentral.blogspot.com/2008/06/how-google-defines-ip-delivery.html.
[29] M. Cutts, How to Report Paid Links, 2007. Retrieved June 10, 2009, from http://www.mattcutts.com/blog/how-to-report-paid-links/.
[30] A. Greenberg, The Saboteurs of Search, 2007. Retrieved June 10, 2009, from http://www.forbes.com/2007/06/28/negative-search-google-tech-ebiz-cx_ag_0628seo.html.
[31] Anonymous, Google Bowling, 2006. Retrieved June 10, 2009, from http://www.watching-paint-dry.com/v7ndotcom-elursrebmem/google-bowling/.
[32] H. Hippsley, Letter to Mr. Gary Ruskin, Executive Director, Commercial Alert, 2002. Retrieved September 3, 2009, from http://www.ftc.gov/os/closings/staff/commercialalertletter.shtm.
[33] BBC, Yahoo 'helped jail China writer', 2007. Retrieved September 4, 2009, from http://news.bbc.co.uk/2/hi/asia-pacific/4221538.stm.

[34] J. Kerstetter, Online Ad Spending Should Grow 20 Percent in 2008, 2008. Retrieved June 10, 2009, from http://news.cnet.com/8301-1023_3-9980927-93.html.
[35] V. Krishnan, R. Raj, Web spam detection with anti-trust rank, in: 2nd Workshop on Adversarial Information Retrieval on the Web, Seattle, WA, August 2006. Retrieved June 10, 2009, from http://i.stanford.edu/~kvijay/krishnan-raj-airweb06.pdf.
[36] M. Kimuara, S. Kazumi, K. Kazuhiro, S. Sato, Detecting Search Engine Spam from a Trackback Network in Blogspace, in: Lecture Notes in Computer Science, Springer, Berlin, 2005, p. 723.
[37] D.J. Kennedy, Google, Yahoo! or MSN—Who Has the Best Cost per Conversion—A Study, 2008. Retrieved June 10, 2009, from http://risetothetop.techwyse.com/pay-per-click-marketing/google-yahoo-or-msn-who-has-the-best-cost-per-conversion-a-study/.

Web Searching and Browsing: A Multilingual Perspective

WINGYAN CHUNG

Department of Operations and Management Information Systems, Leavey School of Business, Santa Clara University, Santa Clara, California, USA

Abstract

Since the publication of "The World Wide Web" in a 1999 volume of *Advances in Computers*, worldwide Internet usage has grown tremendously, with the most rapid growth in some non-English-speaking regions. A widening gap exists between the surging demand for non-English Web content and the availability of non-English resources. This chapter reviews previous works on computer-mediated information seeking on the Web, computing technologies that support Web searching and browsing, and Web portals in several major regions and languages of the world. We introduce a general framework for supporting Web searching and browsing in a multilingual world. Three Web portals were developed to support searching, browsing, and postretrieval analysis of Chinese business intelligence, Spanish business intelligence, and Arabic medical intelligence. Results of experiments involving 67 native speakers of the three languages confirm the usability and benefits of the portals and support the applicability of the framework. The review, framework, and findings presented in this chapter contribute to the fields of Web analysis, text mining, business and medical informatics, and human–computer interaction.

1. Introduction 42
2. Literature Review 43
   2.1. Information Seeking on the Web 43
   2.2. Technologies and Approaches to Support Web Searching and Browsing . . . 46
   2.3. Web Portals in a Multilingual World 51
3. A Multilingual Perspective 53

    3.1. Domain Analysis 54
    3.2. Collection Building and Metasearching 54
    3.3. Analysis Modules 56
    3.4. Web Directory Building 57
4. Case Studies 57
    4.1. CBizPort: Supporting Metasearching and Data
        Analysis of Chinese Business Web Pages 58
    4.2. SBizPort: Bridging Cross-regional Web Usage
        in the Spanish Business Domain 59
    4.3. AMedDir: Facilitating Web Browsing of Arabic Medical Resources 61
5. Summary and Future Directions 63
    5.1. Summary of Findings 65
    5.2. Limitations 65
    5.3. Future Directions 65
    Acknowledgments 66
    References 66

1. Introduction

The World Wide Web is characterized by its global nature and the use of different languages. It is estimated that 23.8% of the world's population uses the Internet [38]. Half or more of the population in some 47 countries use the Internet in a variety of languages. The most popular languages used on the Internet are English, Chinese, Spanish, Arabic, Japanese, French, Portuguese, German, Russian, and Korean. Despite these many languages, computers were originally designed to facilitate input and processing using English. Because interconnected computers form the backbone of the Internet, users must rely primarily on English to access the Internet.

Since the publication of an article entitled "The World Wide Web" in 1999 describing the technologies, social issues, and Web behavior [2], there has been a widening gap between the surging demand for non-English Web content and the availability of non-English resources. According to a report published in 2009, Internet usage has grown tremendously during 2000–2008, with the most rapid growths in non-English-speaking regions [38]. For example, the numbers of Internet users in the Middle East and in Latin America have soared by 1296.2% and 860.9%, respectively. China has more than 298 million Internet users, which has been growing faster than the number of available IP addresses. The number of Web sites in China had reached 2.878 million, surging by 91.4% between 2007 and

2008 [16]. Fueled by the large user base, Arabic Web content is estimated to double every year. This growth contributes to the soaring demands for better Web searching and browsing in many different languages other than English [1]. There is a growing need for Web portal developers and researchers to meet the demands.

This chapter has three goals. First, it provides a review of previous work on computer-mediated information seeking on the Web, computing technologies that support Web searching and browsing, and Web search portals in several major regions and languages of the world. Second, this chapter describes a general framework for supporting Web searching and browsing in a multilingual world. Third, the chapter provides three case studies of supporting Web searching and browsing in a multilingual world. Overall, the chapter contributes to describing the state of the art in Web searching and browsing in a multilingual world and to informing practitioners, researchers, and educators on the relevant topics.

2. Literature Review

As the Web is increasingly used in non-English-speaking regions, the use of non-English languages has been gaining popularity. Online users rely on the Web to seek for information and to conduct commercial activities. Search and browse tools emerge to support these activities. Understanding the theoretical and technical aspects of Web searching and browsing in a multilingual world will enable researchers and developers to design better technologies and tools and to create better online experience for a multitude of online users around the world. Therefore, we review below the related issues, including information-seeking processes on the Web, technologies for Web searching and browsing, and Web portals and search engines in different languages.

2.1 Information Seeking on the Web

2.1.1 Information-Seeking Models

Information seeking in electronic environments (such as the Web) has been studied widely in previous work. Kuhlthau [26] developed a six-stage model of the information-seeking process that includes initiation, problem definition, source selection, formulating queries, examining results, and extracting useful information. Marchionini [33] describes a more detailed process that involves different cognitive and behavioral steps such as recognizing and accepting an information problem, defining and understanding the problem, choosing a search system, formulating a query, semantic, and action mapping, executing a search, examining results, making relevance judgments, extracting information, reading, scanning, listening, classifying, copying, storing information, and reflecting/iterating/stopping. Sutcliffe and Ennis [54] summarized succinctly four main activities in their process model of information searching: problem identification, need articulation, query formulation, and results evaluation. They considered "information searching" to be a range of behaviors from goal-directed information searching to more exploratory information browsing. In directed searching, the user first decomposes his goal into smaller problems, then expresses his needs as concepts and higher-level semantics, formulates queries using such supports as Boolean query languages and syntax-directed editors, and finally evaluates the results by serial search or systematic sampling. In exploratory browsing, the user first transforms his general information need into a problem. He then articulates his needs as search terms or hyperlinks that appear on the system interface, searches using those terms or explores hyperlinks using such browse supports as automatic summarization, clustering and visualization tools, and Web directories, and finally evaluates the results by scanning through them. Due to its comprehensiveness, Sutcliffe and Ennis's process model can be used as a cognitive framework for designing a framework for Web information seeking in a multilingual world. In addition, human information seeking has been described as a behavior that includes questions, dialogue, and social and cognitive situations, associated with a user's interaction with an information retrieval system [25, 27]. The information-seeking process involves user judgments, search tactics or moves, interactive feedback loops, and cycles [52]. Previous research has dealt with issues relating to user's cognitive structure [22] and factors affecting the user-intermediary interaction process [47]. However, relatively little research studying perception by information seekers has been done in the context of non-English Web searching.

2.1.2 Searching and Browsing

Searching and browsing on the Internet are two major information-seeking activities. While researchers have used the term "searching" to refer to a range of information-seeking behaviors from goal-directed searching to serendipity browsing (e.g., [54]), we distinguish "searching" from "browsing" in terms of the extent to which the information seeking is directed to a certain goal. As for the meaning of "searching," Cove and Walsh's [17] three-stage model involves three information-seeking tasks: directed searching, general browsing, and serendipity browsing in which "searching" has a higher degree of goal-directedness. Broder's [3] taxonomy of Web search distinguishes among informational, navigational, and transactional Web searches and "searching" refers largely to informational Web searches. From these works, searching is goal-directed information seeking that involves answering "yes/no" questions or giving a specific answer.

Apart from searching, Internet users frequently engage in browsing. Marchionini and Shneiderman [34] defined browsing as "an exploratory, information seeking strategy that depends upon serendipity." Chang and Rice [5] stated that browsing is a direct application of human perception to information seeking. Spence [50] defined "browse" as the registration of content into a human mental model. Having compared various definitions, Chung et al. [14] defined "browsing" as an exploratory information-seeking process characterized by the absence of planning, with a view to forming a mental model of the content being browsed.

2.1.3 Regional Impacts and Information Quality

Web searching and browsing involve the use of languages. As a language can be used in more than one place or country, regional impacts arise because cultural, social, and economic environments differ. For example, Spanish is widely used in Europe, North America, and South America where the economic environments differ vastly. Chinese is used in mainland China, Taiwan, Hong Kong, and a number of Asian countries where the cultures are very different. Spink et al. [51] compared the searching behaviors of a search engine (FAST) whose users are largely European with those of another search engine (Excite) whose users are largely American and found that FAST users input queries more frequently while Excite users focused more on e-commerce topics. These results suggest regional differences on the Web, arising from possible cultural and social differences. However, these studies focused only on query and topic differences and did not reveal differences in search engine effectiveness. Chinese is another non-English language that is gaining popularity on the Web. However, because Chinese is mainly used in three closely located geographic regions (mainland China, Hong Kong, and Taiwan), its regional impact is much less than that of Spanish, which is used across continents and widely separated geographic regions. Unfortunately, there has been little research on the cross-regional impacts of Spanish search engines, although evaluation of them should improve understanding of optimal design of search engines and portals.

Information quality, a multifaceted concept considered an important aspect of evaluating the quality of a Web site [32], has been explored in previous research [56]. To evaluate information quality, a set of 16 dimensions was developed [56] and tested by Pipino et al. [44]. These dimensions were for the most part used in evaluating the quality of information about organizations or companies, not the quality of information obtained from search engines. Although Marsico and Levialdi [36] have developed a Web site evaluation methodology that considers a site's information quality, their methodology was designed for evaluating general Web sites (e.g., travel information Web sites) and does not consider the special requirements of non-English Web searching.

2.2 Technologies and Approaches to Support Web Searching and Browsing

As the Internet evolves as a major information-seeking platform, prior works have developed technologies to support searching and browsing on the Web. Information retrieval system (most notably search engines) is a major technology used to support searching. Web directories are developed to support browsing. Semantic networks are developed to support searching and deeper understanding of Web content. A review of these technologies and approaches follows.

2.2.1 Web Searching Technologies

2.2.1.1 Components of Search Engines.
Typical search engines operate by collecting and indexing resources on the Internet. A gathering program, often called a Web spider or crawler, explores hyperlinked pages of the Web and fetches them to index. From these pages, an indexer extracts keywords, phrases, and other entities (such as hyperlinks) that are stored in a database or repository. Finally, a retrieval program takes a user query and retrieves a ranked list of relevant Web documents from its database.

Because different search engines have different methods of page collecting, indexing, and ranking, they may include systematic bias in their search results. Such bias has been termed "indexical bias" (bias in the selection of items) and has been detected in major search engines such as Google, Yahoo, and AltaVista [39]. Mowshowitz and Kawaguchi [39] concluded from their study that the only realistic way to counter the adverse effects of search engine bias is to use a number of alternative search engines.

2.2.1.2 Metasearching.
Instead of relying on one search engine, metasearching uses a number of search engines for Internet searching and has been shown to be a highly effective method of resource discovery and collection on the Web. By sending queries to multiple search engines and collating the set of top-ranked results from each of them, metasearch engines can greatly reduce bias in search results and improve coverage. Chen et al. [7] showed that integrating metasearching with textual clustering tools achieved high precision in searching the Web.

Some domain-specific search portals also have employed metasearching to search the Internet. MedTextus [60] and HelpfulMed [9] are two Internet search portals that serve medical professionals by providing searches of several popular and authoritative online databases of medical literature, including MedLine, Merck, ACP, NGC, and DARE. They fetch related documents returned by those databases and filter duplicates. NanoPort [6] is an Internet search portal that allows nanotechnology researchers and professionals to search online journals (e.g., Science, MIT Technology Review), literature databases (e.g., ScienceDirect, RaDiUS), and search engines (e.g., Scirus, NanoSpot, AltaVista) related to the field.

Many commercial metasearch engines also are available and provide analysis functions. Originated from Carnegie Mellon University and now operated by Vivisimo, Clusty (http://www.clusty.com/) searches Bing, Ask, Open Directory, Yahoo, DMOZ, etc., and automatically clusters the search results into groups [42]. Powered by InfoSpace, Inc. MetaCrawler (http://www.metacrawler.com/) and Excite (http://www.excite.com/) metasearches a variety of search engines, including Google, Yahoo!, Bing, and Ask.com. They also provide other services such as Web directory, multimedia search, and news search.

2.2.1.3 Postretrieval Analysis.

Postretrieval analysis provides added values regarding the Web pages returned by search engines. Previews and overviews of retrieved Web pages are important elements in postretrieval analysis. A preview is extracted from, and acts as a surrogate for, a single object of interest [20]. Grounded in the information retrieval field, document summarization techniques provide previews of individual Web pages in the form of summaries. Such different kinds of summaries as indicative summaries [19], query-biased summaries [55], and generic summaries [37] have been proposed.

An overview is constructed from and represents a collection of objects of interest [20]. Document categorization techniques have been used to provide overviews of retrieved Web pages. Chen et al. [10] proposed a self-organizing approach to categorizing and searching the Internet. A multilayered neural network clustering algorithm employing the Kohonen self-organizing feature map (SOM) [24] was used to categorize about 10,000 Internet homepages related to entertainment. Experimental results show that the category hierarchies created could serve to partition the Internet into subject-specific categories and improve keyword searching and browsing. In another research using 110,000 entertainment-related homepages as a testbed, the SOM category map was shown to benefit broad browsing tasks. An automatically generated thesaurus (called a concept space) also was shown to improve keyword-based searching tasks [8].

Apart from document categorization techniques, document visualization techniques also have been used to provide overviews of retrieved Web pages. To facilitate Internet browsing, Yang and Lee [59] compared fisheye and fractal views built upon a SOM category map and found that both techniques could significantly increase the effectiveness of visualizing the map. Reiterer et al. [45] proposed an information assistant called INSYDER for seeking business information on the Internet. The system provides four visualization views (result list, scatterplot, barchart, and tilebars) to facilitate browsing and it was found that result list achieved the highest average effectiveness and efficiency. To discover business intelligence (BI) on the Web, Chung et al. [14] proposed a system called "Business Intelligence Explorer (BIE)" that employed visualization techniques to help reduce information overload in business analysis. Two browsing methods—Web community and knowledge map—which were based on a genetic algorithm hierarchical display and a multidimensional scaling display, respectively, were developed in BIE and were shown to help users browse business information more effectively and efficiently than a result list display. Despite the potential advantages of metasearching and information previews and overview, they are rarely applied to developing non-English search engines.

2.2.2 Web Browsing Technologies

Web directories are developed to support information seeking on the Web via browsing. Previous work in developing Web directories falls into two categories: (1) extensive manual identification and categorization of Web resources and (2) automatic construction of directories using machine learning or Web mining techniques.

2.2.2.1 Manual Categorization.

Manual identification and categorization have been used in various domains, ranging from general search engines to domain-specific Web portals. The Open Directory Project, also known as Directory Mozilla (DMOZ) (http://dmoz.org), is constructed and maintained by a large, global community of volunteer editors. As of December 2009, DMOZ has over 84,626 human editors and lists more than 4,523,956 sites classified into over 590,000 categories. The rationale of DMOZ is to use extensive human work to combat growth of human-created Web resources, which often grow with the size of the online population. Another well-known human-created Web directories is the Yahoo! Directory (http://dir.yahoo.com/), which is built and maintained by a team of paid editors who organize Web sites into categories and subcategories. The hierarchy has 4–5 levels and contains 14 main categories (e.g., Arts & Humanities, Business & Economy, Computers & Internet, Education, etc.), each having around 20–45 subcategories. Apart from Yahoo! Directory, the Librarian's Index to the Internet (LII, http://lii.org/) provides a searchable, annotated subject directory of more than 12,000 Internet resources selected and evaluated by librarians for their usefulness to users of public libraries. Over 100 contributors from libraries in California and Washington State participate in building and maintaining the index. The process of building and updating the directory is facilitated by a Web-based system, through which human indexers can edit existing records, create new records, and preview the edited records [31]. In the biomedical domain, the UMLS Semantic network is one of three UMLS Knowledge Sources being developed by the United States National Library of Medicine (http://www.nlm.nih.gov/pubs/factsheets/umlssemn.html). Based on 134 semantic types and 54 links, the network provides a categorization of all concepts represented in the UMLS Metathesaurus and represents important relationships in the biomedical domain.

In the above four examples, the quality of the directories constructed depends highly on the human editors' domain knowledge, which usually varies from person to person. Two other problems are scalability and inability to meet the needs of less dominant communities on the Web.

2.2.2.2 Automated Approaches.

Other than manual methods, automatic approaches to constructing directory and ontology have been proposed in previous research. Sato and Sato [48] developed an automated editing system that generates a Web directory from a given category word without human intervention. Based on the category term, the system gathers instance names belonging to this category. For every instance name, it uses two search engines, goo and InfoSeek, to collect related Web pages and then removes duplicated contents. Then the system generates a Web directory organized according to geographic regions. Although the system is efficient for generating a directory from a category label and runs without human intervention, the generated directory has only one level that restricts its use in more complicated browse tasks. In another research, Chuang and Chien [11] propose a query-categorization approach to facilitate the construction of Web directories. Obtained from search engine log files, a total of 18,017 query terms were categorized by using a hierarchical agglomerative clustering algorithm into a predefined two-level hierarchical structure, consisting of 14 major categories together with 100 subcategories. The approach requires search engine log data that are typically not accessible by outsiders. The predefined directory structure also does not suit domains of the non-English Web that have relatively smaller coverage on the Web.

To automatically generate a Web directory and identify directory labels, a self-organizing map (SOM) approach was proposed that built up the relationships among Web pages and extracted category labels [59]. The approach recursively generated superclusters via congregating neighboring neurons, then created the hierarchical structure of the Web directories. However, the directory generated by this approach tends to include noisy content. Without precise filtering and editing, the directory is less reasonable and logical. Variations of the approach have been proposed, such as Refs. [10] and [35].

Kumar [28] presents a two-phase, semiautomatic approach to directory construction. The approach combines human knowledge with search engine efficiency.

But the quality of results depends highly on the ontologist's limited knowledge. Moreover, searching only the *Clever* system (based on HITS algorithm [23]) limited the coverage of results.

Stamou et al. [53] developed an approach for automatically assigning Web pages to a directory framework based on the linguistic information in Web textual data. The approach leveraged a variety of lexical resources such as WordNet [18], suggested upper merged ontology (SUMO) [43], Google directory (http://dir.google.com/), and WordNet Domains (http://wndomains.itc.it/) to build a subject hierarchy and to define concepts in the hierarchy. However, the highly predefined nature of the hierarchy combined with the unsatisfactory categorization accuracy makes the approach not promising for constructing Web directories, especially for the non-English Web, which has a rapidly growing content.

2.2.3 Semantic Network Technologies

A semantic network is a method of knowledge representation that represents semantic relations between concepts using a directed or undirected graph consisting of vertices (indicating concepts) and edges (indicating relations) [49]. Richens and Booth [46] first introduced the concept of using semantic unit in machine translation, forming the notion of using human languages to represent concepts and their connections. This work laid a foundation for machine translation that would be useful for enhancing multilingual application on the Web and Web searching and browsing. Using semantic networks to support finding information on the Web, Woods proposed annotating documents with natural language descriptions of their content and taking account of the conceptual structures of those descriptions [57]. Paths connecting concepts in a search request with related concepts can be found, and specific information can be located in response to specific requests. His technique provides a way for people to find answers to specific questions, but it requires humans to evaluate the retrieved answers [58].

Semantic network technologies have been applied to multilingual concept retrieval. For example, Li and Yang [29] developed an algorithmic approach to generate a robust knowledgebase based on statistical correlation analysis of the semantics (knowledge) embedded in the bilingual (English/Chinese) press release corpus obtained from the Web. Li et al. [30] developed a schema design method to support multilingual markup of XML based on two attributes introduced for every element to be transformed, that is, locID and attrList. The approach is language-independent, ensuring all XML instances to be understood by automated program. While semantic network is used extensively in knowledge representation and cross-lingual applications, its application to Web searching and browsing in a multilingual world has not been widely studied.

2.3 Web Portals in a Multilingual World

As more non-English-speaking people use the Internet to search and browse information, major search engines have attempted to extend their services to non-English speakers. Regional search engines that provide more localized searching have begun to emerge. In addition to English, these search engines typically accept queries in a user's native language and return pages from the regions being served. We survey below major Web portals and search engines in Chinese, Spanish, and Arabic, three of the most widely used languages on the Web.

2.3.1 Chinese-Speaking Regions

Chinese is the primary language used by people in mainland China, Taiwan, and Hong Kong. Language encoding, vocabularies, economies, and societies of the three regions differ significantly. In mainland China, Baidu.com is one of the largest search engines and serves many large enterprises including Dell (China), Lenovo, and Yahoo! China. It collected over 1 billion Chinese Web pages from mainland China, Hong Kong, Taiwan, and other regions, and its collection grows by several hundreds of thousands of Web pages per day. Another major Web portal in China, Sina.com.cn provides comprehensive services such as Web searching, email, news, business directory, entertainment, weather forecast, etc. Leveraging on its rich content and a large user base, Sina has its own search engine called iAsk.com that uses both Web content and usage information to rank Web pages. Other search engines in mainland China include Sogou.com and Zhongsou.com. In Taiwan, the two major search portals are Openfind and Yam. Openfind.com.tw, established in 1998, suggests relevant terms to refine users' search queries and allows users to find other related items from each search result. Established in 1995, Yam.com provides comprehensive online services, allowing users to search various media: Web sites, Web pages, news, Internet forum messages, and activities (in major Taiwan cities or regions). Since 2000, Yam.com has been partnering with Google to provide search service. In Hong Kong, due to its bilingual culture, people rely on both English and Chinese when searching the Web. Yahoo! Hong Kong (http://hk.yahoo.com/) returns results in different categories, Web sites, Web pages, and news. Established in 1997, Timway.com searches more than 30,000 Hong Kong Web sites categorized into over 3000 groups and attracts 2.6 million visits per months.

2.3.2 Spanish-Speaking Regions

Spanish is the second most popular language in the United States and the primary language for Spain and some 22 Latin American countries, where regional search engines provide search and browse services. Having 19 regional sites, Terra.com offers its services to more than 3.1 million Internet users in the United States, Spain, and Latin America. A Gallup poll in 2002 reported Terra to be the most popular search engine in Spain while Orange.es (formerly Wanadoo), a subsidiary of France Telecom, was rated second. Yahoo! Telemundo (Spain, http://telemundo.yahoo.com/), the Spanish version of Yahoo! serving the United States and Latin America, provides a Web directory compiled by human editors who categorized millions of listed sites. Yahoo! Telemundo also supplements its results with those from Inktomi and Google. Established in 1995, BIWE.com is one of the earliest search engines for searching Spanish information on the Web and provides a variety of services including a Web directory, email, entertainment, and market information for Hispanics. Headquartered in the United States, Quepasa.com is a bilingual Web portal (Spanish/English) serving Hispanic populations in the United States and Latin America.

The following Spanish search engines primarily serve their own or adjacent regions. Launched in 1998, Ahijuna (Argentina, http://www.ahijuna.com.ar/) provides searching of Argentina Web sites and other Spanish Web sites. It contains a Web directory with 14 categories having a total of 7578 hyperlinks. Based in Venezuela, Auyantepui (http://www.auyantepui.com/) provides a searchable Web directory of Spanish sites. It grew from 14 categories listing 117 Web sites in 1996 to 550 categories with over 18,000 Web sites in 2002. Launched in 1998, Conexcol (Colombia, http://www.conexcol.com/) provides a searchable Web directory containing 14 categories having 400 subcategories and 13,214 Web sites' URLs. With more than 150,000 unique visitors per month, it is one of the four most often visited sites in Colombia. Bacan (Ecuador, http://www.bacan.com/), which began its operations in 1996, provides services such as news, email, online chat, entertainment, and shopping guides. Every month Bacan has 80,000 individual visitors and generates more than 2 million hits. Ascinsa Internet (http://www.ascinsa.com/) is widely used in Peru and contains Web sites from Latin American countries and the United States. It provides services such as Internet access, email, Web page design, domain registration, and Web hosting, among others. It also contains a directory listed by countries and then by domains.

2.3.3 Arabic-Speaking Regions

Arabic is spoken by more than 284 million people in about 22 countries, most of which are developing countries. Although Arabic is the fifth most frequently spoken language in the world, the Arabic Web is still in its infancy, constituting less than 1% of the total Web content and having a low 2.2% penetration rate [1]. The cross-regional use of Arabic and the exponential growth of Arabic Web [40] have nevertheless highlighted the necessity of providing better Web searching and browsing.

Four major search engines offer the Arab World comprehensive services and extensive content coverage. Ajeeb (http://www.ajeeb.com/) is a bilingual Web portal (English/Arabic) launched in 2000 by Sakhr Software Company, a Middle East-based software company specializing on Arabic language processing and content management. Sakhr's database contains over 1 million searchable Arabic Web pages, which can be translated to English using the online version of its machine-translation software. Having a multilingual dictionary, Ajeeb is known for its large Web directory, "Dalil Ajeeb," which Sakhr claims is the world's largest online Arabic directory. Ajeeb's tool called "Johaina" automatically gathers news from many Middle Eastern and worldwide news agencies. Using Sakhr's "Idrisi" Arabic search engine, Johaina gathers mainly Middle East-related news and categorizes them into primary and secondary topic categories. Albawaba.com (http://www.albawaba.com/) is a comprehensive Web portal offering such services as news, sports, entertainment, email, and online chatting. Albawaba supports searching for both Arabic and English pages and the results are classified according to language and relevancy. It provides metasearching of other search engines (Google, Yahoo, Excite, Alltheweb, Dogpile) and a comprehensive directory of all Arab countries. UAE-based Albahhar (http://www.albahhar.com/), launched in 2000, provides a wide range of online services such as searching, news, online chatting, and entertainment. Albahhar searches its 1.25 million Arabic Web pages and provides a wide range of other Arabic online services like news, chat, and entertainment. With offices in UAE, United States, and Lebanon, Ayna (http://www.ayna.com/) is a Web portal providing an Arabic Web directory, an Arabic search engine, and other services such as a bilingual (English/Arabic) email system, chat, greeting cards, personal homepage hosting, and personal commercial classifieds. In 2008, Ayna launched an online map service covering 26 cities in 17 Middle-Eastern countries. In 2007, Ayna began an online recruitment system at AynaBeirut (its Lebanon branch). Targeting Arabic audience, the AynaAD network consists of a group of prescreened Arabic content publishers who place banners and text advertisements on their sites. Due to Ayna's popularity, Alexa Research ranks it among the top three leading Web sites in the Arab World.

3. A Multilingual Perspective

Existing search portals in Chinese, Spanish, and Arabic typically present results as long lists of textual items. While such presentation is convenient for users to view, it may limit users' ability to understand and to analyze these results. The collections searched by these search engines are often region-specific, so they do not provide a comprehensive understanding of the environment where they are operating. Major English search engines such as Google support searching of non-English resources but fall short of covering domain- and region-specific information. There is a need for better supporting Web searching in some emerging non-English languages. Our research proposed a framework for Web searching and browsing in a multilingual world and studied the way this framework helped develop Web search portals that serve the needs of users speaking those languages. Figure 1 shows the framework that consists of domain collection building, metasearching, Web directory building, statistical language processing, Web page summarization, categorization, and visualization. In the following, we describe the components of the framework.

3.1 Domain Analysis

Because of regional and language differences, a careful domain analysis must be conducted before building a Web portal in a particular language. To ensure a comprehensive coverage, the analysis involves surveying existing Web portals and technologies, studying the characteristics of the language, and selecting an area or theme that significant Web resources in the language have been developed. In applying the framework to developing Web portals, we have reviewed regional search engines, government and business Web sites, and news Web sites to select relevant Web content for building a domain-specific collection or for metasearching. Important keywords and URLs relevant to the chosen domain are gathered as seeds to build the collection.

3.2 Collection Building and Metasearching

To provide high-quality information, the existing information sources in the chosen domains must be analyzed and selected carefully. For example, key business categories such as e-commerce, international business, and competitive intelligence were searched to obtain seed URLs for domain spidering/collecting of Web pages. A Web crawler then followed these URLs to collect pages automatically. The pages were then automatically indexed and stored in a database. In addition to domain spidering, we performed metaspidering of major search engines using queries translated from English queries that were used to build an English business intelligence search portal [35]. The search engines must contain rich content in the chosen domain. Metasearching is used as well to provide online searching of resources.

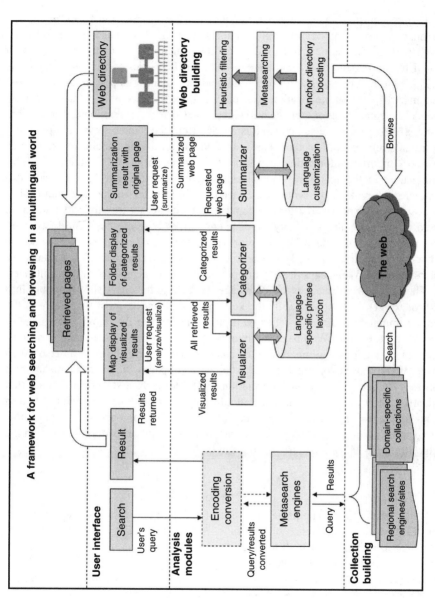

FIG. 1. A general framework for Web searching and browsing in a multilingual world.

3.3 Analysis Modules

Different postretrieval analysis techniques are used in the framework to support summarization, categorization, and visualization of the Web pages retrieval by the Web portals. The summarizer used in building our portals was modified from an English summarizer that uses sentence-selection heuristics to rank text segments [37]. These heuristics strive to reduce redundancy of information in a query-based summary [4]. The summarization takes place in three main steps: (1) sentence evaluation, (2) segmentation or topic identification, and (3) segment ranking and extraction. First, a Web page to be summarized is fetched from the remote server and parsed to extract its full text. All sentences are extracted by identifying punctuation serving as periods. Important information such as presence of cue phrases (e.g., "therefore," "in summary" in the respective languages), sentence lengths and positions are also extracted for ranking the sentences. Second, we use the text-tiling algorithm [21] to analyze the Web page and determine topic boundaries. A Jaccard similarity function is used to compare the similarity of different blocks of sentences. Third, we rank document segments identified in the previous step according to the ranking scores obtained in the first step and key sentences are extracted as summary. The summarizer can summarize Web pages flexibly in a pop-up window, using three or five sentences.

The categorizer organizes the Web pages (related to the query shown on top) into 20 (or fewer) folders labeled by the key phrases appearing most frequently in the page summaries or titles. It relies on a phrase lexicon in the relevant language to extract phrases from Web page summaries obtained from metasearching or searching our collections. To create the lexicons, we collected a large number of Web pages in the chosen domain. From each collection of pages, we extracted meaningful phrases by using the mutual information (MI) approach, a statistical method that identifies significant patterns as meaningful phrases from a large amount of text in any language [41]. The MI algorithm is used in the approach to compute how frequently a pattern appears in the corpus, relative to its subpatterns. Based on the algorithm, the MI of a pattern c (MI_c) can be found by

$$\text{MI}_c = \frac{f_c}{f_{\text{left}} + f_{\text{right}} - f_c},$$

where f stands for the frequency of a set of words. Intuitively, MI_c represents the probability of co-occurrence of pattern c, relative to its left subpattern and right subpattern. Phrases with high MI are likely to be extracted and used in automatic indexing. In addition, we employed an updateable PAT-tree data structure developed by Ong and Chen [41] that supports online frequency update after removing extracted patterns to facilitate subsequent extraction.

The visualizer uses a Kohonen SOM algorithm [24] to categorize and place Web pages onto a two-dimensional jigsaw map [35]. SOM is a neural networks algorithm that has been used in image processing and pattern recognition applications. When applied to automatic categorization and visualization of Web pages, SOM assigns similar pages to adjacent regions with each region labeled by the most frequently occurring phrases extracted by the MI approach described. The larger the size of a region on the map, the more the Web pages are assigned to it. Users can click on a region to see a list of pages on the right and can open pages by clicking the link-embedded titles.

3.4 Web Directory Building

Our framework includes steps for directory building to support Web browsing. Three steps are involved. First, an existing Web directory is identified as the anchor directory and its category labels are modified to suit the chosen domain. This anchor directory should be chosen based on the comprehensiveness of domain coverage and richness of Web content. The directory labels collectively serve as a base framework for further modification, which includes changing category labels and enriching the domain-specific content. The result of this step is a directory framework that consists of topical labels organized in a hierarchical structure.

Second, metasearching is used to collect directory items (Web site URLs) automatically to fill in the directory framework (obtained from the first step). The set of search engines to be used as metasearchers should be chosen based on a comprehensive review of Web search engines. Category labels of the directory framework are used as input queries for metasearching. Because search engines typically return a large number of duplicating results, only top-ranked results should be used (with duplicates filtered) to limit the scope of coverage.

Third, human domain knowledge is used to enhance the quality of the automatically generated directory (obtained from the previous step). A number of heuristic rules must be established to ensure consistency in the work. Both general rules and domain-specific rules should be developed to remove nonrelevant items and to include items that might have been missed in the metasearching process. The rules also help to maintain scalability of the framework in constructing Web directories in different domains.

4. Case Studies

This section provides three case studies of applying the framework to building three Web portals in the Chinese business, Spanish business, and Arabic medical domains. Results of the experiments were summarized as well.

4.1 CBizPort: Supporting Metasearching and Data Analysis of Chinese Business Web Pages

Because Chinese business information sources are numerous, diverse, and have varying quality, information overload becomes an issue. Users are more concerned with BI than business information. BI is obtained through the acquisition, interpretation, collation, assessment, and exploitation of information in the business domain [14]. Professionals such as business consultants, marketing executives, and financial analysts are heavily involved in the discovery of BI. The quality of their work relies mainly on the capability of the tools they use to obtain business information.

Developed based on the aforementioned framework, the Chinese Business Intelligence Portal (CBizPort) is a metasearch portal for business information of greater China–Mainland China, Hong Kong, and Taiwan. User interfaces in Simplified Chinese and Traditional Chinese were developed with the same look and feel. Each version uses its own character encoding when processing queries. The encoding converter relies on a conversion dictionary with 6737 Chinese characters in each of the two encodings (Big5 and GB2312). The dictionary includes the most commonly used characters in the Chinese language. Encoding conversion is performed when the portal sends out queries to other search engines having encoding different from its own or when the portal collates results from those search engines. The eight information sources selected for metasearching are major Chinese search engines or business-related portals from the three regions (see Table I).

TABLE I
INFORMATION SOURCES OF CBIZPORT

Region	Information source	Description
Mainland China	Baidu	A general search engine for mainland China
	China Security Regulatory Commission	A portal containing news and financial reports of the listed companies in mainland China
Hong Kong	Yahoo Hong Kong	A general search engine for Hong Kong
	Hong Kong Trade Development Council	A business portal providing information about local companies, products, trading opportunities
	Hong Kong Government Information Center	A portal with government publications, services and policies, business statistics, etc.
Taiwan	Yam	A general search engine for Taiwan
	PCHome	An IT news portal with hundreds of online publications in business and IT areas
	Taiwan Government Information Office	A government portal with business and legal information

The CBizPort summarizer was developed using the same approach mentioned above, but needed cue phrases translated in Chinese for sentence ranking. Sentences were extracted by identifying Chinese punctuations acting as periods such as "。," "、," "!," and "?." The CBizPort categorizer relies on a Chinese phrase lexicon to extract phrases from Web page summaries obtained from the eight search engines or portals. The lexicon was created using the aforementioned MI algorithm. For creating the Simplified Chinese lexicon, over 100,000 Web pages in GB2312 encoding were collected from major business portals such as Sohu.com, Sina Tech, and Sina Finance in mainland China. For creating the Traditional Chinese lexicon, over 200,000 Web pages in Big5 encoding were collected from major business or news portals in Hong Kong and Taiwan (e.g., HKTDC, HK Government, Taiwan United Daily News Finance Section, Central Daily News). The Simplified Chinese lexicon has about 38,000 phrases and the Traditional Chinese lexicon has about 22,000 phrases. Using the Chinese phrase lexicon, the categorizer performed full-text indexing on the title and summary of each result (or Web page) and extracted the top 20 (or fewer) phrases from the results. Phrases occurring in the text of more Web pages were ranked higher. A folder then was used to represent a phrase and the categorizer assigned the Web pages to respective folders based on the occurrences of the phrase in the text. A Web page can be assigned to more than one folder if it contains more than one of the extracted phrases. Figure 2 provides screen shots of CBizPort, where the user was searching and browsing about trading issues among Hong Kong, Taiwan, and mainland China.

In the CBizPort experiment [15] with 30 Chinese subjects from mainland China, Hong Kong, and Taiwan and three Chinese business academics and practitioners serving as experts, we found that the effectiveness of existing Chinese search engines could increase significantly by augmenting it with CBizPort. However, no significant difference was found in the effectiveness of using summarizer or categorizer of the Chinese portal and in user satisfaction rating. Despite this, 11 subjects commented that the summarizer and categorizer could facilitate their understanding and searching of results. These results indicate that CBizPort could augment these search engines in searching and browsing Chinese business information but its summarizer and categorizer need further improvement in precision and browse support.

4.2 SBizPort: Bridging Cross-regional Web Usage in the Spanish Business Domain

Given the growing Spanish-speaking populations in the United States, Spain, and Latin America, businesses actively expand their opportunities by seeking information on the Web. The Spanish Business Intelligence Portal (SBizPort) was developed

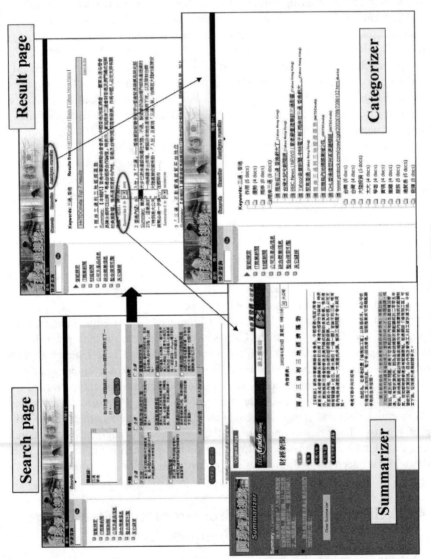

Fig. 2. Screen shots of CBizPort.

to address these growing needs. BI is presented in the forms of collated search results obtained from various high-quality information sources, Web page summaries, categorized search results, and visual maps showing clusters of Web pages.

Key business categories in Spanish including e-commerce, international business, and competitive intelligence were searched to obtain over 183 seed URLs (translated into English) that were used for domain spidering/collecting of Web pages. A Web crawler then followed these URLs to collect pages automatically. The resulting Spanish business collection contained more than 476,084 Web pages covering more than 22 countries. The SBizPort categorizer uses a lexicon built by the MI algorithm that extracted 19,417 phrases from the Spanish business collection. In addition, the SBizPort provides a visualizer that uses Kohonen SOM to visualize retrieved Web pages. A summarizer is included as well. Figure 3 provides screen shots of SBizPort, where the user searches and browses about electronic commerce.

In the SBizPort experiment [12] with 19 Spanish subjects from six countries and a veteran Spanish business consultant serving as the expert, we found the SOM visualizer achieved significantly better browse effectiveness than BIWE, the benchmark search engine, showing that the tool can help alleviate information overload and support browsing. The use of domain-specific collection achieved higher mean accuracy and search efficiency than not using it, but the differences were not significant. Subjects rated SBizPort significantly better than BIWE, citing precision and relevancy of returned results as major reasons. These results indicate the information visualization tool can be an alternative to presenting search results as a textual list.

4.3 AMedDir: Facilitating Web Browsing of Arabic Medical Resources

The growing Arabic online population and medical professionals seek a comprehensive, one-stop Web portal through which to communicate medical information among different Arab regions. We developed the Arabic Medical Web Directory (AMedDir) to support browsing online Arabic medical resources. We followed three steps to build the directory. First, we identified the DMOZ directory (http://www.dmoz.org/) as the anchor directory because of its comprehensiveness in the English medical domain. We removed 46 nodes (due to irrelevant content) from the original 356 nodes of DMOZ's medical subdirectory, leaving 310 nodes in the directory. Then, 11 nodes were manually added by including cultural-specific items such as Islamic medicine, resulting in a 321-node Arabic medical directory framework. Second, we filled in the directory framework with items obtained by metasearching

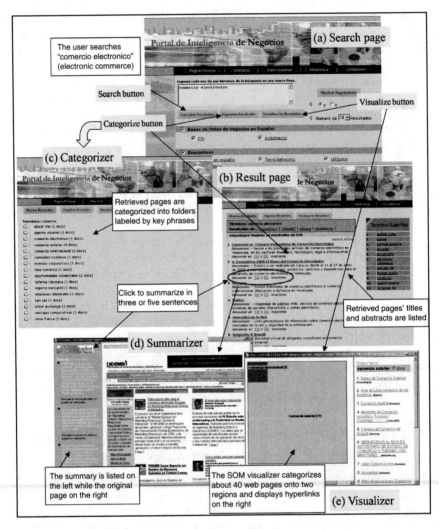

FIG. 3. Screen shots of SBizPort.

six major search engines (Ba7th, Arabmedmag, Google, Ayna, Sehha, and ArabVista) using the 321 category labels of the directory framework as input queries. Running an automatic metasearch program with category labels as query terms, we obtained 8040 unique URLs related to 292 category labels (nonempty nodes) out of the 321 nodes. The maximum depth of the resulting directory was 5. Third, we developed a number of

TABLE II
SUMMARY STATISTICS OF THE AMED WEB DIRECTORY

Statistics	AMed directory
Total number of categories	232
Total number of Web pages	5107
Average number of pages per category	22.1
Maximum depth	5

rules to filter and enhance the directory. URLs were removed if they were not relevant to the topic or to the Arabic medical domain. Empty nodes were removed. Subtopics of deleted nodes were removed as well. Web sites that contained too few links and pages (typically fewer than 10) were removed. Duplicated category labels were consolidated into one label. The statistics of the resulting directory are shown in Table II shows screen shots of the directory (Fig. 4).

In the AMedDir experiment [13] with a total of 18 Arab subjects from six countries and a final-year Arab medical student serving as the expert, we found that AMedDir achieved a significantly higher efficiency than the benchmark directories (Albawaba and Ajeeb). The participants used on average 1 min less using AMedDir to complete a task than benchmark directories. The participants rated the information quality of AMedDir to be significantly higher than that of the benchmark directories, showing that the information provided by AMedDir enabled them to perform the tasks more effectively. AMedDir achieved significantly better ratings on "system usefulness" and "information display and interface design" than Albawaba, and significantly better ratings on "system usefulness," "ease of use," and "information display and interface design." The participants felt that the AMed directory was helpful to browsing different topics in the Arabic medical domain. We believe that the design and development of the directory contributed to the higher usefulness and quality of display in participants' task performance.

5. Summary and Future Directions

The surging demand for non-English Web content and the growing online population in many non-English-speaking regions point to a need to better support Web searching and browsing in a multilingual world. In this chapter, we review issues, technologies, and research related to information seeking on the Web, with a particular emphasis on a non-English environment. We describe a general

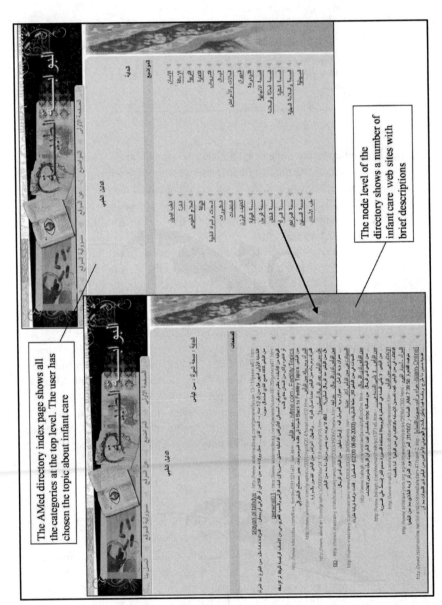

Fig. 4. Screen shots of AMedDir.

framework for developing portals to support Web searching and browsing in a multilingual world and summarize results of using the framework to develop three Web portals.

5.1 Summary of Findings

The experimental results demonstrate that the framework could be used to support Web searching and browsing in a multilingual world. Postretrieval analysis techniques such as summarization and visualization were found to alleviate information overload but the extent of such improvement varies across the domains we studied. The CBizPort study highlighted the importance of summarizing business information in Web search tasks. In the SBizPort study, information visualization achieved significant performance improvement in browsing Web search results. We believe the capability to visualize a large number of search results was most essential for good performance in the portals we studied. In the AMedDir study, we found significant benefits of using our framework to organize online materials to facilitate Web browsing in the Arabic medical domain. We therefore believe that the framework provides useful support to enhance Web searching and browsing in a multilingual world.

5.2 Limitations

Our research prototype portals have speed and stability that are generally not as good as those of commercial search engines like some of the chosen benchmarks. Some subjects complained about the slow responses of our systems. We also were limited by the scarcity of prior work on non-English Web searching, which prevented a more comprehensive review of a topic that possibly would offer better criteria for designing our approach. As for the user study, we had difficulty recruiting native speakers as our subjects. Future work should consider expanding the sample size to establish a higher statistical confidence in the experimental results.

5.3 Future Directions

As the Web continues to be shaped by the growing non-English-speaking population and their needs, we are developing scalable techniques to collect and analyze information in different languages meaningfully to relate diverse content to produce intelligence. For instance, multinational corporations (MNCs) typically provide Web site information in different languages. Analyzing MNC's relationships with their multinational stakeholders could help provide a holistic picture of how they stand in the international arena. The resulting BI from stakeholders will serve to guide global development strategies. Another challenging area is the digital archiving of multilingual data from heterogeneous sources—often scattered in different regions. We will investigate techniques and methods to facilitate such a process and better support non-English Web searching. Furthermore, we will develop and validate new visualization techniques to support browsing and comprehending massive multilingual information on the Web. Other browsing support techniques will be studied as well.

ACKNOWLEDGMENTS

The author thanks the editor and reviewers for their work and their useful comments.

REFERENCES

[1] R. Abbi, Internet in the Arab World, UNESCO Observatory on the Information Society, 2002, p. 3.
[2] H. Berghel, D. Blank, The World Wide Web, in: M.V. Zelkowitz (Ed.), Advances in Computers, Academic Press, New York, NY, 1999, pp. 178–218.
[3] A. Broder, A taxonomy of Web search, ACM SIGIR Forum 36 (2) (2002).
[4] J. Carbonell, J. Goldstein, The use of MMR: diversity-based reranking for reordering documents and producing summaries, in: Proceedings of the 21st Annual International ACM-SIGIR Conference on Research and Development in Information Retrieval (Melbourne, Australia), ACM Press, New York, NY, 1998, pp. 335–336.
[5] S.J. Chang, R.E. Rice, Browsing: a multidimensional framework, in: M.E. Williams (Ed.), Annual Review of Information Science and Technology, Information Today, Inc., Medford, NJ, 1993, pp. 231–276.
[6] M. Chau, H. Chen, J. Qin, Y. Zhou, W.-K. Sung, Y. Chen, Y. Qin, D. McDonald, A. Lally, M. Landon, NanoPort: a Web portal for nanoscale science and technology, in: Proceedings of the 2nd ACM/IEEE-CS Joint Conference on Digital Libraries (Portland, OR), ACM/IEEE-CS, New York, NY, 2002, p. 373.
[7] H. Chen, H. Fan, M. Chau, D. Zeng, MetaSpider: meta-searching and categorization on the Web, J. Am. Soc. Inform. Sci. Technol. 52 (13) (2001) 1134–1147.
[8] H. Chen, A. Houston, R. Sewell, B. Schatz, Internet browsing and searching: user evaluation of category map and concept space techniques, J. Am. Soc. Inform. Sci. [Special Issue on AI Techniques for Emerging Information Systems Applications] 49 (7) (1998) 582–603.
[9] H. Chen, A. Lally, B. Zhu, M. Chau, HelpfulMed: intelligent searching for medical information over the Internet, J. Am. Soc. Inform. Sci. Technol. 54 (7) (2003) 683–694.
[10] H. Chen, C. Schuffels, R. Orwig, Internet categorization and search: a self-organizing approach, J. Visual Commun. Image Representation 7 (1) (1996) 88–102.
[11] S.-L. Chuang, L.-F. Chien, Enriching Web taxonomies through subject categorization of query terms from search engine logs, Decision Support Syst. 35 (1) (2003) 113–127.
[12] W. Chung, Studying information seeking in the non-English Web: an experiment on a Spanish business Web portal, Int. J. Hum. Comput. Stud. 64 (9) (2006) 811–829.
[13] W. Chung, H. Chen, Browsing the underdeveloped Web: an experiment on the Arabic Medical Web Directory, J. Am. Soc. Inform. Sci. Technol. 60 (3) (2009) 595–607.

[14] W. Chung, H. Chen, J.F. Nunamaker, A visual framework for knowledge discovery on the Web, J. Manage. Inform. Syst. 21 (4) (2005) 57–84.
[15] W. Chung, Y. Zhang, Z. Huang, G. Wang, T.-H. Ong, H. Chen, Internet searching and browsing in a multilingual world: an experiment on the Chinese Business Intelligence Portal (CBizPort), J. Am. Soc. Inform. Sci. Technol. 55 (9) (2004) 818–831.
[16] CNNIC, The 23rd Statistical Survey Report on the Internet Development in China, 2009, http://www.cnnic.cn/uploadfiles/pdf/2009/3/23/153540.pdf.
[17] J.F. Cove, B.C. Walsh, Online text retrieval via browsing, Inform. Process. Manage. 24 (1) (1988) 31–37.
[18] C. Fellbaum, WordNet: An Electronic Lexical Database, MIT Press, Cambridge, MA, 1998.
[19] T. Firmin, M.J. Chrzanowski, An evaluation of automatic text summarization systems, MIT Press, Cambridge, MA, 1999.
[20] S. Greene, G. Marchionini, C. Plaisant, B. Shneiderman, Previews and overviews in digital libraries: designing surrogates to support visual information seeking, J. Am. Soc. Inform. Sci. 51 (4) (2000) 380–393.
[21] M.A. Hearst, Multi-paragraph segmentation of expository text, in: Proceedings of the 32nd Annual Meeting of the Association for Computational Linguistics (Las Cruces, NM), 1994, pp. 9–16.
[22] P. Ingwersen, Information Retrieval Interaction, Taylor Graham, London, 1992.
[23] J. Kleinberg, Authoritative sources in a hyperlinked environment, J. Assoc. Comput. Mach. 46 (5) (1999) 604–632.
[24] T. Kohonen, Self-Organizing Maps, Springer-Verlag, Berlin, 1995.
[25] C. Kuhlthau, A principle of uncertainty for information seeking, J. Doc. 49 (4) (1993) 339–355.
[26] C. Kuhlthau, Longitudinal case studies of the information search process of users in libraries, Libr. Inform. Sci. Res. 10 (3) (1998) 257–304.
[27] C. Kuhlthau, A. Spink, C. Cool, Exploration into stages in the information search process in on-line IR: communication between users and intermediaries, Proc. Annu. Meet. Am. Soc. Inform. Sci. 29 (1992) 67–71.
[28] R. Kumar, P. Raghavan, S. Rajagopalan, A. Tomkins, On semi-automated Web taxonomy construction, in: Proceedings of the 4th International Workshop on the Web and Databases (Santa Barbara, CA), ACM Press, New York, NY, 2001, pp. 91–96.
[29] K.W. Li, C.C. Yang, Automatic crosslingual thesaurus generated from the Hong Kong SAR Police Department Web corpus for crime analysis, J. Am. Soc. Inform. Sci. Technol. 56 (3) (2005) 272–282.
[30] N. Li, Y.-a. Tian, Q. Liang, Making XML schema to support multilingual markup, in: Proceedings of the 4th International Conference on Computer Science & Education (Beijing, China), 2009, pp. 1029–1034.
[31] Librarians' Internet Index, About LII—Overview, 2006, http://lii.org/pub/htdocs/about_overview.htm.
[32] E. Loiacono, WebQual™: a Web site quality instrument, in: Proceedings of International Conference on Information Systems (ICIS) Doctoral Consortium (Charlotte, NC), 2002.
[33] G. Marchionini, Information seeking in electronic environments, Cambridge University Press, New York, NY, 1995.
[34] G. Marchionini, B. Shneiderman, Finding facts vs. browsing knowledge in hypertext systems, IEEE Comput. 21 (1) (1988) 70–80.
[35] B. Marshall, D. McDonald, H. Chen, W. Chung, EBizPort: collecting and analyzing business intelligence information, J. Am. Soc. Inform. Sci. Technol. 55 (10) (2004) 873–891.
[36] M.D. Marsico, S. Levialdi, Evaluating Web sites: exploiting user's expectations, Int. J. Hum. Comput. Stud. 60 (3) (2004) 381–416.

[37] D. McDonald, H. Chen, Using sentence selection heuristics to rank text segments in TXTRACTOR, in: Proceedings of the 2nd ACM/IEEE-CS Joint Conference on Digital Libraries (Portland, OR), ACM/IEEE-CS, New York, NY, 2002, pp. 28–35.
[38] Miniwatts International Internet Usage Statistics—The Internet Big Picture, 2009 (updated on March 31, 2009), http://www.internetworldstats.com/stats.htm.
[39] A. Mowshowitz, A. Kawaguchi, Bias on the Web, Commun. ACM 45 (9) (2002) 56–60.
[40] L. Norton, The Expanding Universe: Internet Adoption in the Arab Region, World Markets Research Centre, Waltham, MA, 2001, p. 3.
[41] T.-H. Ong, H. Chen, Updateable PAT-array approach for Chinese key phrase extraction using mutual information: a linguistic foundation for knowledge management, in: Proceedings of the 2nd Asian Digital Library Conference (Taipei, Taiwan), 1999, pp. 63–84.
[42] C. Palmer, J. Pesenti, R. Valdes-Perez, M. Christel, A. Hauptmann, D. Ng, H. Wactlar, Demonstration of hierarchical document clustering of digital library retrieval results, in: Proceedings of the 1st ACM/IEEE Joint Conference on Digital Libraries (Roanoke, VA), 2001.
[43] A. Pease, I. Niles, J. Li, The suggested upper merged ontology: a large ontology for the semantic Web and its applications, in: Working Notes of the AAAI-2002 Workshop on Ontologies and the Semantic Web (Edmonton, Canada), 2002.
[44] L.L. Pipino, Y.W. Lee, R.Y. Wang, Data quality assessment, Commun. ACM 45 (4) (2002) 211–218.
[45] H. Reiterer, G. MuBler, T.M. Mann, S. Handschuh, INSYDER—an information assistant for business intelligence, in: Proceedings of the 23rd Annual International ACM SIGIR Conference on Research and Development in Information Retrieval (Athens, Greece), ACM Press, New York, NY, 2000, pp. 112–119.
[46] R.H. Richens, A.D. Booth, Some methods of mechanized translation, in: W.N. Locke, A.D. Booth (Eds.), Machine Translation of Languages: Fourteen Essays, MIT Press, Cambridge, MA, 1955.
[47] T. Saracevic, Modeling interaction in IR. Review and proposal, Proc. Annu. Meet. Am. Soc. Inform. Sci. (1996) 3–9.
[48] S. Sato, M. Sato, Automatic generation of Web directories for specific categories, in: Proceedings of the AAAI Workshop on Intelligent Information Systems (Orlando, FL), AAAI Press, Orlando, FL, 1999.
[49] J.F. Sowa, Semantic networks, in: S.C. Shapiro (Ed.), Encyclopedia of Artificial Intelligence, John Wiley & Sons, New York, NY, 1992.
[50] R. Spence, A framework for navigation, Int. J. Hum. Comput. Stud. 51 (5) (1999) 919–945.
[51] A. Spink, S. Ozmutlu, H.C. Ozmutlu, B.J. Jansen, U.S. versus European Web searching trends, SIGIR Forum 36 (2) (2002) 32–38.
[52] A. Spink, T. Saracevic, Interaction in IR: selection and effectiveness of search terms, J. Am. Soc. Inform. Sci. 48 (8) (1997) 741–761.
[53] S. Stamou, V. Krikos, P. Kokosis, A. Ntoulas, D. Christodoulakis, Web directory construction using lexical chains, in: Proceedings of the 10th International Conference on Applications of Natural Language to Information Systems (Alicante, Spain), Springer-Verlag, Berlin, 2005.
[54] A.G. Sutcliffe, M. Ennis, Towards a cognitive theory of information retrieval, Interact. Comput. [Special Edition on HCI & Information Retrieval] 10 (1998) 321–351.
[55] A. Tombros, M. Sanderson, Advantages of query biased summaries in information retrieval, in: Proceedings of the 21st Annual International ACM-SIGIR Conference on Research and Development in Information Retrieval (Melbourne, Australia), ACM Press, New York, NY, 1998, pp. 2–10.
[56] R.Y. Wang, D.M. Strong, Beyond accuracy: what data quality means to data consumers, J. Manage. Inform. Syst. 12 (4) (1996) 5–34.

[57] W.A. Woods, Finding information on the Web: a knowledge representation approach, in: Proceedings of the 4th International World Wide Web Conference (Boston, MA), 1995.
[58] W.A. Woods, Searching vs. finding, Queue 2 (2) (2004) 26–35.
[59] H.-C. Yang, C.-H. Lee, A text mining approach on automatic generation of Web directories and hierarchies, in: Proceedings of the IEEE/WIC International Conference on Web Intelligence (Halifax, Canada), IEEE Computer Society, Los Alamitos, CA, 2003.
[60] B. Zhu, G. Leroy, H. Chen, Y. Chen, MedTextus: an intelligent Web-based medical meta-search system, in: Proceedings of the 2nd ACM/IEEE-CS Joint Conference on Digital Libraries (Portland, OR), ACM/IEEE-CS, New York, NY, 2002, p. 386.

Features for Content-Based Audio Retrieval

DALIBOR MITROVIĆ

Vienna University of Technology, Vienna, Austria

MATTHIAS ZEPPELZAUER

Vienna University of Technology, Vienna, Austria

CHRISTIAN BREITENEDER

Vienna University of Technology, Vienna, Austria

Abstract

Today, a large number of audio features exists in audio retrieval for different purposes, such as automatic speech recognition, music information retrieval, audio segmentation, and environmental sound retrieval. The goal of this chapter is to review latest research in the context of audio feature extraction and to give an application-independent overview of the most important existing techniques. We survey state-of-the-art features from various domains and propose a novel taxonomy for the organization of audio features. Additionally, we identify the building blocks of audio features and propose a scheme that allows for the description of arbitrary features. We present an extensive literature survey and provide more than 200 references to relevant high-quality publications.

1. Introduction 72
2. Background 74
   2.1. A Brief Overview on Content-Based Audio Retrieval 74
   2.2. Architecture of a Typical Audio Retrieval System 75
   2.3. Objective Evaluation of Audio Retrieval Techniques 81
   2.4. Attributes of Audio 81

3. Audio Feature Design 84
   3.1. Properties of Audio Features 84
   3.2. Building Blocks of Features 86
   3.3. Challenges in Features Design 89
4. A Novel Taxonomy for Audio Features 94
5. Audio Features 99
   5.1. Overview 99
   5.2. Temporal Features 109
   5.3. Physical Frequency Features 112
   5.4. Perceptual Frequency Features 116
   5.5. Cepstral Features 124
   5.6. Modulation Frequency Features 126
   5.7. Eigendomain Features 131
   5.8. Phase Space Features 132
6. Related Literature 133
   6.1. Application Domains 133
   6.2. Literature on Audio Features 135
   6.3. Relevant Published Surveys 137
7. Summary and Conclusions 139
   Acknowledgments 139
   References 139

1. Introduction

The increasing amounts of publicly available audio data demand for efficient indexing and annotation to enable access to the media. Consequently, content-based audio retrieval has been a growing field of research for several decades. Today, content-based audio retrieval systems are employed in manifold application domains and scenarios such as music retrieval, speech recognition, and acoustic surveillance.

A major challenge during the development of an audio retrieval system is the identification of appropriate content-based features for the representation of the audio signals under consideration. The number of published content-based audio features is too large for quickly getting an overview of the relevant ones. This

FEATURES FOR CONTENT-BASED AUDIO RETRIEVAL 73

chapter tries to facilitate feature selection by organizing the large set of available features into a novel structure.

Audio feature extraction addresses the analysis and extraction of meaningful information from audio signals to obtain a compact and expressive description that is machine-processable. Audio features are usually developed in the context of a specific task and domain. Popular audio domains include audio segmentation, automatic speech recognition, music information retrieval, and environmental/general-purpose sound recognition, see Section 6.1. We observe that features originally developed for a particular task and domain are later often employed for other tasks in other domains. A good example are cepstral coefficients, such as Mel-frequency cepstral coefficients (MFCCs, see Section 5.5.1). MFCCs have originally been employed for automatic speech recognition and were later used in other domains such as music information retrieval and environmental sound retrieval as well. Based on these observations, we conclude that audio features may be considered independently from their original application domain.

This chapter provides a comprehensive survey on content-based audio features. It differs from other surveys in audio retrieval in the fact that it does not restrict itself to a particular application domain. We bring together state-of-the-art and traditional features from various domains and analyze and compare their properties.

It is nearly impossible to give a complete overview of audio features since they are widely distributed across the scientific literature of several decades. We survey publications in high-quality audio- and multimedia-related journals and conference proceedings. The resulting literature survey covers more than 200 relevant publications. From these publications, we select a manifold set of state-of-the-art features. Additionally, we include traditional features that are still competitive. The major criterion for selection is the maximization of heterogeneity between the features in relation to what information they carry and how they are computed. The result is a selection of more than 70 audio features together with references to the relevant literature. We direct the chapter toward researchers in all domains of audio retrieval and developers of retrieval systems.

The presented set of audio features is heterogeneous and has no well-defined structure. We develop a taxonomy in order to structure the set of audio features into meaningful groups. The taxonomy groups the audio features by properties, such as the domain they live in, perceptual properties, and computational similarities. It organizes the entire set of selected features into a single structure that is independent of any application domain. This novel organization groups features with similar characteristics from different application domains. The taxonomy represents a toolkit that facilitates the selection of features for a particular task. It further enables the comparison of features by formal and semantic properties.

This chapter is organized as follows. We give background information on audio retrieval in Section 2. Characteristics of audio features and the challenges in feature design are discussed in Section 3. Section 4 introduces a novel taxonomy for audio features. We summarize the features in Section 5. Section 6 is devoted to related literature. Finally, we summarize the chapter and draw conclusions in Section 7.

2. Background

This section covers different aspects that may allow for better understanding of the authors' view on content-based audio retrieval and its challenges.

2.1 A Brief Overview on Content-Based Audio Retrieval

There are different fields of research in content-based audio retrieval, such as segmentation, automatic speech recognition, music information retrieval, and environmental sound retrieval which we list in the following. *Segmentation* covers the distinction of different types of sound such as speech, music, silence, and environmental sounds. Segmentation is an important preprocessing step used to identify homogeneous parts in an audio stream. Based on segmentation, the different audio types are further analyzed by appropriate techniques.

Traditionally, *automatic speech recognition* focuses on the recognition of the spoken word on the syntactical level [1]. Additionally, research addresses the recognition of the spoken language, the speaker, and the extraction of emotions.

In the last decade *music information retrieval* became a popular domain [2]. It deals with retrieval of similar pieces of music, instruments, artists, musical genres, and the analysis of musical structures. Another focus is music transcription which aims at extracting pitch, attack, duration, and signal source of each sound in a piece of music [3].

Environmental sound retrieval comprises all types of sound that are neither speech nor music. Since this domain is arbitrary in size, most investigations are restricted to a limited domain of sounds. A survey of techniques for feature extraction and classification in the context of environmental sounds is given in Ref. [4].

One major goal of content-based audio retrieval is the identification of perceptually similar audio content. This task is often trivial for humans due to powerful mechanisms in our brain. The human brain has the ability to distinguish between a wide range of sounds and to correctly assign them to semantic categories and previously heard sounds. This is much more difficult for computer systems, where an audio signal is simply represented by a numeric series of samples without any semantic meaning.

Content-based audio retrieval is an ill-posed problem (also known as inverse problem). In general, an ill-posed problem is concerned with the estimation of model parameters by the manipulation of observed data. In case of a retrieval task, model parameters are terms, properties, and concepts that may represent class labels (e.g., terms like "car" and "cat," properties like "male" and "female," and concepts like "outdoor" and "indoor").

The ill-posed nature of content-based retrieval introduces a *semantic gap*. The semantic gap refers to the mismatch between high-level concepts and low-level descriptions. In content-based retrieval, the semantic gap is positioned between the audio signals and the semantics of their contents. It refers to the fact that the same media object may represent several concepts. For example, a recording of Beethoven's Symphony No. 9 is a series of numeric values (samples) for a computer system. On a higher semantic level the symphony is a sequence of notes with specific durations. A human may perceive high-level semantic concepts like musical entities (motifs, themes, movements) and emotions (excitement, euphoria).

Humans bridge the semantic gap based on prior knowledge and (cultural) context. Machines are usually not able to complete this task. Today, the goal of the research community is to narrow the semantic gap as far as possible.

2.2 Architecture of a Typical Audio Retrieval System

A content-based (audio) retrieval system consists of multiple parts, illustrated in Fig. 1. There are three modules: the input module, the query module, and the retrieval module. The task of the input module is to extract features from audio objects stored in an *audio database* (e.g., a music database). *Feature extraction* aims at reducing the amount of data and extracting meaningful information from the signal for a particular retrieval task. Note that the amount of raw data would be much too big for direct processing. For example, an audio signal in standard CD quality consists of 44,100 samples per second for each channel. Furthermore, a lot of information (e.g., harmonics and timbre) is not apparent in the waveform of a signal. Consequently, the raw waveform is often not adequate for retrieval.

The result of feature extraction are parametric numerical descriptions (features) that characterize meaningful information of the input signals. Features may capture audio properties, such as the fundamental frequency and the loudness of a signal. We discuss fundamental audio attributes in Section 2.4. Feature extraction usually reduces the amount of data by several orders of magnitude. The features are extracted once from all objects in the database and stored in a *feature database*.

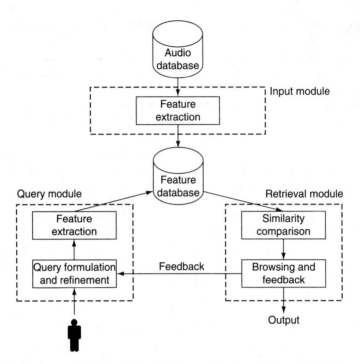

FIG. 1. The components of a typical content-based audio retrieval system and their relations.

The user communicates with the retrieval system by formulating queries. There are different types of queries. Usually, the user provides the system with a query that contains one or more audio objects of interest (query by example). Other possibilities are query by humming (QBH) and query by whistling which are often applied in music retrieval systems. In these approaches, the user has to hum or whistle a melody which is then used as a query object. In both cases, the user asks the system to find objects with similar content as that of the query object(s).

After formulation of a query, features are extracted from the query object(s). This is the same procedure as in the input module. The resulting features have to be compared to the features stored in the feature database in order to find objects with similar properties. This is the task of the retrieval module.

The crucial step in the retrieval module is *similarity comparison* which estimates the similarity of different feature-based media descriptions. Similarity judgments usually base on distance measurements. The most popular approach in this context is the *vector space model* [5]. The basic assumption of this model is that the numeric values of a feature may be regarded as a vector in a high-dimensional space. Consequently, each feature vector denotes one position in this vector space. Distances between feature vectors may be measured by metrics (e.g., Euclidean metric). Similarity measurement is performed by mapping distances in the vector space to similarities. We expect that similar content is represented by feature vectors that are spatially close in the vector space while dissimilar content will be spatially separated.

Similarity measures derived from distance metrics are only appropriate to a certain degree, since mathematical metrics usually do not fully match the human perception of similarity. The mismatch between perceived similarity and computed similarity often leads to unexpected retrieval results.

After similarity comparison the audio objects that are most similar to the query object(s) are returned to the user. In general, not all returned media objects satisfy the query. Additionally, the query may be imperfect, for example in a QBH application. Consequently, most retrieval systems offer the user the opportunity to give feedback based on the output of the retrieval process. The user may specify which of the returned objects meet her expectations and which do not (relevance feedback) [6]. This information may be used to iteratively refine the original query. Iterative refinement enables the system to improve the quality of retrieval by incorporating the user's knowledge.

In the following, we mainly focus on the process of feature extraction. Feature extraction is a crucial step in retrieval since the quality of retrieval heavily relies on the quality of the features. The features determine which audio properties are available during processing. Information not captured by the features is unavailable to the system.

For successful retrieval, it is necessary that those audio properties are extracted from the input signals that are significant for the particular task. In general, features should capture audio properties that show high variation across the available (classes of) audio objects. It is not reasonable to extract features that capture invariant properties of the audio objects, since they do not produce discriminatory information. Furthermore, in some applications, for example, automatic speech recognition the features should reflect perceptually meaningful information. This enables similarity comparisons that imitate human perception. In most applications, the features should be robust against signal distortions and interfering noise and should filter components of the signal that are not perceivable by the human auditory system.

In the following, we present three example sound clips together with different (feature) representations to show how different features capture different aspects of the signals and how features influence similarity measurements. The three example sounds are all one second long and originate from three different sound sources all playing the musical note A4 (440 Hz). The sources are a tuning fork, a flute, and a violin. Figure 2A–C shows plots of the sounds' amplitudes over

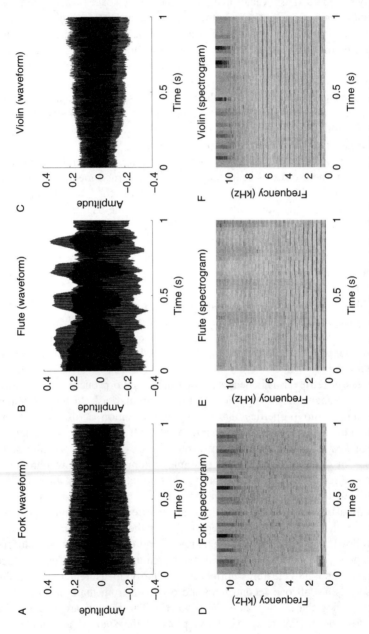

FIG. 2. Three example sounds from different sound sources: tuning fork, flute, and violin. The first row (A)–(C) shows their waveforms and the second row (D)–(F) shows their spectrograms.

time (also called waveforms). The sound produced by the tuning fork has higher amplitude at the beginning and lower amplitude at the end because it dies out slowly after striking the tuning fork. The flute's sound (hereafter flute) exhibits higher variation of the amplitude because it contains tremolo. The amplitude of the violin's sound (hereafter violin) slowly increases toward the end. Except for the similar range of values the waveforms are not similar at all. Signal properties and similarities can hardly be derived from the waveforms. A much more expressive visualization of sounds is the spectrogram which reveals the distribution of frequencies over time. The spectrogram of the fork sound in Fig. 2D contains only one strong frequency component at 440 Hz. The spectrograms of flute (Fig. 2E) and violin (Fig. 2F) are similar to each other. They exhibit strong frequency components at 440 Hz and contain a large number of harmonics (multiples of the fundamental frequency). In the spectrogram of flute, we further observe that the periodic change in amplitude is accompanied by a change in the frequency distribution.

We present two different feature representations of the example sounds and the similarities they reveal in Fig. 3. Figure 3A–C depicts the content-based feature *pitch* which is an estimate of the fundamental frequency of a sound (see Sections 2.4 and 5.4.4). The values of the pitch feature are almost identical for all sounds (approximately at 440 Hz). Considering pitch, the three sounds are extremely similar and cannot be discriminated. However, the three sounds have significantly differing acoustic colors (timbre, see Section 2.4). Consequently, a feature that captures timbral information may be better suited to discriminate between the different sound sources. Figure 3D–F shows visualizations of the first 13 MFCCs which coarsely represent the spectral envelope of the signals for each frame (see Section 5.5.1). We observe that the three plots vary considerably. For example, the violin's sound has much higher values in the third and fifth MFCC than the fork and the flute. Under consideration of this feature all three sounds are different from each other.

This example demonstrates that different content-based features represent different information and that the retrieval task determines which information is necessary for measuring similarities. For example, pitch is suitable to determine the musical note from a given audio signal (e.g., for automatic music transcription). Classification of sound sources (e.g., different instruments) requires a feature that captures timbral characteristics such as MFCCs.

We conclude that the selection and design of features is a nontrivial task that has to take several aspects into account, such as the particular retrieval task, available data, and physical and psychoacoustic properties. We summarize aspects of feature design in Section 3.

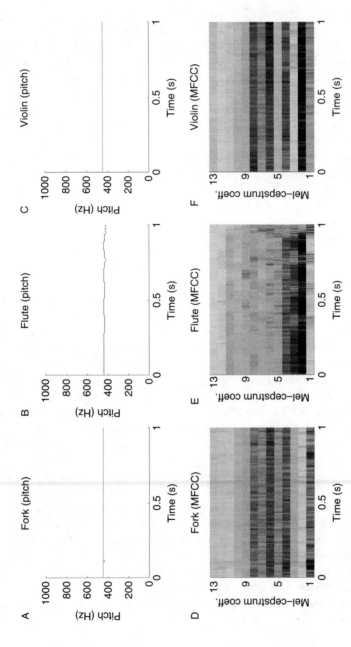

FIG. 3. Two features [pitch in the first row (A)–(C) and MFCCs in the second row (D)–(F)] for the tuning fork, flute, and violin. While all three sounds have similar pitch, their representations in terms of MFCCs differ considerably.

2.3 Objective Evaluation of Audio Retrieval Techniques

An open issue is the evaluation of content-based audio retrieval systems. The results of a retrieval system depend heavily on the input data. Hence, it may happen that a retrieval system is optimized for a specific data set. This may degrade the objectivity of the retrieval results.

The lack of readily available ground truths is an underestimated challenge. There is a need for standardized ground truths to objectively evaluate the performance of different retrieval systems. Currently, ground truths are mostly available in the domains of music information retrieval and automatic speech recognition. Due to legal and economic reasons they frequently are not for free. For speech data, high costs are introduced through the necessary transcription by humans. In the domain of music, copyrights constrain the availability of free data. The situation for environmental sounds is even worse. Due to the infinite range of environmental sounds, it is difficult to build a representative ground truth. Furthermore, the partition of environmental sounds into distinct classes is much more demanding than in the domains of speech and music due to the vast amount of possible sound sources.

Recently, there have been attempts to standardize data and evaluation metrics for music retrieval, for example the audio description contest at the International Conference on Music Information Retrieval in 2004 [7] and the Music Information Retrieval Evaluation eXchange [8]. These contests provide ground truths for free to the participants. According to the authors' knowledge there are no such efforts in the context of environmental sound recognition.

We believe that a set of freely available benchmarking databases and well-defined performance metrics would promote the entire field of audio retrieval. Additionally, independent domain experts should be employed in the process of building ground truths due to their unbiased view. Even though this leads to a decrease of performance, the objectivity and comparability of the results would improve. Although there are efforts in this direction, more attention has to be turned to standardized and easily available ground truths.

2.4 Attributes of Audio

Audio features represent specific properties of audio signals. Hence, we should briefly discuss the different types of audio signals and the general attributes of audio prior to studying audio features.

Generally, we distinguish between tones and noise. Tones are characterized by the fact that they are "capable of exciting an auditory sensation having pitch" [9] while noise not necessarily has a pitch (see below). Tones may be *pure tones* or *complex tones*. A pure tone is a sound wave where "the instantaneous sound pressure of which is a simple sinusoidal function in time" while a complex tone contains "sinusoidal components of different frequencies" [9].

Complex tones may be further distinguished into *harmonic complex tones* and *inharmonic complex tones*. Harmonic complex tones comprise of partials with frequencies at integer multiples of the fundamental frequency (so-called harmonics). Inharmonic complex tones consist of partials whose frequencies significantly differ from integer multiples of the fundamental frequency.

There are different types of noise, distinguished by their temporal and spectral characteristics. Noise may be stationary or nonstationary in time. Stationary noise is defined as "noise with negligibly small fluctuations of level within the period of observation" while nonstationary noise is "noise with or without audible tones, for which the level varies substantially during the period of observation" [9].

The spectral composition of noise is important for its characterization. We distinguish between *broadband noise* and *narrow-band noise*. Broadband noise usually has no pitch while narrow-band noise may stimulate pitch perception. Special types of noise are for example *white noise*, which equally contains all frequencies within a band, and *colored noise* where the spectral power distribution is a function of frequency (e.g., pink ($1/f$) noise).

From a psychoacoustic point of view, all types of audio signals may be described in terms of the following attributes: duration, loudness, pitch, and timbre.

Duration is the time between the start and the end of the audio signal of interest. The temporal extent of a sound may be divided into attack, decay, sustain, and release depending on the envelope of the sound. Not all sounds necessarily have all four phases. Note that in certain cases silence (absence of audio signals) may be of interest as well.

Loudness is an auditory sensation mainly related to sound pressure level (SPL) changes induced by the producing signal. Loudness is commonly defined as "that attribute of auditory sensation in terms of which sounds can be ordered on a scale extending from soft to loud" with the unit *sone* [9].

The American Standards Association defines (spectral) *pitch* as "that attribute of auditory sensation in terms of which sounds may be ordered on a scale extending from low to high" with the unit *mel* [9]. However, pitch has several meanings in literature. It is often used synonymously with the fundamental frequency. In speech processing pitch is linked to the glottis, the source in the source and filter model of speech production. In psychoacoustics, pitch mainly relates to the frequency of a sound but also depends on duration, loudness, and timbre. In the context of this chapter, we refer to the psychoacoustic definition.

Additionally, to spectral pitch, there is the phenomenon of *virtual pitch*. The model of virtual pitch has been introduced by Terhardt [10]. It refers to the ability of auditory perception to reproduce a missing fundamental of a complex tone by its harmonics.

An attribute related to pitch is *pitch strength*. Pitch strength is the "subjective magnitude of the auditory sensation related to pitch" [9]. For example, a pure tone produces a stronger pitch sensation than high-pass noise [11]. Generally, the spectral shape determines the pitch strength. Sounds with line spectra and narrow-band noise evoke larger pitch strength than signals with broader spectral distributions.

The most complex attribute of sounds is *timbre*. According to the ANSI standard timbre is "that attribute of auditory sensation which enables a listener to judge that two non-identical sounds, similarly presented and having the same loudness and pitch, are dissimilar" [9]. For example, timbre reflects the difference between hearing sensations evoked by different musical instruments playing the same musical note (e.g., piano and violin).

In contrast to the aforementioned attributes, it has no single determining physical counterpart [12]. Due to the multidimensionality of timbre, objective measurements are difficult. Terasawa et al. [13] propose a method to compare model representations of timbre with human perception.

Timbre is a high-dimensional audio attribute and is influenced by both stationary and nonstationary patterns. It takes the distribution of energy in the critical bands into account (e.g., the tonal or noise-like character of sound and its harmonics structure). Furthermore, timbre perception involves any aspect of sound that changes over time (changes of the spectral envelope and temporal characteristics, such as attack, decay, sustain, and release). Preceding and following sounds influence timbre as well.

Each of the attributes duration, loudness, pitch, and pitch strength generally allow for ordering on a unidimensional scale. From a physical point of view, one may be tempted to consider them as independent. Unfortunately, the sensations of these attributes are not independent. In the following, we summarize some relations to illustrate the complexity of auditory perception.

Pitch perception is affected not only by the frequency content of a sound, but also by the sound pressure and the waveform [9, 14]. For example, the perceived pitch of sounds with frequencies above approximately 2 kHz increases with rising amplitudes, while sounds below 2 kHz are perceived to have lower pitch when the amplitude increases. Pitch is usually measured using models of the human perception. Evaluation is performed by comparison of the automatic measurements with human assessments.

There are only few sounds that do not have a pitch at all, such as broadband noise. Nonpitched sounds, are for example, produced by percussive instruments. Byrd and Crawford [15] list nonpitched sounds as one of the current real-world problems in music information retrieval.

Pitch strength is related to duration, amplitude, and frequency of a signal. For example, in case of pure tones the pitch strength increases with both the amplitude and the duration. Additionally, it reaches a maximum in the frequency range between 1 and 3 kHz for pure sounds [11].

Loudness is a subjective sensation that does not only relate to the sound pressure but also to the frequency content and the waveform of a signal as well as its duration [9]. Sounds with durations below 100 ms appear less loud than the same sounds with longer durations [11]. Furthermore, loudness sensation varies with the frequency. This relation is described by equal-loudness contours (see Section 3.3.1).

Generally, audio features describe aspects of the aforementioned audio attributes. For example there is a variety of features that aim at representing pitch and loudness. Other features capture particular aspects of timbre, such as sharpness, tonality, and frequency modulations. We present the overview of audio features in Section 5.

3. Audio Feature Design

Feature design is an early conceptual phase in the process of feature development. During this process, we first determine what aspects of the audio signal the feature should capture. This is performed in the context of the application domain in question and the specific retrieval task. The next step is the development of a technical solution that fulfills the specified requirements and the implementation of the feature.

In this section, we investigate properties of content-based audio features. Additionally, we analyze the fundamental building blocks of features from a mathematically motivated point of view. Finally, we summarize important challenges and problems in feature design.

3.1 Properties of Audio Features

Content-based audio features share several structural and semantical properties that help in classifying the features. In Table I, we summarize properties of audio features that are most frequently used in literature.

A basic property of a feature is the *audio representation* it is specified for. We distinguish between two groups of features: features based on linear-coded signals and features that operate on lossily compressed (subband-coded) audio signals. Most feature extraction methods operate on linear-coded signals. However, there has been some research on lossily compressed-domain audio features, especially for MPEG audio encoded signals due to their wide distribution. Lossy audio compression transforms the signal into a frequency representation by employing psychoacoustic

FEATURES FOR CONTENT-BASED AUDIO RETRIEVAL 85

TABLE I
THE FORMAL PROPERTIES OF AUDIO FEATURES AND THEIR POSSIBLE VALUES

Property	Values
Signal representation	Linear coded, lossily compressed
Domain	Temporal, frequency, correlation, cepstral, modulation frequency, reconstructed phase space, eigendomain
Temporal scale	Intraframe, interframe, global
Semantic meaning	Perceptual, physical
Underlying model	Psychoacoustic, nonpsychoacoustic

models which remove information from the signal that is not perceptible to human listeners (e.g., due to masking effects). Although lossy compression has different goals than feature extraction, features may benefit from the psychoacoustically preprocessed signal representation, especially for tasks in which the human perception is modeled. Furthermore, compressed-domain features may reduce computation time significantly if the source material is already compressed. Wang et al. [16] provide a survey of compressed-domain audio features. We focus on features for linear-coded audio signals, since they are most popular and form the basis for most lossily compressed-domain audio features.

Another property is the *domain* of an audio feature. This is the representation a feature resides in after feature extraction. The domain allows for the interpretation of the feature data and provides information about the extraction process and the computational complexity. For example, a feature in *temporal* domain directly describes the waveform while a feature in *frequency* domain represents spectral characteristics of the signal. It is important to note that we only consider the final domain of a feature and not the intermediate representations during feature extraction. For example, MFCCs are a feature in *cepstral* domain, regardless of the fact that the computation of MFCCs first takes place in frequency domain. We summarize the different domains in Section 3.2.

Another property is the *temporal scale* of a feature. In general, audio is a nonstationary time-dependent signal. Hence, various feature extraction methods operate on short frames of audio where the signal is considered to be locally stationary (usually in the range of milliseconds). Each frame is processed separately (eventually by taking a small number of neighboring frames into account, such as spectral flux—SF) which results in one feature vector for each frame. We call such features *intraframe* features because they operate on independent frames. Intraframe features are sometimes called frame-level, short-time, and steady features [17]. A well-known example for an intraframe feature are MFCCs which are frequently extracted for frames of 10–30 ms length.

In contrast, *interframe* features describe the temporal change of an audio signal. They operate on a larger temporal scale than intraframe features to capture the dynamics of a signal. In practice, interframe features are often computed from intraframe representations. Examples for interframe features are features that represent rhythm and modulation information (see Section 5.6). Interframe features are often called long-time features, global features, dynamic features, clip-level features, and contour features [17, 18].

In addition to interframe and intraframe features, there are *global* features. According to Peeters a global feature is computed for the entire audio signal. An example is the attack duration of a sound. However, a global feature does not necessarily take the entire signal into account [19].

The *semantic interpretation* of a feature indicates whether or not the feature represents aspects of human perception. *Perceptual* features approximate semantic properties known by human listeners, for example, pitch, loudness, rhythm, and harmonicity [20]. Additionally to perceptual features, there are *physical* features. Physical features describe audio signals in terms of mathematical, statistical, and physical properties without emphasizing human perception in the first place (e.g., Fourier transform coefficients and the signal energy).

We may further distinguish features by the type of the *underlying model*. In recent years, researchers incorporated psychoacoustic models into the feature extraction process in order to improve the information content of the features and to approximate human similarity matching [21]. Psychoacoustic models for example incorporate filter banks that simulate the frequency resolution of the human auditory system. Furthermore, these models consider psychoacoustic properties, such as masking, specific loudness sensation, and equal-loudness contours (see Section 3.3.1). Investigations show that retrieval results often benefit from features that model psychoacoustical properties [21–24]. In the context of this work, we distinguish between *psychoacoustic* and *nonpsychoacoustic* features.

Each audio feature can be characterized in terms of the aforementioned properties. We employ several of these properties in the design of the taxonomy in Section 4.

3.2 Building Blocks of Features

In this section, we analyze the mathematical structure of selected features and identify common components (building blocks). This approach offers a novel perspective on content-based audio features that reveals their structural similarities.

We decompose audio features into a sequence of basic mathematical operations similarly to Mierswa and Morik [25]. We distinguish between three basic groups of functions: transformations, filters, and aggregations. *Transformations* are functions that map data (numeric values) from one domain into another domain. An example for a transformation is the discrete Fourier transform that maps data from temporal domain into frequency domain and reveals the frequency distribution of the signal. It is important that the transformation from one domain into the other changes the interpretation of the data. The following domains are frequently used in audio feature extraction.

Temporal domain. The temporal domain represents the signal changes over time (the waveform). The abscissa of a temporal representation is the sampled time domain and the ordinate corresponds to the amplitude of the sampled signal. While this domain is the basis for feature extraction algorithms the signals are often transformed into more expressive domains that are better suited for audio analysis.

Frequency domain. The frequency domain reveals the spectral distribution of a signal and allows for example the analysis of harmonic structures, bandwidth, and tonality. For each frequency (or frequency band) the domain provides the corresponding magnitude and phase. Popular transformations from time to frequency domain are Fourier (DFT), Cosine (DCT), and Wavelet transform. Another widely used way to transform a signal from temporal to frequency domain is the application of banks of band-pass filters with, for example, Mel- and Bark-scaled filters to the time domain signal. Note that Fourier, Cosine, and Wavelet transforms may also be considered as filter banks.

Correlation domain. The correlation domain represents temporal relationships between signals. For audio features especially the *auto*correlation domain is of interest. The autocorrelation domain represents the correlation of a signal with a time-shifted version of the same signal for different time lags. It reveals repeating patterns and their periodicities in a signal and may be employed, for example for the estimation of the fundamental frequency of a signal.

Cepstral domain. The concept of cepstrum has been introduced by Bogert et al. [26]. A representation in cepstral domain is obtained by taking the Fourier transform of the logarithm of the magnitude of the spectrum. The second Fourier transform may be replaced by the inverse DFT, DCT, and inverse DCT. The Cosine transform better decorrelates the data than the Fourier transform and thus is often preferred. A cepstral representation is one way to compute an approximation of the shape (envelope) of the spectrum. Hence, cepstral features usually capture timbral information [13]. They are frequently applied in automatic speech recognition and audio fingerprinting.

Modulation frequency domain. The modulation frequency domain reveals information about the temporal modulations contained in a signal. A typical representation is the joint acoustic and modulation frequency graph which represents the temporal structure of a signal in terms of low-frequency amplitude modulations [24]. The abscissa represents modulation frequencies and the ordinate corresponds to acoustic frequencies. Another representation is the modulation spectrogram introduced by Greenberg and Kingsbury [27] which displays the distribution of slow modulations across time and frequency. Modulation information may be employed for the analysis of rhythmic structures in music [28] and noise-robust speech recognition [27, 29].

Reconstructed phase space. Audio signals such as speech and singing may show nonlinear (chaotic) phenomena that are hardly represented by the domains mentioned so far. The nonlinear dynamics of a system may be reconstructed by embedding the signal into a phase space. The reconstructed phase space is a high-dimensional space (usually $d > 3$), where every point corresponds to a specific state of the system. The reconstructed phase space reveals the attractor of the system under the condition that the embedding dimension d has been chosen adequately. Features derived from the reconstructed phase space may estimate the degree of chaos in a dynamic system and are often applied in automatic speech recognition for the description of phonemes [30, 31].

Eigendomain. We consider a representation to be in eigendomain if it is spanned by eigen- or singular vectors. There are different transformations and decompositions that generate eigendomains in this sense, such as principal component analysis (PCA) and singular value decomposition (SVD). The (statistical) methods have in common that they decompose a mixture of variables into some canonical form, for example uncorrelated principal components in case of the PCA. Features in eigendomain have decorrelated or even statistically independent feature components. These representations enable easy and efficient reduction of data (e.g., by removing principal components with low eigenvalues).

Additionally to transformations, we define *filters* as the second group of operators. In the context of this chapter, we define a filter as a mapping of a set of numeric values into another set of numeric values residing in the *same* domain. In general, a filter changes the values of a given numeric series but not their number. Note that this definition of the term filter is broader than the definition usually employed in signal processing.

Simple filters are, for example, scaling, normalization, magnitude, square, exponential function, logarithm, and derivative of a set of numeric values. Other filters are quantization and thresholding. These operations have in common that they reduce the range of possible values of the original series.

We further consider the process of windowing (framing) as a filter. Windowing is simply the multiplication of a series of values with a weighting (window) function where all values inside the window are weighted according to the function and the values outside the window are set to zero. Windowing may be applied for (non) uniform scaling and for the extraction of frames from a signal (e.g., by repeated application of hamming windows).

Similarly, there are low-pass, high-pass, and band-pass filters. Filters in the domain of audio feature extraction are often based on Bark- [32], ERB- [33], and Mel-scale [34]. We consider the application of a filter (or a bank of filters) as a filter according to our definition, if the output of each filter is again a series of values (the subband signal). Note that a filter bank may also represent a transformation. In this case, the power of each subband is aggregated over time, which results in a spectrum of a signal. Consequently, a filter bank may be considered as both, a filter and a transformation, depending on its output.

The third category of operations are *aggregations*. An aggregation is a mapping of a series of values into a single scalar. The purpose of aggregations is the reduction of data, for example, the summarization of information from multiple subbands. Typical aggregations are mean, variance, median, sum, minimum, and maximum. A more comprehensive aggregation is a histogram. In this case each bin of the histogram corresponds to one aggregation. Similarly, binning of frequencies (e.g., spectral binning into Bark and Mel bands) is an aggregation.

A subgroup of aggregations are *detectors*. A detector reduces data by locating distinct points of interest in a value series, for example, peaks, zero crossings, and roots.

We assign each mathematical operation that occurs during feature extraction to one of the three proposed categories (see Section 5.1). These operations form the building blocks of features. We are able to describe the process of computation of a feature in a very compact way, by referring to these building blocks. As we will see, the number of different transformations, filters, and aggregations employed in audio feature extraction is relatively low, since most audio features share similar operations.

3.3 Challenges in Features Design

The task of feature design is the development of a feature for a specific task under consideration of all interfering influences from the environment and constraints defined by the task. Environmental influences are interfering noise, concurrent sounds, distortions in the transmission channel, and characteristics of the signal source. Typical constraints are, for example, the computational complexity, dimension, and statistical properties and the information carried by the feature. Feature design poses various challenges to the developer. We distinguish between psychoacoustic, technical, and numeric challenges.

3.3.1 *Psychoacoustic Challenges*

Psychoacoustics focuses on the mechanisms that process an audio signal in a way that sensations in our brain are caused. Even if the human auditory system has been extensively investigated in recent years, we still do not fully understand all aspects of auditory perception.

Models of psychoacoustic functions play an important role in feature design. Audio features incorporate psychoacoustic properties to simulate human perception. Psychoacoustically enriched features enable similarity measurements that correspond to some degree to the human concepts of similarity.

We briefly describe the function of the human ear, before we present some aspects of psychoacoustics. The human ear comprises of three sections: the outer ear, the middle ear, and the inner ear. The audio signal enters the outer ear at the pinna, travels down the auditory canal, and causes the ear drum to vibrate. The vibrations of the ear drum are transmitted to the three bones of the middle ear (Malleus, Incus, and Stapes) which in turn transmit the vibrations to the cochlea. The cochlea in the inner ear performs a frequency-to-place conversion. A specific point on the basilar membrane inside the cochlear is excited, depending on the frequency of the incoming signal. The movement of the basilar membrane stimulates the hair cells which are connected to the auditory nerve fibers. The inner hair cells transform the hydromechanical vibration into action potentials while the outer hair cells actively influence the vibrations of the basilar membrane. The outer hair cells receive efferent activity from the higher centers of the auditory system. This feedback mechanism increases the sensitivity and frequency resolution of the basilar membrane [35]. In the following, we summarize important aspects of auditory perception that are often integrated into audio features.

Frequency selectivity. The frequency resolution of the basilar membrane is higher at low frequencies than at high frequencies. Each point on the basilar membrane may be considered as a band-pass filter (auditory filter) with a particular bandwidth (critical bandwidth) and center frequency. We refer the reader to Refs. [11, 35] for a comprehensive introduction to the frequency selectivity of the human auditory system.

In practice, a critical-band spectrum is obtained by the application of logarithmically scaled band-pass filters where the bandwidth increases with center frequency. Psychoacoustical scales, such as Bark- and ERB-scale are employed to approximate the frequency resolution of the basilar membrane [32, 36].

Auditory masking. Masking is "the process by which the threshold of hearing for one sound is raised by the presence of another (masking) sound" [9]. The amount of masking is expressed in decibels. We distinguish between simultaneous masking and temporal masking. Simultaneous masking is related to the frequency selectivity of the human auditory system. One effect is that when two spectral components of similar frequency occur simultaneously in the same critical band, the louder sound may mask the softer sound [37]. Spectral masking effects are implemented for the computation of loudness for example in Ref. [28].

In temporal masking, the signal and the masker occur consecutively in time. This means for example that a loud (masking) sound may decrease the perceived loudness of a preceding sound. We distinguish between forward masking (also poststimulus masking) which refers to a "condition in which the signal appears after the masking sound" and backward masking (also prestimulus masking) where the signal appears before the masking sound [9].

Loudness levels. The loudness of sinusoids is not constant over all frequencies. The loudness of two tones of same SPL but different frequency varies [38]. Standardized equal-loudness contours relate tones of different frequencies and SPL to loudness levels (measured in phon) [39]. Figure 4 shows equal-loudness contours for different loudness levels. Pfeiffer [40] presents a method to approximate loudness by incorporating equal-loudness contours.

Psychophysical power law. According to Stevens [41], the loudness is a power function of the physical intensity. A tenfold change in intensity (interval of 10 phons) approximately results in a twofold change in loudness. The unit of

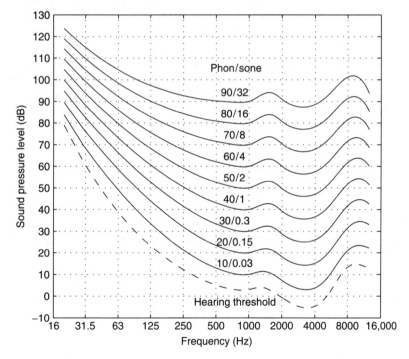

FIG. 4. The solid lines are the equal-loudness contours for 10–90 phons as specified by the ISO 226 standard. Additionally, the corresponding sone values are given. The dashed line is the threshold of hearing. We are most sensitive to frequencies around 2 and 5 kHz.

loudness is sone, where 1 sone is defined as the loudness of a pure 1000 Hz tone at 40 dB SPL (40 phon). Figure 4 shows the phon and corresponding sone values of several equal-loudness contours.

In many domains psychoacoustically motivated features have shown to be superior to features that do not simulate auditory perception, for example in automatic speech recognition [22], fingerprinting [24], and audio classification [21].

3.3.2 Technical Challenges

An audio signal is usually exposed to distortions, such as interfering noise and channel distortions. Techniques robust to a wide range of distortions have been proposed, for example, in Refs. [42, 43]. Important factors are:

Noise. Noise is present in each audio signal and is usually an unwanted component that interferes with the signal. Thermal noise is always introduced during capturing and processing of signals by analog devices (microphones, amplifiers, recorders) due to thermal motion of charge carriers. In digital systems additional noise may be introduced through sampling and quantization. These types of noise are often neglected in audio retrieval.

More disturbing are background noise and channel distortions. Some characteristics of noise have already been summarized in Section 2.4. Additionally, noise may be characterized by the way it is embedded into the signal. The simplest case is additive noise. A more complicated case is convolutional noise, usually induced by the transmission channel. Generally, noise is considered to be independent from the signal of interest; however, this is not true in all situations. Noise robustness is one of the main challenges in audio feature design [21, 44, 45].

Sound pressure level variations. For many retrieval tasks, it is desired that an audio feature is invariant to the SPL of the input signal (except for features that are explicitly designed to measure loudness, see Section 5.4.3). For example, in automatic speech recognition, an utterance at different SPLs should ideally yield the same feature-based representation.

Tempo variations. In most application domains uncontrolled tempo variations decrease retrieval performance. For example, in music similarity retrieval one is interested in finding all interpretations of a piece of music independent of their respective tempos. A challenge in feature design is to create audio descriptions that are invariant against temporal shifts and distortions. Therefore, it is important to maintain the original frequency characteristics [24, 46].

Concurrency. Concurrent audio signals (background noise and reverberation) pose problems to feature extraction. In many situations the audio signal contains components of more than one signal source, for example, multiple instruments or a mixture of environmental sounds. It is difficult (and generally impossible) to filter all unwanted portions from the composite signal.

Available resources. Finally, the computational complexity of an audio feature is a critical factor especially in real-time applications. While feature extraction on standard PCs is often possible in real-time applications on mobile devices, such as PDAs and mobile phones pose novel challenges to efficient feature extraction.

3.3.3 Numeric Challenges

The result of feature extraction is a numeric feature vector that represents particular aspects of the underlying signal. The feature vector should fulfill a number of statistical and numeric requirements depending on the employed classifier and similarity/distance measure. In the following, we summarize the most important statistical and numeric properties.

Compactness. This property refers to the dimensionality of the feature vector. A compact representation is desired to decrease the computational complexity of subsequent calculations.

Numeric range. The components of a feature vector should be in the same numeric range to allow for comparisons of the components. Different numeric ranges of components in the same vector may lead to unwanted bias in following similarity judgments (depending on the employed classifier and distance metric). Therefore, normalization may be applied after feature extraction.

Completeness. A feature should be able to completely cover the range of values of the property it describes. For example, a feature that describes the pitch of an audio signal should cover the entire range of possible pitches.

Redundancy. The correlation between components of a feature vector is an indicator for its quality. The components of a feature vector should be decorrelated to maximize the expressive power. We find features with decorrelated components especially in the cepstral- and eigendomain (see Sections 5.5 and 5.7).

Discriminant power. For different audio signals, a feature should provide different values. A measure for the discriminant power of a feature is the variance of the resulting feature vectors for a set of input signals. Given different classes of similar signals, a discriminatory feature should have low variance inside each class and high variance over different classes.

Sensitivity. An indicator for the robustness of a feature is the sensitivity to minor changes in the underlying signal. Usually, low sensitivity is desired to remain robust against noise and other sources of irritation.

In general, it is not possible to optimize all mentioned properties simultaneously, because they are not independent from each other. For example, with increasing discriminant power of a feature, its sensitivity to the content increases as well which in turn may reduce noise robustness. Usually, tradeoffs have to be found in the context of the particular retrieval task.

4. A Novel Taxonomy for Audio Features

Audio features describe various aspects and properties of sound and form a versatile set of techniques that has no inherent structure. One goal of this chapter is to introduce some structure into this field and to provide a novel, holistic perspective. Therefore, we introduce a taxonomy that is applicable to general-purpose audio features independent from their application domain.

A taxonomy is an organization of entities according to different principles. The proposed taxonomy organizes the audio features into hierarchical groups with similar characteristics. There is no single, unambiguous and generally applicable taxonomy of audio features, due to their manifold nature. A number of valid and consistent taxonomies exist. Usually, they are defined with particular research fields in mind. Hence, most of them are tailored to the needs of these particular fields which diminishes their general applicability.

We want to point out some issues related to the design of a taxonomy by discussing related approaches. Tzanetakis [18] proposes a categorization for audio features in the domain of music information retrieval. The author employs two organizing principles. The first principle corresponds to computational issues of a feature, for example, *Wavelet transform features, short-time Fourier transform (STFT)-based features*. The second principle relates to qualities like texture, timbre, rhythm, and pitch. This results in groups of features that either are computed similarly or describe similar audio qualities.

Two groups in this categorization are remarkable. There is a group called *other features* that incorporates features that do not fit into any other group. This reflects the difficulties associated with the definition of a complete and clear taxonomy. The other remarkable group is the one named *musical content features*. This group contains combinations of features from the other groups and cannot be regarded to be on the same structural level as the other groups. Tzanetakis' categorization is appropriate for music information retrieval [47]. However, it is too coarse for a general application in audio retrieval.

Peeters [48] promotes four organizing principles for the categorization of audio features. The first one relates to the *steadiness or dynamicity* of a feature. The second principle takes the *time extent* of a feature into account. The third principle is the *abstractness* of the representation resulting from feature extraction. The last organizing principle is the *extraction process* of the feature. Peeters describes an organization that is better suited for general use, though we believe a more systematic approach is needed.

We have identified several principles that allow for classification of audio features inspired by existing organizations and the literature survey presented in Section 6. Generally, these principles relate to feature properties, such as the domain, the carried information (semantic meaning), and the extraction process. The selection of organizing principles is crucial to the worth of a taxonomy. There is no broad consensus on the allocation of features to particular groups, for example, Lu et al. [49] regard zero crossing rate (ZCR) as a perceptual feature, whereas Essid et al. [50] assign ZCR to the group of temporal features. This lack of consensus may stem from the different viewpoints of the authors.

Despite the aforementioned difficulties, we propose a novel taxonomy that aims at being generally applicable. The taxonomy follows a method-oriented approach that reveals the internal structure of different features and their similarities. Additionally, it facilitates the selection of features for a particular task. In practice, the selection of features is driven by factors such as computational constraints (e.g., feature extraction on (mobile) devices with limited capabilities) or semantic issues (e.g., features describing rhythm). The proposed taxonomy is directed toward these requirements.

We believe that a taxonomy of features has to be as fine-grained as possible in order to maximize the degree of introduced structure. However, at the same time the taxonomy should maintain an abstract view to provide groups with semantic meaning. We aim at providing a tradeoff between these conflicting goals in the proposed taxonomy.

We assign features to groups in a way that avoids ambiguities. However, we are aware that even with the proposed organizing principles, certain ambiguities will remain. Generally, the number of computationally and conceptually valid views of features, renders the elimination of ambiguities impossible.

The proposed taxonomy has several levels. On the highest level, we distinguish features by their *domain* as specified in Section 3.1. This organizing principle is well suited for the taxonomy, since each feature resides in one distinct domain. The domains employed for the taxonomy are presented in Section 3.2.

Figure 5 depicts the groups of the first level of the taxonomy. Note that we group features from frequency domain and from autocorrelation domain into the same group of the taxonomy (named frequency domain) since both domains represent similar information. The frequency domain represents the frequency distribution of a signal while the autocorrelation domain reveals the same frequencies (periodicities) in terms of time lags.

The domain a feature resides in reveals the basic meaning of the data represented by that feature, for example, whether or not it represents frequency content.

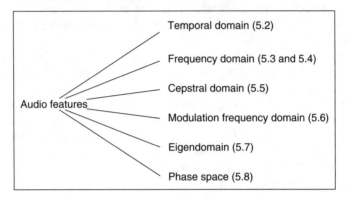

Fig. 5. The first level of the proposed taxonomy. The organizing principle is the domain the features reside in. In brackets a reference to the section containing the corresponding features is given.

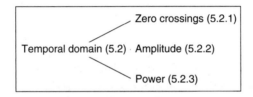

Fig. 6. The organization of features in the temporal domain relates to physical properties of the signal. In brackets a reference to the section containing the corresponding features is given.

Additionally, it allows the user to coarsely estimate the computational complexity of a feature. It further provides information on the data quality, such as statistical independence of the feature components.

On the next level, we apply organizing principles based on computational and semantic concepts. Inside one domain we consistently categorize features according to the property that structures them best. The structure of the *temporal domain* bases on what aspect of the signal the feature represents. In the temporal domain, depicted in Fig. 6, we distinguish between three groups of features: amplitude-, power-, and zero crossing-based features. Each group contains features related to a particular physical property of the waveform.

For the frequency domain we propose a deeper hierarchy due to the diversity of the features that live in it. We introduce a semantic layer that divides the set of features into two distinct groups. One group are *perceptual* features and the other group are *physical* features. Perceptual features represent information that has a semantic meaning to a human listener, while physical features describe audio signals in terms of mathematical, statistical, and physical properties of the audio signal (see Section 3.1). We believe that this layer of the taxonomy supports clarity and practicability.

We organize the perceptual features according to semantically meaningful aspects of sound. These aspects are brightness, chroma, harmonicity, loudness, pitch, and tonality. Each of these properties forms one subgroup of the perceptual frequency features (see Fig. 7). This structure facilitates the selection of audio features for particular retrieval tasks. For example, if the user needs to extract harmonic content, the taxonomy makes identification of relevant features an easy task.

Note that we do not employ timbre as a semantic category in the taxonomy because of its versatile nature. Its many facets would lead to an agglomeration of diverse features into this group. Many audio features represent one or more facets of timbre. In this taxonomy features that describe timbral properties are distributed over several groups.

A semantic organization of the physical features in the frequency domain is not reasonable, since physical features do not explicitly describe semantically meaningful aspects of audio. We employ a mathematically motivated organizing principle

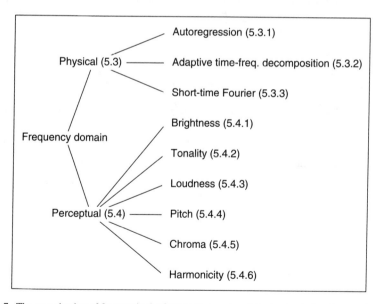

FIG. 7. The organization of features in the frequency domain relates to physical and semantic properties of the signal. In brackets a reference to the section containing the corresponding features is given.

for physical features. The features are grouped according to their extraction process. We distinguish between features that are based on autoregression, adaptive time-frequency decomposition (e.g., Wavelet transform), and STFT. Features that base on STFT may be further separated into features that take the complex part into account (phase) and features that operate on the real part (envelope) of the spectrum.

Similarly to the physical features in the frequency domain, we organize the features in the cepstral domain. Cepstral features have in common that they approximate the spectral envelope. We distinguish between cepstral features by differences in their extraction process.

Figure 8 illustrates the structure of the cepstral domain. The first group of cepstral features employs critical-band filters, features in the second group incorporate advanced psychoacoustic models during feature extraction and the third group applies autoregression.

Modulation frequency features carry information on long-term frequency modulations. All features in this domain employ similar long-term spectral analyses. A group of features we want to emphasize are rhythm-related features, since they represent semantically meaningful information. Consequently, these features form a subgroup in this domain (see Fig. 9).

The remaining domains of the first level of the taxonomy are *eigendomain* and *phase space*. We do not further subdivide these domains, since the taxonomy does

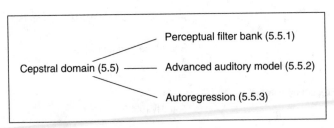

FIG. 8. The organization of features in the cepstral domain relates to the computational properties of the features. In brackets a reference to the section containing the corresponding features is given.

Modulation frequency domain (5.6) —— Rhythm (5.6.1)

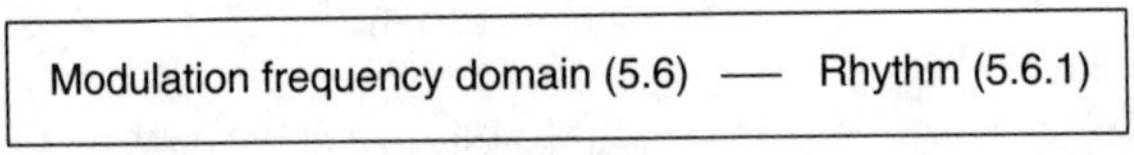

FIG. 9. The organization of features in the modulation frequency domain. We group features that relate to rhythmic content into a separate semantic group. In brackets a reference to the section containing the corresponding features is given.

not profit from further subdivision. A further partition of the domains would decrease the general applicability of the taxonomy.

The taxonomy allows for the selection of features by the information the features carry (e.g., harmonic and rhythm-related features) as well as by computational criteria (e.g., temporal features). We believe that the taxonomy groups features in a way that makes it generally applicable to all areas of audio retrieval and demands only a small number of tradeoffs.

5. Audio Features

In the previous section, we have introduced a taxonomy that represents a hierarchy of feature groups that share similar characteristics. We investigate more than 70 state-of-the-art and traditional audio features from an extensive literature survey. In the following, we briefly present each audio feature in the context of the taxonomy. The sections and subsections reflect the structure of the taxonomy. We describe important characteristics of the features and point out similarities and differences. Before we describe the features in more detail, we give an overview of all covered features and introduce a compact notation for describing the feature extraction process. We compactly present properties of the features, such as the extraction process, domain, temporal structure, application domain, complexity, etc. A tabular representation gives the reader the opportunity to structurally compare and survey all features.

5.1 Overview

Before we present the tables containing the properties of the features, we introduce a notation that allows for the compact representation of the extraction process of a feature. In Section 3.2, we have introduced three groups of mathematical operations that are usually employed in audio feature extraction: transformations, filters, and aggregations. We identify the most important operators belonging to these categories by analyzing the features covered in this chapter. The resulting sets of transformations, filters, and aggregations are listed in Tables II–IV. We arrange similar operations into groups by horizontal bars in order to improve readability.

In the tables, we assign a character to each operation as an abbreviation. Transformations are abbreviated by uppercase Latin characters and filters by lowercase Latin characters. We assign Greek characters (lower and upper case) to aggregations. We observe that the number of identified operations (building blocks) is relatively small, considering that they originate from the analysis of more than 70 different audio features.

TABLE II
FREQUENT TRANSFORMATIONS EMPLOYED IN AUDIO
FEATURES AND THEIR SYMBOLS (UPPERCASE LETTERS, LEFT)

	Transformations
A	Autocorrelation
R	Crosscorrelation
B	Band-pass filter bank
F	Discrete Fourier transform (DFT)
C	(Inverse) discrete cosine transform (DCT/IDCT)
Q	Constant Q transform (CQT)
M	Modulated complex lapped transform (MCLT)
V	Adaptive time-frequency transform (ATFT)
W	Discrete wavelet (packet) transform (DW(P)T)
E	Phase space embedding
I	Independent component analysis (ICA)
P	(Oriented) principal component analysis ((O)PCA)
S	Singular value decomposition (SVD)

TABLE III
FREQUENT FILTERS EMPLOYED IN AUDIO FEATURES AND
THEIR SYMBOLS (LOWERCASE LETTERS, LEFT)

	Filters
b	Band-pass filter (bank)
c	Comb filter (bank)
o	Low-pass filter
f	Framing/windowing
w	(Non)linear weighting function
d	Derivation, difference
e	Energy spectral density
g	Group delay function
l	Logarithm
x	Exponential function
n	Normalization
a	Autoregression (linear prediction analysis)
r	Cepstral recursion formula

The process of computation of a feature may be described as a sequence of the identified operations. We introduce a *signature* as a compact representation that summarizes the computational steps of the extraction process of a feature.

FEATURES FOR CONTENT-BASED AUDIO RETRIEVAL 101

TABLE IV
FREQUENT AGGREGATIONS EMPLOYED IN AUDIO FEATURES AND THEIR SYMBOLS
(GREEK LETTERS, LEFT)

	Aggregations and detectors
χ	Maximum
ι	Minimum
μ	Mean (weighted, arithmetic, geometric)
ϕ	Median
Σ	Sum, weighted sum
σ	Deviation, sum of differences
ϖ	Root-mean-square
ω	Power (mean square)
H	Entropy
π	Percentile
ρ	Regression
Λ	Histogram
β	Spectral binning
κ	Peak detection
ψ	Harmonic peak detection
θ	Polynomial root finding
ζ	Zero/level-crossing detector

The subgroup of detectors are summarized at the bottom of the table.

A signature is a sequence of transformations, filters, and aggregations represented by the previously assigned symbols in Tables II–IV. The characters are arranged from left to right in the order the corresponding operations are performed during feature extraction.

We demonstrate the composition of a signature by means of the well-known MFCC feature [51]. MFCCs are usually computed as follows. At first the Fourier transform of the windowed input signal is computed (a STFT). Then a Mel-filter bank, consisting of logarithmically positioned triangular band-pass filters is applied. After taking the logarithm of the magnitude of the band-pass filtered amplitudes, the Cosine transform is taken to obtain MFCCs.

We can easily construct the corresponding signature for MFCCs by selecting the necessary building blocks from Tables II to IV. First, a single frame ("f") of the input signal is extracted and a Fourier transform ("F") is performed. Then spectral binning of the Fourier coefficients is performed to obtain the responses of the Mel-filters ("β"). Taking the logarithm corresponds to "l" and the completing Cosine transform matches "C." The resulting sequence for the MFCC feature is "f F βl C."

Additionally to transformations, filters, and aggregations, the signatures may contain two structural elements: parentheses and brackets. Parentheses indicate optional operations. We apply parentheses in cases where different definitions of a feature exist to express that more than one computation is possible. Brackets label operations that are repeated for several (two or more) audio frames. For example, in the signature of MPEG-7 temporal centroid "[f ϖ] μ" the brackets indicate that the mean operator is applied to several root-mean-squared frames.

We construct signatures for all features to enable a *structural* comparison of the features and present them together with other properties in Tables V–VII. The tables organize the features according to the taxonomy. The first column presents the domain of the features (which is the first level of the taxonomy). The second column contains references to the sections where the corresponding features are presented (each section covers a subgroup of the taxonomy).

For each feature we specify its temporal scale: "I," "X," and "G" denote intraframe, interframe, and global features, respectively (see Section 3.1). "Y" and "N" in column "perceptual" indicate whether or not a feature is perceptual. The same is done in the column "psychoacoustic model." Furthermore, we rate the computational complexity of each feature ("L," "M," and "H" denote low, medium, and high, respectively). The next column lists the proposed dimension of the feature vectors. The character "V" indicates that the dimension of a feature is parameterized (variable). Additionally, we list the "application domain" where the feature is mostly used. The abbreviation "ASR" stands for automatic speech recognition, "ESR" is environmental sound recognition, "MIR" is music information retrieval, "AS" is audio segmentation, "FP" is fingerprinting, and "VAR" indicates that the feature is applied across several application domains.

The benefit of the signatures in Tables V–VII is not only the compact representation of the extraction process. More important is the ability to identify structurally similar features by comparing rows in the tables. Note that this may be done very quickly without decoding the signatures. Additionally to structural similarities, we may identify preferred operations for particular tasks (e.g., time-to-frequency transformation, analysis of harmonic structures), typical combinations of building blocks and coarsely estimate the complexity of a feature.

In the following, we summarize some observations from the signatures in Tables V–VII. We observe that framing ("f") is part of almost every audio feature independent from the temporal scale. Most of the features are intraframe features, which means that the feature generates one vector for every frame (see Section 3.1). Features that contain brackets in their signature are most often interframe features, for example modulation frequency domain features. These features incorporate information from several frames and represent long-term properties, such as rhythm and tempo.

The signatures reveal the usage and distribution of mathematical transformations among the audio features. Most features employ the (short-time) Fourier transform

TABLE V
An Overview of Temporal and Frequency Domain Features

Domain	Section	Feature name	Temporal scale	Perceptual	Psychoacoustic model	Complexity	Dimension	Application domain	Signature
Temporal	5.2.1	Zero crossing rate (ZCR)	I	N	N	L	1	VAR	f ζ
		Linear prediction ZCR	I	N	N	L	1	ASR	f ζ a ζ
		Zero crossing peak amplitudes (ZCPA)	I	N	Y	M	V	ASR	f b ζ κ 1 Λ Σ
		Pitch synchronous ZCPA	I	N	Y	M	V	ASR	f b A χ ζ κ 1 Λ Σ
	5.2.2	MPEG-7 audio waveform	I	N	N	L	2	—	f χ ι
		Amplitude descriptor	I	N	N	L	9	ESR	f μ σ ζ μ σ
	5.2.3	Short-time energy, MPEG-7 audio power	I	N	N	L	1	VAR	f ω
		Volume	I	N	N	L	1	VAR	f ϖ
		MPEG-7 temporal centroid	X	N	N	L	1	MIR	[f ϖ] μ
		MPEG-7 log attack time	G	N	N	L	1	MIR	[f ϖ] κ 1
Frequency— physical	5.3.1	Linear predictive coding	I	N	N	L	V	ASR	f(b)a(F)
		Line spectral frequencies	I	N	N	M	V	VAR	f a θ

(*continued*)

TABLE V (*Continued*)

Domain	Section	Feature name	Temporal scale	Perceptual	Psychoacoustic model	Complexity	Dimension	Application domain	Signature
Frequency—physical	5.3.2	Daubechies wavelet coefficient histogram	I	N	N	M	28	MIR	f W Λ
		Adaptive time-frequency transform	G	N	N	M	42	MIR	V Λ
	5.3.3	Subband energy ratio	I	N	N	L	V	VAR	f F β e n
		Spectral flux	I	N	N	L	1	VAR	[f F] d $\sum$
		Spectral slope	I	N	N	L	4	VAR	f F ρ
		Spectral peaks (Modified)	X	N	N	L	V	MIR	[f F χ] d
		group delay	I	N	N	M	V	ASR	f F(o) g(C)
Frequency—perceptual	5.4.1	MPEG-7 spectral centroid	G	Y	N	L	1	MIR	F μ
		MPEG-7 audio spectrum centroid	I	Y	Y	M	1	VAR	f F β1 μ
		Spectral centroid	I	Y	N	L	1	VAR	f F(β)(l) μ
		Sharpness	I	Y	Y	M	1	VAR	f F β w w μ
		Spectral center	I	Y	N	L	1	MIR	f F e ϕ

For each feature, we list the domain, a reference to the describing section, temporal scale, whether or not the feature is perceptual and employs psychoacoustic models, the complexity, dimension, application domain, and signature.

TABLE VI
AN OVERVIEW OF FREQUENCY DOMAIN PERCEPTUAL FEATURES

Domain	Section	Feature name	Temporal scale	Perceptual	Psychoacoustic model	Complexity	Dimension	Application domain	Signature
Frequency—perceptual	5.4.2	Bandwidth	I	Y	N	L	1	VAR	f F β(l) σ
		MPEG-7 audio spectrum spread	I	Y	Y	M	1	VAR	f F β l σ
		Spectral dispersion	I	Y	N	L	1	MIR	f F e ϕ σ
		Spectral rolloff	I	Y	N	L	1	VAR	f F π
		Spectral crest	I	Y	N	L	V	FP	f F β χ $\mu(l)$
		Spectral flatness	I	Y	N	M	V	FP	f F β $\mu(l)$
		Subband spectral flux	I	Y	N	M	8	ESR	f F 1 n β d μ
		(Multiresolution) entropy	I	Y	N	M	V	ASR	f F n β H
	5.4.3	Sone	I	Y	Y	H	V	MIR	f F β o l w
		Integral loudness	I	Y	Y	H	1	MIR	f F l $\sum$ w x $\sum$
	5.4.4	Pitch (dominant frequency)	I	Y	N	L	1	VAR	f A χ
		MPEG-7 audio fundamental frequency	I	Y	N	L	2	VAR	f A χ
		Pitch histogram	X	Y	N	M	V	MIR	[f A κ] Λ ($\sum$)
		Psychoacoustical pitch	I	Y	Y	H	V	VAR	b b w A $\sum$
	5.4.5	Chromagram	I	Y	N	M	12	MIR	f F l $\sum$
		Chroma CENS features	I	Y	N	M	12	MIR	f B $\sum$ n o
		Pitch profile	I	Y	N	H	12	MIR	f Q κ $\sum$ χ Λ χ χ $\sum$

(continued)

TABLE VI (Continued)

Domain	Section	Feature name	Temporal scale	Perceptual	Psychoacoustic model	Complexity	Dimension	Application domain	Signature
Frequency—perceptual	5.4.6	MPEG-7 audio harmonicity	I	Y	N	M	2	VAR	$fA\chi$
		Harmonic coefficient	I	Y	N	L	1	AS	$fA\chi$
		Harmonic prominence	I	Y	N	M	1	ESR	$fA\psi$
		Inharmonicity	I	Y	N	M	1	MIR	$fA\psi\sigma$
		MPEG-7 harmonic spectral centroid	I	Y	N	M	1	MIR	$fF\psi\mu$
		MPEG-7 harmonic spectral deviation	I	Y	N	M	1	MIR	$fF\psi\mu1\sigma$
		MPEG-7 harmonic spectral spread	I	Y	N	M	1	MIR	$fF\psi\sigma$
		MPEG-7 harmonic spectral variation	I	Y	N	M	1	MIR	$[fF\psi]R$
		Harmonic energy entropy	I	Y	N	M	1	MIR	$fF\psi H$
		Harmonic concentration	I	Y	N	M	1	MIR	$fF\psi e\sum$
		Spectral peak structure	I	Y	N	M	1	MIR	$fF\psi d\Lambda H$
		Harmonic derivate	I	Y	N	M	V	MIR	$fF1d$

For each feature, we list the domain, a reference to the describing section, temporal scale, whether or not the feature is perceptual and employs psychoacoustic models, the complexity, dimension, application domain, and signature.

TABLE VII
An Overview of Features in Cepstral Domain, Modulation Frequency Domain, Eigendomain, and Phase Space

Domain	Section	Feature name	Temporal scale	Perceptual	Psychoacoustic model	Complexity	Dimension	Application domain	Signature
Cepstral	5.5.1	Mel-scale frequency cepstral coefficients	I	N	Y	H	V	VAR	f F β 1 C
		Bark-scale frequency cepstral coefficients	I	N	Y	H	V	VAR	f F β 1 C
		Autocorrelation MFCCs	I	N	Y	H	V	ASR	f A o F β 1 C
	5.5.2	Noise-robust auditory feature	I	N	Y	H	256	ESR	f B w d o 1 C
	5.5.3	Perceptual linear prediction (PLP)	I	N	Y	H	V	ASR	f F β w w C a r
		Relative spectral PLP	I	N	Y	H	V	ASR	f F β 1 b w w x C a r
		Linear prediction cepstral coefficients	I	N	N	M	V	ASR	f (b) a r
Modulation frequency	5.6	Auditory filter bank temporal envelopes	I	N	Y	M	62	MIR	f b b e $\sum$
		Joint acoustic and modulation frequency features	X	N	Y	H	V	VAR	[f F βo]W$\sum$
		4-Hz modulation harmonic coefficient	X	N	N	M	1	AS	[f A χ]C b
		4-Hz modulation energy	X	N	Y	M	1	AS	[f F β]b e n $\sum$

(*continued*)

TABLE VII (*Continued*)

Domain	Section	Feature name	Temporal scale	Perceptual	Psychoacoustic model	Complexity	Dimension	Application domain	Signature
Modulation frequency	5.6.1	Band periodicity	X	Y	N	M	4	AS	[f b A χ]∑
		Pulse metric	I	Y	N	M	1	AS	f b κ A κ
		Beat spectrum (beat spectrogram)	X	Y	N	H	V	MIR	[f F1o]R A
		Cyclic beat spectrum	X	Y	N	H	V	MIR	o[f F d ∑]c o ∑ κ
		Beat tracker	X	Y	N	H	1	MIR	[f b o d c ∑]κ
		Beat histogram	X	Y	N	M	6	MIR	[f W o ∑ A κ]Λ
		DWPT-based rhythm feature	X	Y	N	M	V	MIR	[f W A κ]Λ
		Rhythm patterns	X	N	Y	H	80	MIR	[[f F β o l w] F w o]φ
Eigendomain	5.7	Rate-scale-frequency features	X	N	Y	H	256	ESR	[f B w d o]W ∑ P
		MPEG-7 audio spectrum basis	X	N	N	H	V	ESR	[f F β l η]S(l)
		Distortion discriminant analysis	X	N	N	H	64	FP	[f M 1 P]P
Phase space	5.8	Phase space features	I	N	N	H	V	ASR	f E

The provided data are organized as in Tables V and VI.

("f F") to obtain a time-frequency representation. We observe that the Cosine transform ("C") is mainly employed for the conversion from frequency to cepstral domain (due to its ability to decorrelate the data). In the set of investigated features, the Wavelet transform ("W") appears rarely compared to the other transformations, although it has better time-frequency resolution than the STFT.

As already mentioned, the features in Tables V–VII are arranged according to the taxonomy (see Section 4). Usually, features from the same group of the taxonomy share similar properties. For example, most harmonicity features share the same building blocks (DFT "F" or autocorrelation "A" followed by a peak detection "h"). Another observation is that pitch and rhythm features make extensive use of autocorrelation.

The identification of building blocks and signatures provides a novel perspective on audio features. Signatures give a compact overview of the computation of a feature and reveal basic properties (e.g., domain, temporal scale, and complexity). Additionally, they enable the comparison of features based on a unified vocabulary of mathematical operations that is independent of any application domain. The literature concerning each feature is listed separately in Section 6.2.

5.2 Temporal Features

The temporal domain is the native domain for audio signals. All temporal features have in common that they are extracted directly from the raw audio signal, without any preceding transformation. Consequently, the computational complexity of temporal features tends to be low.

We partition the group of temporal features into three groups, depending on what the feature describes. First, we investigate features that are based on zero crossings; then, we survey features that describe the amplitude and the energy of a signal, respectively. Figure 10 depicts the groups of the taxonomy.

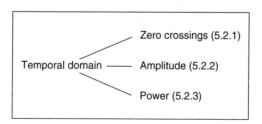

FIG. 10. The organization of features in the temporal domain relates to the captured physical properties of the signal. In brackets a reference to the section containing the corresponding features is given.

5.2.1 Zero Crossing-Based Features

Zero crossings are a basic property of an audio signal that is often employed in audio classification. Zero crossings allow for a rough estimation of dominant frequency and the spectral centroid (SC) [52].

Zero crossing rate. One of the cheapest and simplest features is the ZCR, which is defined as the number of zero crossings in the temporal domain within one second. According to Kedem [53], the ZCR is a measure for the dominant frequency in a signal. ZCR is a popular feature for speech/music discrimination [54, 55] due to its simplicity. However, it is extensively used in a wide range of other audio application domains, such as musical genre classification [56], highlight detection [57], speech analysis [58], singing voice detection in music [59], and environmental sound recognition [60].

Linear prediction zero crossing ratio (LP-ZCR). LP-ZCR is the ratio of the zero crossing count of the waveform and the zero crossing count of the output of a linear prediction analysis filter [52]. The feature quantifies the degree of correlation in a signal. It helps to distinguish between different types of audio, such as (higher correlated) voiced speech and (lower correlated) unvoiced speech.

Zero crossing peak amplitudes (ZCPA). The ZCPA feature has been proposed by Kim et al. [61, 62] for ASR in noisy environments. The ZCPA technique extracts frequency information and corresponding intensities in several psychoacoustically scaled subbands from time domain zero crossings. Information from all subbands is accumulated into a histogram where each bin represents a frequency. The ZCPA feature is an approximation of the spectrum that is directly computed from the signal in temporal domain and may be regarded as a descriptor of the spectral shape. Kim et al. [62] show that ZCPA outperforms linear prediction cepstral coefficients (LPCCs; see Section 5.5.3) under noisy conditions for ASR.

Pitch synchronous zero crossing peak amplitudes (PS-ZCPA). PS-ZCPA is an extension of ZCPA that additionally takes pitch information into account [63]. Small peak amplitudes, which are prone to noise, are removed by synchronizing the ZCPA with the pitch. Ghulam et al. [63] show that the resulting feature is more robust to noise than ZCPA. They further increase the performance of PS-ZCPA by taking auditory masking effects into account in Ref. [64].

5.2.2 Amplitude-Based Features

Some features are directly computed from the amplitude (pressure variation) of a signal. Amplitude-based features are easy and fast to compute but limited in their expressiveness. They represent the temporal envelope of the audio signal.

MPEG-7 audio waveform (AW). The audio waveform descriptor gives a compact description of the shape of a waveform by computing the minimum and maximum samples within nonoverlapping frames. The AW descriptor represents the (downsampled) waveform envelope over time. The purpose of the descriptor is the display and comparison of waveforms rather than retrieval [65].

Amplitude descriptor (AD). The amplitude descriptor has been developed for the recognition of animal sounds [66]. The descriptor separates the signal into segments with low and high amplitude by an adaptive threshold (a level-crossing operation). The duration, variation of duration, and energy of these segments make up the descriptor. AD characterizes the waveform envelope in terms of quiet and loud segments. It allows us to distinguish sounds with characteristic waveform envelopes.

5.2.3 Power-Based Features

The energy of a signal is the square of the amplitude represented by the waveform. The power of a sound is the energy transmitted per unit time (second) [35]. Consequently, power is the mean-square of a signal. Sometimes the root of power (root-mean-square, RMS) is used in feature extraction. In the following, we summarize features that represent the power of a signal (short-time energy—STE, volume) and its temporal distribution (temporal centroid, log attack time—LAT).

Short-time energy. STE describes the envelope of a signal and is extensively used in various fields of audio retrieval (see Table IX for a list of references). We define STE according to Zhang and Kuo [20] as the mean energy per frame (which actually is a measure for power). The same definition is used for the *MPEG-7 audio power descriptor* [65]. Note that there are varying definitions for STE that take the *spectral* power into account [49, 67].

Volume. Volume is a popular feature in audio retrieval, for example in silence detection and speech/music segmentation [54, 68]. Volume is sometimes called loudness, as in Ref. [69]. We use the term loudness for features that model human sensation of loudness (see Section 5.4.3). Volume is usually approximated by the RMS of the signal magnitude within a frame [70]. Consequently, volume is the square root of STE. Both volume and STE reveal the magnitude variation over time.

MPEG-7 temporal centroid. The temporal centroid is the time average over the envelope of a signal in seconds [65]. It is the point in time where most of the energy of the signal is located in average. Note that the computation of temporal centroid is equivalent to that of spectral centroid (Section 5.4.1) in the frequency domain.

MPEG-7 log attack time. LAT characterizes the attack of a sound. LAT is the logarithm of the time it takes from the beginning of a sound signal to the point in time where the amplitude reaches a first significant maximum [65]. The attack characterizes the beginning of a sound, which can be either smooth or sudden. LAT may be employed for classification of musical instruments by their onsets.

5.3 Physical Frequency Features

The group of frequency domain features is the largest group of audio features. All features in this group have in common that they live in frequency or autocorrelation domain. From the signatures in Tables V–VII, we observe that there are several ways to obtain a representation in these domains. The most popular methods are the Fourier transform and the autocorrelation. Other popular methods are the Cosine transform, Wavelet transform, and the constant Q transform. For some features the spectrogram is computed by directly applying a bank of band-pass filters to the temporal signal followed by framing of the subband signals.

We divide frequency features into two subsets: *physical* features and *perceptual* features. See Section 3.1 for more details on these two properties. In this section, we focus on physical frequency features. These features describe a signal in terms of its physical properties. Usually, we cannot assign a semantic meaning to these features. Figure 11 shows the corresponding groups of the taxonomy.

5.3.1 Autoregression-Based Features

Autoregression analysis is a standard technique in signal processing where a linear predictor estimates the value of each sample of a signal by a linear combination of previous values. Linear prediction analysis has a long tradition in audio retrieval and signal coding [71, 72].

Linear predictive coding (LPC). LPC is extensively used in ASR since it takes into account the source-filter model of speech production (by employing an all-pole filter) [71]. The goal of LPC is to estimate basic parameters of a speech signal, such as formant frequencies and the vocal tract transfer function. LPC is applied in other

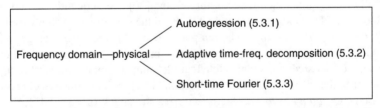

FIG. 11. The organization of physical features in the frequency domain. In brackets a reference to the section containing the corresponding features is given.

domains as well, such as audio segmentation and general-purpose audio retrieval where the LPC spectrum is used as an approximation of the spectral envelope [73–75].

In practice, the cepstral representation of LPC coefficients is mostly used due to their higher retrieval efficiency [76]. We address LPCCs in Section 5.5.3.

Line spectral frequencies (LSF). LSF (also called line spectral pairs) are an alternative representation of linear prediction coefficients. LSF are obtained by decomposing the linear prediction polynomial into two separate polynomials. LSF are at the roots of these two polynomials [77].

LSF characterize the resonances of the linear prediction polynomial together with their bandwidths [78]. While LSF describe equivalent information to LPC coefficients, they have statistical properties that make them better suited for pattern recognition applications [79]. LSF are employed in various application domains, such as in speech/music discrimination [52], instrument recognition [78], and speaker segmentation [80].

5.3.2 Adaptive Time-Frequency Decomposition-Based Features

STFT is widely used in audio feature extraction for time-frequency decomposition. This can be observed from the signatures in Tables V–VII. However, STFT provides only a suboptimal tradeoff between time and frequency resolution since the frequency resolution of the STFT is the same for all locations in the spectrogram. The advantage of adaptive time-frequency decompositions, like the Wavelet transform is that they provide a frequency resolution that varies with the temporal resolution.

This group of the taxonomy comprises features that employ Wavelet transform and related transformations for time-frequency decomposition. Features in this group are based on the transform coefficients. For example, Khan and Al-Khatib [74] successfully employ the variance of Haar Wavelet coefficients over several frames for speech/music discrimination. We consider such features as physical features since they do not have a semantic interpretation.

Daubechies Wavelet coefficient histogram features (DWCHs). DWCHs have been proposed by Li et al. [47] for music genre classification. The authors decompose the audio signal by Daubechies Wavelets and build histograms from the Wavelet coefficients for each subband. The subband histograms provide an approximation of the waveform variation in each subband. The first three statistical moments of each coefficient histogram together with the energy per subband make up the feature vector. Li et al. [47] show that DWCHs improve efficiency in combination with traditional features for music genre classification. Further studies on DWCHs in the fields of artist style identification, emotion detection, and similarity retrieval may be found in Refs. [81, 82].

Adaptive time-frequency transform (ATFT) features. The ATFT investigated by Umapathy et al. [83] is similar to the Wavelet transform. The signal is decomposed into a set of Gaussian basis functions of several scales, translations, and center frequencies. The scale parameter varies with the waveform envelope of the signal and represents for example rhythmic structures. It shows that the scale parameter contains discriminatory information for musical genres.

5.3.3 Short-Time Fourier Transform-Based Features

In this section, we group physical frequency features that employ the STFT for computation of the spectrogram. The STFT yields real and complex values. The real values represent the distribution of the frequency components while the complex values carry information on the phase of the components. Consequently, we distinguish between features that rely on the frequency distribution (spectral envelope) and features that evaluate the phase information. First, we present features that capture basic properties of the spectral envelope: subband energy ratio, spectral flux, spectral slope, and spectral peaks. Then, we focus on phase-based features, such as the (modified) group delay function (GDF).

Subband energy ratio. The subband energy ratio gives a coarse approximation of the energy distribution of the spectrum. There are slightly different definitions concerning the selection of the subbands. Usually, four subbands are used as in Ref. [75]. However, Cai et al. [60] divide the spectrum into eight Mel-scaled bands. The feature is extensively used in audio segmentation [68, 84] and music analysis [85]. See Table IX for further references.

Spectral flux. The SF is the 2-norm of the frame-to-frame spectral amplitude difference vector [86]. It quantifies (abrupt) changes in the shape of the spectrum over time. Signals with slowly varying (or nearly constant) spectral properties (e.g., noise) have low SF, while signals with abrupt spectral changes (e.g., note onsets) have high SF.

A slightly different definition is provided by Lu et al. [87] where the authors compute SF based on the logarithm of the spectrum. Similarly to SF, the cepstrum flux is defined in Ref. [76]. SF is widely used in audio retrieval, for example, in speech/music discrimination [68, 73, 74], music information retrieval [81, 88], and speech analysis [89].

Spectral slope. The spectral slope is a basic approximation of the spectrum shape by a linear regression line [85]. It represents the decrease of the spectral amplitudes from low to high frequencies (the spectral tilt) [48]. The slope, the y-intersection, the maximum and median regression error may be used as features. Spectral slope/tilt may be employed for discrimination of voiced and unvoiced speech segments.

Spectral peaks. Wang [90, 91] introduces features that allow for a very compact and noise-robust representation of an audio signal. The features are part of an audio search engine that is able to identify a piece of music by a short segment captured by a mobile phone.

The author first computes the Fourier spectrogram and detects local peaks. The result is a sparse set of time-frequency points—the constellation map. From the constellation map, pairs of time-frequency points are formed. For each pair, the two frequency components, the time difference, and the time offset from the beginning of the audio signal are combined into a feature. Each piece of music is represented by a large number of such time-frequency pairs. An efficient and scalable search algorithm proposed by Wang allows for efficiently searching large databases built from these features. The search system is best described in Ref. [90].

The proposed feature represents a piece of music in terms of spatiotemporal combinations of dominant frequencies. The strength of the technique is that it solely relies on the salient frequencies (peaks) and rejects all other spectral content. This preserves the main characteristics of the spectrum and makes the representation highly robust to noise since the peak frequencies are usually less influenced by noise than the other frequencies.

Group delay function. The features mentioned above take the real part (magnitude) of the Fourier transform into account. Only a few features describe the phase information of the Fourier spectrum.

Usually, the phase is featureless and difficult to interpret due to polarity and wrapping artifacts. The GDF is the negative derivative of the unwrapped Fourier transform phase [92]. The GDF reveals meaningful information from the phase, such as peaks of the spectral envelope.

The GDF is traditionally employed in speech analysis, for example for the determination of significant excitations [93]. A recent approach applies the GDF in music analysis for rhythm tracking [94]. Since the GDF is not robust against noise and windowing effects, the *modified* GDF is often employed instead [95].

Modified group delay function (MGDF). The MGDF algorithm applies a low-pass filter (cepstral smoothing) to the Fourier spectrum prior to computing the GDF [92]. Cepstral smoothing removes artifacts contributed by noise and windowing, which makes the MGDF more robust and better suited to speech analysis than the GDF [95]. The MGDF is employed in various subdomains of speech analysis, such as speaker identification, phoneme recognition, syllable detection, and language recognition [96–99]. Murthy et al. [100] show that the MGDF robustly estimates formant frequencies.

5.4 Perceptual Frequency Features

So far we have focused on physical frequency features that have no perceptual interpretation. In this section, we cover features that have a semantic meaning in the context of human auditory perception. In the following, we group the features according to the auditory quality that they describe (see Fig. 12).

5.4.1 Brightness

Brightness characterizes the spectral distribution of frequencies and describes whether a signal is dominated by low or high frequencies, respectively. A sound becomes brighter as the high-frequency content becomes more dominant and the low-frequency content becomes less dominant. Brightness is often defined as the *balancing point* of the spectrum [75, 86]. Brightness is closely related to the sensation of sharpness [11].

Spectral centroid. A common approximation of brightness is the SC (or frequency centroid). It is defined as the center of gravity of the magnitude spectrum (first moment) [88, 101]. The SC determines the point in the spectrum where most of the energy is concentrated and is correlated with the dominant frequency of the signal. A definition of spectral centroid in logarithmic frequency can be found in Ref. [102]. Furthermore, SC may be computed for several frequency bands as in Ref. [103].

The MPEG-7 standard provides further definitions of SC [65]. The *MPEG-7 audio spectrum centroid (ASC)* differs from the SC in that it employs a power spectrum in the octave-frequency scale. The ASC approximates the perceptual sharpness of a sound [104]. Another definition of SC is the *MPEG-7 spectral centroid*. The difference to SC is that MPEG-7 spectral centroid is defined for entire

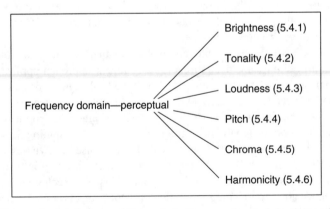

FIG. 12. The organization of perceptual features in the frequency domain. In brackets a reference to the section containing the corresponding features is given.

signals instead of single frames and that the power spectrum is used instead of the magnitude spectrum. The different definitions of spectral centroid are very similar, as shown by the signatures in Table V.

Sharpness. Sharpness is closely related to brightness. Sharpness is a dimension of timbre that is influenced by the center frequency of narrow-band sounds. Sharpness grows with the strength of high frequencies in the spectrum [11]. It may be computed similarly to the spectral centroid but based on the specific loudness instead of the magnitude spectrum. A mathematical model of sharpness is provided by Zwicker and Fastl [11]. Sharpness is employed in audio similarity analysis in Refs. [48, 105].

Spectral center. The spectral center is the frequency where half of the energy in the spectrum is below and half is above that frequency [94]. It describes the distribution of energy and is correlated with the spectral centroid and thus with the dominant frequency of a signal. Sethares et al. [94] employ spectral center together with other features for rhythm tracking.

5.4.2 Tonality

Tonality is the property of sound that distinguishes noise-like from tonal sounds [11]. Noise-like sounds have a continuous spectrum while tonal sounds typically have line spectra. For example, white noise has a flat spectrum and consequently a minimum of tonality while a pure sine wave results in high tonality. Tonality is related to the pitch strength that describes the strength of the perceived pitch of a sound (see Section 2.4). Sounds with distinct (sinusoidal) components tend to produce larger pitch strength than sounds with continuous spectra.

We distinguish between two classes of features that (partially) measure tonality: *flatness measures* and *bandwidth measures*. In the following, we first describe bandwidth measures (bandwidth, spectral dispersion, and spectral rolloff point) and then we focus on flatness measures (spectral crest, spectral flatness, subband spectral flux, and entropy).

Bandwidth. Bandwidth is usually defined as the magnitude-weighted average of the differences between the spectral components and the spectral centroid [69]. The bandwidth is the second-order statistic of the spectrum. Tonal sounds usually have a low bandwidth (single peak in the spectrum) while noise-like sounds have high bandwidth. However, this is not the case for more complex sounds. For example in music we find broadband signals with tonal characteristics. The same applies to complex tones with a large number of harmonics that may have a broadband line spectrum. Consequently bandwidth may not be a sufficient indicator for tonality for particular tasks. Additional features (e.g., harmonicity features, see Section 5.4.6 and flatness features, see below) may be necessary to distinguish between tonal and noise-like signals.

Bandwidth may be defined in the logarithmized spectrum or the power spectrum [49, 102, 106]. Additionally, it may be computed within one or more subbands of the spectrum [103, 107].

In the MPEG-7 standard the measure for bandwidth is called *spectral spread* [65, 104]. Similarly to the bandwidth measures above, the MPEG-7 audio spectrum spread (ASS) is the RMS deviation from the spectrum centroid (MPEG-7 ASC descriptor, see Section 5.4.1). Measures for bandwidth are often combined with that of spectral centroid in literature since they represent complementary information [49, 103, 107].

Spectral dispersion. The spectral dispersion is a measure for the spread of the spectrum around its spectral center [94]. See Section 5.4.1 for a description of spectral center. In contrast to bandwidth, the computation of spectral dispersion takes the spectral center into account instead of the spectral centroid.

Spectral rolloff point. The spectral rolloff point is the $N\%$ percentile of the power spectral distribution, where N is usually 85% or 95% [86]. The rolloff point is the frequency below which $N\%$ of the magnitude distribution is concentrated. It increases with the bandwidth of a signal. Spectral rolloff is extensively used in music information retrieval [85, 108] and speech/music segmentation [86].

Spectral flatness. Spectral flatness estimates to which degree the frequencies in a spectrum are uniformly distributed (noise-like) [109]. The spectral flatness is the ratio of the geometric and the arithmetic mean of a subband in the power spectrum [103]. The same definition is used by the MPEG-7 standard for the *audio spectrum flatness* descriptor [65]. Spectral flatness may be further computed in decibel scale as in Refs. [110, 111]. Noise-like sounds have a higher flatness value (flat spectrum) while tonal sounds have lower flatness values. Spectral flatness is often used (together with spectral crest factor) for audio fingerprinting [111, 112].

Spectral crest factor. The spectral crest factor is a measure for the "peakiness" of a spectrum and is inversely proportional to the flatness. It is used to distinguish noise-like and tone-like sounds due to their characteristic spectral shapes. Spectral crest factor is the ratio of the maximum spectrum power and the mean spectrum power of a subband. In Ref. [111], the spectral crest factor is additionally logarithmized. For noise-like sounds the spectral crest is lower than for tonal sounds. A traditional application of spectral crest factor is fingerprinting [103, 111, 112].

Subband spectral flux (SSF). The SSF has been introduced by Cai et al. [60] for the recognition of environmental sounds. The feature is a measure for the portion of prominent partials ("peakiness") in different subbands. SSF is computed from the logarithmized short-time Fourier spectrum. For each subband the SSF is the accumulation of the differences between adjacent frequencies in that subband. SSF is low for flat subbands and high for subbands that contain distinct frequencies. Consequently, SSF is inversely proportional to spectral flatness.

Entropy. Another measure that correlates with the flatness of a spectrum is entropy. Usually, Shannon and Renyi entropy are computed in several subbands [103]. The entropy represents the uniformity of the spectrum. A *multiresolution entropy* feature is proposed by Misra et al. [113, 114]. The authors split the spectrum into overlapping Mel-scaled subbands and compute the Shannon entropy for each subband. For a flat distribution in the spectrum the entropy is low while a spectrum with sharp peaks (e.g., formants in speech) has high entropy. The feature captures the "peakiness" of a subband and may be used for speech/silence detection and automatic speech recognition.

5.4.3 Loudness

Loudness features aim at simulating the human sensation of loudness. Loudness is "that attribute of auditory sensation in terms of which sounds may be ordered on a scale extending from soft to loud" [9]. The auditory system incorporates a number of physiological mechanisms that influence the transformation of the incoming physical sound intensity into the sensational loudness [11]. See Section 3.3 for a summary of important effects.

Specific loudness sensation (sone). Pampalk et al. [28] propose a feature that approximates the specific loudness sensation per critical band of the human auditory system. The authors first compute a Bark-scaled spectrogram and then apply spectral masking and equal-loudness contours (expressed in phon). Finally, the spectrum is transformed to specific loudness sensation (in sone). The feature is the basis for rhythm patterns (see Section 5.6.1). The representation in sone may be applied to audio retrieval as in Refs. [85, 115].

Integral loudness. The specific loudness sensation (sone) gives the loudness of a single sine tone. A spectral integration of loudness over several frequencies enables the estimation of the loudness of more complex tones [11]. Pfeiffer proposes an approach to compute the integral loudness by summing up the loudness in different frequency groups [40]. The author empirically shows that the proposed method closely approximates the human sensation of loudness. The integral loudness feature is applied to foreground/background segmentation in Ref. [116].

5.4.4 Pitch

Pitch is a basic dimension of sound, together with loudness, duration, and timbre. The hearing sensation of pitch is defined as "that attribute of auditory sensation in terms of which sounds may be ordered on a scale extending from low to high" [9]. The term pitch is widely used in literature and may refer to both, a stimulus parameter (fundamental frequency or frequency of glottal oscillation) and an auditory sensation (the perceived frequency of a signal), depending on the application domain.

In this section, we first focus on features that capture the fundamental frequency and then present a technique that models the psychoacoustic pitch. Features that describe pitch are correlated to chroma and harmonicity features (see Sections 5.4.5 and 5.4.6).

Fundamental frequency. The fundamental frequency is the lowest frequency of a harmonic series and is a coarse approximation of the psychoacoustic pitch. Fundamental frequency estimation employs a wide range of techniques, such as temporal autocorrelation, spectral, and cepstral methods and combinations of these techniques. An overview of techniques is given in Ref. [117].

The MPEG-7 standard proposes a descriptor for the fundamental frequency (*MPEG-7 audio fundamental frequency*) which is defined as the first peak of the local normalized spectrotemporal autocorrelation function [65, 118]. Fundamental frequency is employed in various application domains [58, 69, 88].

Pitch histogram. The pitch histogram describes the pitch content of a signal in a compact way and has been introduced for musical genre classification in Refs. [18, 88]. In musical analysis pitch usually corresponds to musical notes. The pitch histogram is a global representation that aggregates the pitch information of several short audio frames. Consequently, the pitch histogram represents the distribution of the musical notes in a piece of music. A similar histogram-based technique is the beat histogram that represents the rhythmic content of a signal (see Section 5.6.1).

Psychoacoustic pitch. Meddis and O'Mard [119] propose a method to model human pitch perception. First the authors apply a band-pass filter to the input signal to emphasize the frequencies relevant for pitch perception. Then the signal is decomposed with a gammatone filter bank that models the frequency selectivity of the cochlea. For each subband an inner hair-cell model transforms the instantaneous amplitudes into continuous firing probabilities. A running autocorrelation function is computed from the firing probabilities in each subband. The resulting autocorrelation functions are summed across the channels to obtain the final feature.

In contrast to other pitch detection techniques, the output of this algorithm is a series of values instead of one single pitch value. These values represent a range of frequencies relevant for pitch perception. Meddis and O'Mard point out that a single pitch frequency is not sufficient for approximation of the pitch perception of complex sounds. Consequently, they employ all values of the feature for matching pitches of different sounds.

5.4.5 Chroma

According to Shepard [120], the sensation of musical pitch may be characterized by two dimensions: *tone height* and *chroma*. The dimension of tone height is partitioned into the musical octaves. The range of chroma is usually divided into

12 pitch classes, where each pitch class corresponds to one note of the 12-tone equal temperament. For example, the pitch class C contains the Cs of all possible octaves ($C_0, C_1, C_2, \ldots$). The pitches (musical notes) of the same pitch class share the same chroma and produce a similar auditory sensation. Chroma-based representations are mainly used in music information analysis and retrieval since they provide an octave-invariant representation of the signal.

Chromagram. The chromagram is a spectrogram that represents the spectral energy of each of the 12 pitch classes [121]. It is based on a logarithmized short-time Fourier spectrum. The frequencies are mapped (quantized) to the 12 pitch classes by an aggregation function. The result is a 12 element vector for each audio frame. A similar algorithm for the extraction of chroma vectors is presented in Ref. [122].

The chromagram maps all frequencies into one octave. This results in a spectral compression that allows for a compact description of harmonic signals. Large harmonic series may be represented by only a few chroma values, since most harmonics fall within the same pitch class [121]. The chromagram represents an octave-invariant (compressed) spectrogram that takes properties of musical perception into account.

Chroma energy distribution normalized statistics (CENS). CENS features are another representation of chroma, introduced for music similarity matching by Müller et al. [46] and Müller [123]. The CENS features are robust against tempo variations and different timbres which makes them suitable for the matching of different interpretations of the same piece of music.

Pitch profile. The pitch profile is a more accurate representation of the pitch content than the chroma features [124]. It takes pitch mistuning (introduced by mistuned instruments) into account and is robust against noisy percussive sounds (e.g., sounds of drums that do not have a pitch). Zhu and Kankanhalli [124] apply the pitch profile in musical key detection and show that the pitch profile outperforms traditional chroma features.

5.4.6 Harmonicity

Harmonicity is a property that distinguishes periodic signals (harmonic sounds) from nonperiodic signals (inharmonic and noise-like sounds). Harmonics are frequencies at integer multiples of the fundamental frequency. Figure 13 presents the spectra of a noise-like (inharmonic) and a harmonic sound. The harmonic spectrum shows peaks at the fundamental frequency and its integer multiples.

Harmonicity relates to the proportion of harmonic components in a signal. Harmonicity features may be employed to distinguish musical instruments. For example harmonic instrument sounds (e.g., violins) have stronger harmonic structure than percussive instrument sounds (e.g., drums). Furthermore, harmonicity may

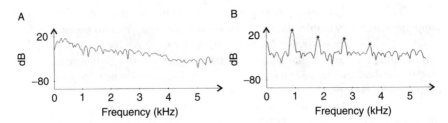

FIG. 13. (A) The spectrum of a noise-like sound (thunder). (B) The spectrum of a harmonic sound (siren). The harmonic sound has peaks at multiples of the fundamental frequencies (the harmonic peaks are marked by asterisks), while the noise-like sound has a flat spectrum.

be useful in ESR to distinguish between harmonic (e.g., bird song) and inharmonic (e.g., dog barks) sounds.

MPEG-7 audio harmonicity. The audio harmonicity descriptor of the MPEG-7 standard comprises two measures. The *harmonic ratio* is the ratio of the fundamental frequency's power to the total power in an audio frame [65, 104]. It is a measure for the degree of harmonicity contained in a signal. The computation of harmonic ratio is similar to that of MPEG-7 audio fundamental frequency, except for the used autocorrelation function.

The second measure in the audio harmonicity descriptor is the *upper limit of harmonicity*. The upper limit of harmonicity is the frequency beyond which the spectrum no longer has any significant harmonic structure. It may be regarded as the bandwidth of the harmonic components. The audio harmonicity descriptor is well suited for the distinction of periodic (e.g., musical instruments, voiced speech) and nonperiodic (e.g., noise, unvoiced speech) sounds.

A similar feature is the *harmonic coefficient* which is defined as the first maximum in the (spectrotemporal) autocorrelation function in Ref. [125]. Note that the definition is nearly equivalent to that of harmonic ratio, except for the employed autocorrelation function.

Inharmonicity measures. Most real-world harmonic signals do not show a perfect harmonic structure. Inharmonicity features measure the difference between observed harmonics and their theoretical (predicted) values which are exactly at integer multiples of the fundamental frequency.

A straightforward cumulative measure for the deviation of the harmonics from their predicted values is presented in Refs. [48, 107]. A more enhanced and more accurate feature is *harmonicity prominence* which additionally takes the energy and the bandwidth of each harmonic component into account [60].

A related feature is *spectral peak structure* which is the entropy of the distances of adjacent peaks in the spectrum. For perfect harmonic sounds, these distances are constant, while for nonharmonic sounds the distances may vary. Consequently, the entropy of the distances is a measure for inharmonicity.

MPEG-7 spectral timbral descriptors. The MPEG-7 standard defines a set of descriptors for the harmonic structure of sounds: *MPEG-7 harmonic spectral centroid (HSC), MPEG-7 harmonic spectral deviation (HSD), MPEG-7 harmonic spectral spread (HSS),* and *MPEG-7 harmonic spectral variation (HSV)* [65, 126]. All descriptors are based on an estimate of the fundamental frequency and the detection of harmonic peaks in the spectrum (see the signatures in Table VI). The descriptors represent statistical properties (moments) of the harmonic frequencies and their amplitudes.

The HSC is the amplitude-weighted average of the harmonic frequencies. Similarly to spectral centroid (see Section 5.4.1) HSC is related to brightness and sharpness [104].

The HSS descriptor is the power-weighted RMS deviation of the harmonic peaks from the HSC. It represents the bandwidth of the harmonic frequencies. HSC and HSS are first and second moment of the harmonic spectrum similarly to spectral centroid and bandwidth (spectral spread) which are first and second moment of the entire spectrum.

HSD measures the amplitude deviation of harmonic peaks from their neighboring harmonic peaks in the same frame. If all harmonic peaks have equal amplitude HSD reaches its minimum. While HSS represents the variation of harmonic frequencies, HSD reflects the variation of harmonics' amplitudes.

The HSV descriptor represents the correlation of harmonic peak amplitudes in two adjacent frames. It represents fast variations of harmonic structures over time. The MPEG-7 spectral timbral descriptors address musical instrument recognition, where the harmonic structure is an important discriminative property [126].

Further harmonicity features. Srinivasan and Kankanhalli [102] introduce harmonicity features for classification of music genre and instrument family. *Harmonic concentration* measures the fraction of energy of the dominant harmonic component of the signal. *Harmonic energy entropy* describes the energy distribution of the harmonic components by computing the entropy of their energies. Finally, Srinivasan and Kankanhalli define the *harmonic derivate* as the difference of the energy of adjacent harmonic frequencies. The feature represents the decay of harmonic energy with increasing frequency.

There is a large number of features that capture harmonic properties in literature. Harmonicity features are related to pitch and chroma features. Additionally, they are correlated to each other to a high degree due to methodological similarities which may be observed from the signatures in Table VI.

5.5 Cepstral Features

The concept of the "cepstrum" has been originally introduced by Bogert et al. [26] for the detection of echoes in seismic signals. In the domain of audio, cepstral features have first been employed for speech analysis [51, 127, 128]. Cepstral features are frequency smoothed representations of the log magnitude spectrum and capture timbral characteristics and pitch. Cepstral features allow for application of the Euclidean metric as distance measure due to their orthogonal basis which facilitates similarity comparisons [127]. Today, cepstral features are widely used in all fields of audio retrieval (speech, music, and environmental sound analysis), for example [19, 129].

We have identified three classes of cepstral features. The first group employs traditional filter banks, such as Mel- and Bark-filters. The second group bases on more elaborate auditory models. The third group are cepstral features that apply autoregression.

5.5.1 Perceptual Filter Bank-Based Features

Bogert et al. [26] define the cepstrum as the FT of the logarithm (log) of the magnitude (mag) of the spectrum of the original signal.

$$\text{signal} \rightarrow \text{FT} \rightarrow \text{mag} \rightarrow \text{log} \rightarrow \text{FT} \rightarrow \text{cepstrum}$$

This sequence is the basis for the cepstral features described in this section. However, in practice the computation slightly differs from this definition. For example, the second Fourier transform is often replaced by a DCT due to its ability to decorrelate output data.

Mel-frequency cepstral coefficients. MFCCs originate from ASR but evolved into one of the standard techniques in most domains of audio retrieval. They represent timbral information (the spectral envelope) of a signal. MFCCs have been successfully applied to timbre measurements by Terasawa et al. [13].

Computation of MFCCs includes a conversion of the Fourier coefficients to Melscale [34]. After conversion, the obtained vectors are logarithmized, and decorrelated by DCT to remove redundant information.

The components of MFCCs are the first few DCT coefficients that describe the coarse spectral shape. The first DCT coefficient represents the average power in the spectrum. The second coefficient approximates the broad shape of the spectrum and is related to the spectral centroid. The higher-order coefficients represent finer spectral details (e.g., pitch). In practice, the first 8–13 MFCC coefficients are used to represent the shape of the spectrum. However, some applications require more higher-order coefficients to capture pitch and tone information. For example, in Chinese speech recognition up to 20 cepstral coefficients may be beneficial [130].

Variations of MFCCs. In the course of time several variations of MFCCs have been proposed. They mainly differ in the applied psychoacoustic scale. Instead of the Mel-scale, variations employ the Bark- [32], ERB- [33], and octave-scale [131]. A typical variation of MFCCs are Bark-frequency cepstral coefficients (BFCCs). However, cepstral coefficients based on the Mel-scale are the most popular variant used today, even if there is no theoretical reason that the Mel-scale is superior to the other scales.

Extensions of MFCCs. A noise-robust extension of MFCCs are *autocorrelation MFCCs* proposed by Shannon and Paliwal [44]. The main difference is the computation of an unbiased autocorrelation from the raw signal. Particular autocorrelation coefficients are removed to filter noise. From this representation more noise-robust MFCCs are extracted.

Yuo et al. [45] introduce two noise-robust extensions of MFCCs, namely RAS-MFCCs and CHNRAS-MFCCs. The features introduce a preprocessing step to the standard computation of MFCCs that filters additive and convolutional noise (cannel distortions) by cepstral mean subtraction.

Another extension of MFCCs is introduced in Ref. [132]. Here, the outputs of the Mel-filters are weighted according to the amount of estimated noise in the bands. The feature improves accuracy of ASR in noisy environments.

Li et al. [133] propose a novel feature that may be regarded as an extension of BFCCs. The feature incorporates additional filters that model the transfer function of the cochlea. This enhances the ability to simulate the human auditory system and improves performance in noisy environments.

5.5.2 Advanced Auditory Model-Based Features

Features in this group base on an auditory model that is designed to closely represent the physiological processes in human hearing.

Noise-robust audio features (NRAF). NRAF are introduced in Ref. [21] and are derived from a mathematical model of the early auditory system [134]. The auditory model yields a psychoacoustically motivated time-frequency representation which is called the auditory spectrum. A logarithmic compression of the auditory spectrum models the behavior of the outer hair cells. Finally, a DCT decorrelates the data. The temporal mean and variance of the resulting decorrelated spectrum make up the components of NRAF. The computation of NRAF is similar to that of MFCCs but it follows the process of hearing in a more precise way. A related audio feature of NRAF are rate-scale-frequency (RSF) features addressed in Section 5.7.

5.5.3 Autoregression-Based Features

Features in this group are cepstral representations that base on linear predictive analysis (see Section 5.3.1).

Perceptual linear prediction (PLP). PLP was introduced by Hermansky [135] in 1990 for speaker-independent speech recognition. It bases on the concepts of hearing and employs linear predictive analysis for the approximation of the spectral shape. In the context of speech PLP represents speaker-independent information, such as vocal tract characteristics. It better represents the spectral shape than conventional linear prediction coding (LPC) by approximating several properties of human hearing. The feature employs Bark-scale as well as asymmetric critical-band masking curves to achieve a higher grade of consistency with human hearing.

Relative spectral—perceptual linear prediction (RASTA-PLP). RASTA-PLP is an extension of PLP introduced by Hermansky and Morgan [23]. The objective of RASTA-PLP is to make PLP more robust to linear spectral distortions. The authors filter each frequency channel with a band-pass filter to alleviate fast variations (frame-to-frame variations introduced by the short-time analysis) and slow variations (convolutional noise introduced by the communication channel). RASTA-PLP better approximates the human abilities to filter noise than PLP and yields a more robust representation of the spectral envelope under noisy conditions.

Linear prediction cepstral coefficients. LPCCs are the inverse Fourier transform of the log magnitude frequency response of the autoregressive filter. They are an alternative representation for linear prediction coefficients and thus capture equivalent information. LPCCs may be directly derived from the LPC coefficients presented in Section 5.3.1 with a recursion formula [136].

In practice, LPCCs have shown to perform better than LPC coefficients, for example, in ASR, since they are a more compact and robust representation of the spectral envelope [137]. In contrast to LPC they allow for the application of the Euclidean distance metric. The traditional application domain of LPCCs is ASR. However, LPCCs may be employed in other domains, such as music information retrieval as well [76].

5.6 Modulation Frequency Features

Modulation frequency features capture low-frequency modulation information in audio signals. A modulated signal contains at least two frequencies: a high carrier frequency and a comparatively low modulation frequency. Modulated sounds cause different hearing sensations in the human auditory system. Low modulation frequencies up to 20 Hz produce the hearing sensation of fluctuation strength [11]. Higher modulation frequencies create the hearing sensation of roughness.

Modulation information is a long-term signal variation of amplitude or frequency that is usually captured by a temporal (interframe) analysis of the spectrogram.

Rhythm and tempo are aspects of sound (especially important in music) that are strongly related to long-time modulations. Rhythmic structures (e.g., sequences of equally spaced beats or pulses) may be revealed by analyzing low-frequency modulations over time. Figure 14 shows a short-time Fourier spectrogram together with the corresponding modulation spectrogram of a piece of music. The spectrogram represents the distribution of acoustic frequencies over time, while the modulation spectrogram shows the distribution of long-term modulation frequencies for each acoustic frequency. In Fig. 14, we observe two strong modulation frequencies at 3 and 6 Hz that are distributed over all critical bands. These frequencies relate to the main- and sub-beats of the song. We discuss features that represent rhythm and tempo-related information in Section 5.6.1.

4-Hz modulation energy. The hearing sensation of fluctuation strength has its peak at 4-Hz modulation frequency (for both amplitude- and frequency-modulated sounds) [138, 139]. This is the modulation frequency that is most often observed in fluent speech, where approximately four syllables per second are produced. Hence, the 4-Hz modulation energy may be employed for distinguishing speech from nonspeech sounds.

Scheirer and Slaney [86] extract the 4-Hz modulation energy by a spectral analysis of the signal. They filter each subband by a 4-Hz band-pass filter along the temporal dimension. The filter outputs represent the 4-Hz modulation energy. A different definition that derives the 4-Hz modulation energy is given in Ref. [70].

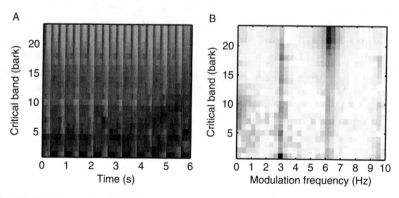

FIG. 14. (A) The spectrogram of a 6 s excerpt of "Rock DJ" by Robbie Williams. (B) The modulation spectrogram reveals modulation frequencies at 3 and 6 Hz. The modulation frequencies relate to the main beat and the sub-beats of the song.

Similarly to 4-Hz modulation frequency, Chou and Gu [125] define the *4-Hz modulation harmonic coefficient* which actually is an estimate of the 4-Hz modulation energy of the fundamental frequency of a signal. The authors report that this feature better discriminates speech from singing than the 4-Hz modulation frequency.

Joint acoustic and modulation frequency features. Sukittanon and Atlas [140] propose a feature for audio fingerprinting that represents the distribution of modulation frequencies in the critical bands. The feature is a time-invariant representation and captures time-varying (nonstationary) behavior of an audio signal.

The authors first decompose the input signal into a Bark-scaled spectrogram. Then they demodulate the spectrogram by extracting frequencies of each subband envelope. A Wavelet transform produces one modulation frequency vector for each subband. The output of this procedure is a matrix (a modulation spectrogram) that contains the modulation frequencies for each acoustic frequency band. The modulation spectrogram is constant in size and time-invariant. Hence, it may be vectorized to build a feature vector (fingerprint) for retrieval.

Sukittanon et al. [24] show that their modulation frequency feature outperforms MFCCs in presence of noise and time-frequency distortions. A similar feature are rhythm patterns which have been developed for music similarity matching. We present rhythm patterns together with other rhythm-related features in Section 5.6.1.

A spectral representation that takes the temporal resolution of modulation information into account is the modulation spectrogram by Greenberg and Kingsbury [27]. In contrast to the features mentioned above, the modulation spectrogram shows the distribution of slow modulations across time *and* frequency. Experiments show that it is more robust to noise than the narrow-band spectrogram.

Auditory filter bank temporal envelopes. McKinney and Breebaart [56] present another approach for the computation of modulation frequency features. They employ logarithmically spaced gamma tone filters for subband decomposition. The resulting subband envelopes are band-pass filtered to obtain modulation information. The feature represents modulation energy for particular acoustic frequency bands similarly to the joint acoustic and modulation frequency features (see above). The features have been successfully employed for musical genre classification and general-purpose audio classification.

5.6.1 Rhythm

Rhythm is a property of an audio signal that represents a change pattern of timbre and energy over time [20]. According to Zwicker and Fastl [11], the hearing sensation of rhythm depends on the temporal variation of loudness. Rhythm is an important element in speech and music. In speech, rhythm relates to stress and pitch and in music it relates to the tempo of a piece of music (in beats per minute). Rhythm may be important for the characterization of environmental sounds as well, for example for the description of footsteps [20].

Rhythm is a property that evolves over time. Consequently, the analysis windows of rhythm features are usually longer than that of other features. Typical analysis windows are in the range of a few seconds (≈ 3–5 s) [88]. Rhythmic patterns are usually obtained by analyzing low-frequency amplitude modulations.

We first present two features that measure the strength of a rhythmic variation in a signal (pulse metric and band periodicity). Then we summarize features that estimate the main- and sub-beats in a piece of music (beat spectrum representations, beat tracker) and finally we address features that globally represent the rhythmic structure of a piece of music (beat histograms and rhythm patterns).

Pulse metric. A measure for the "rhythmicness" of sound is proposed by Scheirer and Slaney [86]. They detect rhythmic modulations by identifying peaks in the autocorrelation function of several subbands. The pulse metric is high when the autocorrelations in all subbands show peaks at similar positions. This indicates a strong rhythmic structure in the sound.

Band periodicity. The band periodicity also measures the strength of rhythmic structures and is similar to pulse metric [87]. The signal is split into subbands and the maximum peak of the subband correlation function is estimated for each analysis frame. The band periodicity for a subband is the mean of the peaks in all frames. It correlates with the rhythm content of a signal, since it captures the strength of repetitive structures over time.

Beat spectrum (beat spectrogram). The beat spectrum represents the self-similarity of a signal for different time lags (similarly to autocorrelation) [141, 142]. The peaks in the beat spectrum indicate strong beats with a specific repetition rate. Hence, this representation allows a description of the rhythm content of a signal. The peaks correspond to note onsets with high periodicity.

The beat spectrum is computed for several audio frames to obtain the beat spectrogram. Each column of the beat spectrogram is the beat spectrum of a single frame. The beat spectrogram shows the rhythmic variation of a signal over time. It is a two-dimensional representation that has the time dimension on the abscissa and the lag time (repetition rate or tempo) on the ordinate. The beat spectrogram visualizes how the tempo changes over time and allows for a detailed analysis of the rhythmic structures and variations.

Note that the beat spectrogram represents similar information as the joint acoustic and modulation frequency feature (see above). Both representations capture rhythmic content of a signal. However, the beat spectrogram represents the variation of tempo over time while the joint acoustic and modulation representation reveals rhythmic patterns independently of time. The difference between both representations is that the beat spectrogram provides temporal information while it neglects the distribution of acoustic frequencies and the modulation spectrogram preserves acoustic frequencies and neglects time. Both complement each other.

The beat spectrum serves as a basis for onset detection and the determination of rhythmically similar music. It may be used for the segmentation of pieces of music into rhythmically different parts, such as chorus and verse.

Cyclic beat spectrum (CBS). A related representation to the beat spectrum is the CBS [143]. The CBS is a compact and robust representation of the *fundamental* tempo of a piece of music. Tempo analysis with the beat spectrum reveals not only the fundamental tempo but also corresponding tempos with a harmonic and subharmonic relationship to the fundamental tempo (e.g., 1/2-, 1/3-, 2-, 3-,...fold tempo). The CBS groups tempos belonging to the same fundamental tempo into one tempo class. This grouping is similar to the grouping of frequencies into chroma classes (see Section 5.4.5).

The CBS is derived from a beat spectrum. Kurth et al. first low-pass filter the signal (to remove timbre information that may be neglected for tempo analysis) and compute a spectrogram by STFT. They derive a novelty curve by summing the differences between adjacent spectral vectors. The novelty curve is then analyzed by a bank of comb filters where each comb filter corresponds to a particular tempo. This analysis results in a beat spectrogram where peaks correspond to dominant tempos. The beat spectrum is divided into logarithmically scaled tempo octaves (tempo classes) similarly to pitch classes in the context of chroma. The CBS is obtained by aggregating the beat spectrum over all tempo classes.

The CBS robustly estimates one or more significant and independent tempos of a signal and serves as a basis for the analysis of rhythmic structures. Kurth et al. [143] employ the beat period (derived from the CBS) together with more complex rhythm and meter features for time-scale invariant audio retrieval.

Beat tracker. An important rhythm feature is Scheirer's [144, 145] beat-tracking algorithm which enables the determination of tempo and beat positions in a piece of music. The algorithm starts with a decomposition of the input signal into subbands. Each subband envelope is analyzed by a bank of comb filters (resonators). The resonators extract periodic modulations from the subband envelopes and are related to particular tempos. The resonator's outputs are summed over all subbands to obtain an estimate for each tempo under consideration. The frequency of the comb filter with the maximum energy output represents the tempo of the signal.

An advantage of using comb filters instead of autocorrelation methods for finding periodic modulations is that they allow for the detection of the beat positions and thus enable beat tracking. Scheirer tracks beat positions by analyzing the phase information preserved by the comb filters. The author empirically shows that the proposed technique approximates the beat-tracking abilities of human listeners.

See Ref. [144] for a comparison of comb filters with autocorrelation methods and more details on the beat-tracking algorithm.

Beat histogram. The beat histogram is a compact global representation of the rhythm content of a piece of music [88, 146]. It describes the repetition rates of main beat and sub-beats together with their strength. Similarly to other rhythm features, the computation is based on periodicity analysis in multiple frequency bands. The authors employ Wavelet transform to obtain an octave-frequency decomposition. They detect the most salient periodicities in each subband and accumulate them into a histogram. This process is similar to that of pitch histograms in Section 5.4.4.

Each bin of the histogram corresponds to a beat period in beats per minute where peaks indicate the main- and sub-beats. The beat histogram compactly summarizes all occurring beat periods (tempos) in a piece of music. The beat histogram is designed for MIR, especially genre classification. A measure for the beat strength may be easily derived from the beat histogram as in Ref. [147]. Grimaldi et al. introduce a derivation of the beat histogram in Ref. [148] that builds upon the discrete Wavelet packet transform (DWPT) [149].

Rhythm patterns. Rhythm patterns are proposed for music similarity retrieval by Pampalk et al. [28]. They build upon the specific loudness sensation in sone (see Section 5.4.3). Given the spectrogram (in specific loudness) the amplitude modulations are extracted by a Fourier analysis of the critical bands over time. The extracted modulation frequencies are weighted according to the fluctuation strength to approximate the human perception [11]. This results in a two-dimensional representation of acoustic versus modulation frequency. A detailed description of the computation is given in Ref. [150]. Note that rhythm patterns are similar to the joint acoustic and modulation frequency features mentioned above.

5.7 Eigendomain Features

Features in this group represent long-term information contained in sound segments that have a duration of several seconds. This leads to large amounts of (redundant) feature data with low expressiveness that may not be suitable for further processing (e.g., classification).

Statistical methods may be applied to reduce the amount of data in a way that preserves the most important information. The employed statistical methods usually decorrelate the feature data by factorization. The resulting representation allows for dimensionality reduction by removing factors with low influence. Methods such as PCA and SVD are standard techniques for this purpose.

Rate-scale-frequency features. Ravindran et al. [21] introduce RSF features for general-purpose sound recognition. The computation of the features relies on a model of the auditory cortex and the early auditory model, used for NRAF (see

NRAF in Section 5.5.2). RSF features describe modulation information for selected frequency bands of the auditory spectrum. Ravindran et al. apply a two-dimensional Wavelet transform to the auditory spectrum in order to extract temporal and spatial modulation information resulting in a three-dimensional representation. They perform PCA for compression and decorrelation of the data to obtain an easily processable fingerprint.

MPEG-7 audio spectrum basis/projection. The MPEG-7 standard defines the combination of audio spectrum basis (ASB) and audio spectrum projection (ASP) descriptors for general-purpose sound recognition [65, 151]. ASB is a compact representation of the short-time spectrogram of a signal. The compression of the spectrogram is performed by SVD. ASB contains the coarse frequency distribution of the *entire* spectrogram. This makes it suitable for general-purpose sound recognition. The ASP descriptor is a projection of a spectrogram against a given ASB. ASP and ASB are usually combined in a retrieval task as described in Ref. [104].

Distortion discriminant analysis (DDA). DDA features are used for noise-robust fingerprinting [152]. Initially, the signal is transformed using a modulated complex lapped transform (MCLT) which yields a time-frequency representation [153]. The resulting spectrogram is passed to a hierarchy of oriented PCAs to subsequently reduce the dimensionality of the spectral vectors and to remove distortions. This hierarchical application of the oriented PCA yields a compact time-invariant and noise-robust representation of the entire sound.

DDA generates features that are robust to several types of noise and distortions, such as time shifts, frequency distortions, and compression artifacts. Burges et al. [43] point out that DDA is even robust against types of noise that are not present in the training set.

5.8 Phase Space Features

In speech production nonlinear phenomena, such as turbulence have been observed in the vocal tract [154]. Features in the domains mentioned so far (temporal, frequency, cepstral, etc.) are not able to capture nonlinear phenomena. The *state space* represents a domain that reveals the nonlinear behavior of a system. However, in general it is not possible to extract the state space for an audio signal, since not all necessary variables may be derived from the audio signal. Alternatively, the *reconstructed phase space*, an approximation that shares important properties with the state space, may be computed. For phase space reconstruction the original audio signal is considered to be a one-dimensional projection of the dynamic system. The reconstructed phase space is built by creating time-lagged versions of the original signal. The original signal is shifted by multiples of a constant time lag. Each dimension of the reconstructed phase space relates to a delayed version of the original signal. The dimension of the reconstructed phase space corresponds to the number of time-lagged versions of the original signal. The critical steps in phase space reconstruction are the determination of embedding dimension and time lag. An extensive description of phase space reconstruction is given in Ref. [30]. The possibly high-dimensional attractor of the system unfolds in the phase space if time lag and embedding dimension are properly selected. Several parameters of the attractor may serve as audio features.

The *Lyapunov exponents* of the attractor measure the "degree of chaos" of a dynamic system. Kokkinos and Maragos [154] employ Lyapunov exponents for the distinction of different phonemes in speech. They observe that phonemes, such as voiced and unvoiced fricatives, (semi)vowels, and stop sounds may be characterized by their Lyapunov exponents due to the different degree of chaos in these phonemes.

Lindgren et al. [31] employ the *natural distribution* of the attractor together with its first derivative as features for phoneme recognition. The natural distribution describes the spatial arrangement of the points of the attractor, that is, the coarse shape of the attractor. The first derivative characterizes the flow or trajectory of the attractor over time.

Further features derived from reconstructed phase space are dimension measures of the attractor, such as *fractal dimension* [154] and *correlation dimension* [155].

Bai et al. [156] show that phase space features are well suited for musical genre classification. They compute the angles between vectors in phase space and employ the variance of these angles as features.

Phase space features capture information that is orthogonal to features that originate from linear models. Experiments show that recognition solely based on phase space features is poor compared to results of standard features, such as MFCCs [31]. Consequently, phase space features are usually combined with traditional features to improve accuracy of recognition.

6. Related Literature

6.1 Application Domains

In the following, we briefly present the application domains that we cover in this chapter together with selected references to relevant publications. The major research areas in audio processing and retrieval are automatic speech recognition, music information retrieval, environmental sound recognition, and audio segmentation. Audio segmentation (often called audio classification) is a preprocessing step in audio analysis that separates different types of sound, for example, speech, music, environmental sounds, silence, and combinations of these sounds [21, 74].

Subdomains of audio segmentation address silence detection [157, 158], the segmentation of speech and nonspeech [159], and the segmentation of speech and music [54].

The segmented audio stream may be further analyzed by more specific analysis methods. ASR is probably the best investigated problem of audio retrieval [1]. However, there is still active research on audio features for ASR [95, 132, 154]. Related fields of research are speaker recognition and speaker segmentation [45, 160]. Speaker recognition deals with the identification of the speaker in an audio stream. Applications of speaker identification are authentication in safety systems and user recognition in dialog systems. Speaker segmentation determines the beginning and end of a speech segment of a particular speaker [80]. Another discipline dealing with speech is language identification where systems automatically predict the language of a speaker [161–164].

Recent approaches aim at the recognition and assessment of stress and other emotions in spoken language which may help to design mood driven human computer interfaces [58, 165–167]. Further domains of speech processing are gender detection and age detection from speech [89, 168]. A novel approach is speech analysis in medical applications for the detection of illnesses that affect human speech [169].

This chapter further focuses on ESR-related techniques. A typical application is the classification of general-purpose sounds, such as dog barks, flute sounds, or applause, which require specialized audio features [75, 170, 171]. Typical ESR tasks are surveillance applications where the environment is scanned for unusual sounds [172]. Furthermore, video analysis and annotation is a popular domain that deals with environmental sounds. Important tasks are violence detection in feature films [173] and highlight detection in video. Highlight detection addresses identification of key scenes in videos, for example, in sports videos [60, 174]. Multimodal approaches improve the detection rate by combining auditory and visual information [57]. Another application is the analysis of affective dimensions in the sound track of feature films (e.g., arousal, valence) [175].

Additionally, ESR covers pattern recognition in bioacoustics. Bioacoustic pattern recognition deals among others with acoustic monitoring of animals in the wild and the discrimination and retrieval of animal sounds, such as bird song and whale sounds [66, 67].

This chapter further addresses features related to MIR. MIR is a rapidly growing field of scientific interest due to the growing number of publicly available music databases. The main research areas of music analysis are recognition of instruments, genres, artists, and singers [59, 81, 85, 88, 102, 176–179]. Music similarity retrieval addresses the identification of pieces of music that sound similar [105, 180–182]. A related task is music identification (or music recognition) where different interpretations or versions of a single piece of music are matched [46, 183]. Furthermore, research focuses on emotion detection in music. The goal of emotion detection is to classify music into categories, such as *cheerful* and *depressive* [182].

A related field is structural music analysis which addresses the extraction of repeated patters, such as chorus and verse of a piece of music [122, 184]. Additionally, the analysis of structures such as rhythm and tempo is a popular task [94, 110]. A related topic is music transcription that deals with the extraction of notes and key(s) from a piece of music [124, 185]. Music summarization and thumbnailing address the extraction of the most significant part(s) in a piece of music [121, 148, 186].

Query by humming is a very popular MIR application. In a QBH application, a user can search for music in a database by humming the melody of the piece of music. The matching between the hummed query and the music database usually employs content-based audio features [21, 187]. Additionally, content-based music visualization, organization and browsing techniques employ audio features for the representation of audio signals [28, 188].

We review a variety of audio features that originate from audio fingerprinting. Audio fingerprinting addresses matching of audio signals based on fingerprints [24, 103]. A fingerprint is a compact numeric representation that captures the most significant information of a signal. A popular application are information systems that retrieve the artist and title of a particular piece of music given only a short clip recorded with a mobile phone.

This chapter covers the most active domains of audio processing and retrieval. We have systematically reviewed the most important conference proceedings and journals that are related to audio retrieval and signal processing. The result of the literature survey is a collection of more than 200 relevant papers that address audio feature extraction.

6.2 Literature on Audio Features

The literature survey yields a large number of publications that deal with feature extraction and audio features. We organize the publications according to the addressed audio features in order to make them manageable for the reader. Tables VIII and IX list relevant publications for each audio feature in alphabetical order and help the reader to get an overview of the literature in the context of an audio feature.

We have tried to identify the base paper for each feature. This is not always possible, since some features do not seem to have a distinct base paper, as in the case of ZCR and STE. In cases where no base paper exists, we have tried to identify an early paper, where the feature is mentioned. Base papers and early papers are printed in boldface.

TABLE VIII
SELECTED REFERENCES FOR EACH AUDIO FEATURE

Audio feature	Selected references
4-Hz modulation energy	[70, **86**, 125]
4-Hz modulation harmonic coefficient	[**125**]
Adaptive time-frequency transform	[**83**]
Amplitude descriptor	[**66**]
Auditory filter bank temporal envelopes	[56]
Autocorrelation MFCCs	[**44**]
Band periodicity	[49, **87**]
Bandwidth	[49, 56, 60, 67, 69, 85, 102, 103, 107]
Bark-scale frequency cepstral coefficients	[22, 85]
Beat histogram	[47, 101, **146**,148, 178, 189]
Beat spectrum (beat spectrogram)	[**141**, **142**,189, 190]
Beat tracker	[**144**, 145]
Chroma CENS features	[**46**]
Chromagram	[115, 121, 122, 181, **191**]
Cyclic beat spectrum	[**143**]
Daubechies wavelet coefficient histogram	[**47**, 81, 82, 182]
Distortion discriminant analysis	[43, **152**]
DWPT-based rhythm feature	[**148**, 178]
(Multiresolution) entropy	[102, 103, 114, 169]
(Modified) group delay	[**92**, 94, 95, 97–99]
Harmonic coefficient	[59, **125**]
Harmonic concentration	[**102**]
Harmonic derivate	[**102**]
Harmonic energy entropy	[**102**]
Harmonic prominence	[**60**]
Inharmonicity	[**12**, 48, 107]
Integral loudness	[**40**, 116]
Line spectral frequencies	[52, 68, 78, 80, 87]
Linear prediction cepstral coefficients	[62, 68, 76, 78, 133, **136**,192, 193]
Linear prediction ZCR	[**52**]
Linear predictive coding	[66, 71, 73–75]
Mel-scale frequency cepstral coefficients	[**51**, 57, 82, 85, 127, 132, 172, 194, 195]
Modulation frequency features	[24, **140**,171, 196, 197]
MPEG-7 audio fundamental frequency	[**65**,104]
MPEG-7 audio harmonicity	[**65**,104]
MPEG-7 audio power	[**65**,104]
MPEG-7 audio spectrum basis	[**65**,104, 151]
MPEG-7 audio spectrum centroid	[**65**, 104]
MPEG-7 audio spectrum spread	[48, **65**, 104, 198]

Base papers and early papers are typeset in bold font.

FEATURES FOR CONTENT-BASED AUDIO RETRIEVAL 137

TABLE IX
SELECTED REFERENCES FOR EACH AUDIO FEATURE

Audio feature	Selected references
MPEG-7 audio waveform	[**65**,104]
MPEG-7 harmonic spectral centroid/deviation/spread/variation	[**65**, 104, 126]
MPEG-7 log attack time	[48, **65**, 104]
MPEG-7 spectral centroid	[**65**, 104]
MPEG-7 temporal centroid	[48, **65**, 104]
Noise-robust auditory feature	[**21**]
Perceptual linear prediction (PLP)	[27, 62, 114, 133, **135**, 192]
Phase space features	[31, 154, 155, 199]
Pitch	[57, 58, 69, 70, 88, 107, 175]
Pitch histogram	[**18**, 47, 88, 101, 200]
Pitch profile	[**124**]
Pitch synchronous ZCPA	[**63**, 64]
Psychoacoustic pitch	[**119**]
Pulse metric	[**86**]
Rate-scale-frequency features	[**21**]
Relative spectral PLP	[**23**]
Rhythm patterns	[**28**, 150]
Sharpness	[**11**, 48, 105]
Short-time energy (STE)	[49, 60, 67, 68, 76, 84, 94, 111, 175]
Sone	[**28**, 85, 150]
Spectral center	[56, 94, 105]
Spectral centroid (SC)	[49, 56, 60, 69, 85, 86, 88, 103, 107]
Spectral crest	[42, 48, 83, 85, 103, 105, 111, 156]
Spectral dispersion	[94]
Spectral flatness	[42, 48, 103, 105, **109**, 111, 201]
Spectral flux	[59, 68, 73, 74, 81, 86, 89, 159]
Spectral peaks	[**90**, 91]
Spectral peak structure	[**102**]
Spectral rolloff	[48, 56, 68, 82, 86, 89, 102, 198]
Spectral slope	[48, 85]
Subband energy ratio	[49, 56, 60, 67, 68, 75, 85, 103]
Subband spectral flux	[60]
Volume	[54, 56, 68, 70, 75, 85, **156**]
Zero crossing peak amplitudes (ZCPA)	[22, **61**, 62]
Zero crossing rate (ZCR)	[52, 54, 57, 60, 68, 74, 84–86, 107]

Base papers and early papers are typeset in bold font.

6.3 Relevant Published Surveys

Audio feature extraction and audio retrieval both have a long tradition. Consequently several surveys have been published that cover these topics. Most related surveys focus on a single application domain, such as MIR or FP and cover a relatively small number of features. In the following, we briefly present important surveys in the field of audio feature extraction.

Lu [202] provides a survey on audio indexing and retrieval techniques. The survey describes a set of traditional time and frequency domain features, such as harmonicity and pitch. The authors focus on feature extraction and classification techniques in the domains of speech and music. Furthermore, the survey discusses concepts of speech and music retrieval systems.

In Ref. [17], the authors present a comprehensive survey of features for multimedia retrieval. The survey covers basic short-time audio features, such as volume, bandwidth, and pitch together with aggregations of short-time features. The authors extract audio features together with video features from a set of TV programs and compute the correlation between the features to show redundancies.

A bibliographical study of content-based audio retrieval is presented in Ref. [203]. The survey covers a set of seven frequently used audio features in detail. The authors perform retrieval experiments to prove the discriminant power of the features.

Tzanetakis [18] surveys a large set of music-related features. The author describes techniques for music analysis and retrieval, such as features for beat tracking, rhythm analysis, and pitch content description. Additionally, the author surveys traditional features that mainly originate from ASR. Finally, the survey presents a set of features that are directly computed from compressed MPEG signals.

Compressed-domain features are also presented in Ref. [16]. The authors discuss features for audio–visual indexing and analysis. The survey analyzes the applicability of traditional audio features and MPEG-7 descriptors in the compressed domain. However, the major part of the chapter addresses content-based video features.

A survey of audio fingerprinting techniques is presented in Ref. [204]. Fingerprints are compact signatures of audio content. The authors review the most important recent feature extraction techniques for fingerprinting.

Peeters [48] summarizes a large set of audio features. The author organizes the features among others in global and frame-based descriptions, spectral features, energy features, harmonic features, and perceptual features. The feature groups in Ref. [48] are similar to the groups of the taxonomy we present in Section 4.

There has been extensive research done in the field of audio feature extraction in recent years. However, we observe that most surveys focus on a small set of widely used traditional features while recent audio features are rarely addressed. In contrast to existing surveys we solely focus on feature extraction which allows us to cover a richer set of features and to introduce some structure in the field. Additionally, the survey presented in this chapter covers a wide range of application domains. The advantage of this approach is that it brings features from different domains together, which facilitates the comparison of techniques with different origins.

7. Summary and Conclusions

This chapter presents a survey on state-of-the-art and traditional content-based audio features originating from numerous application domains. We select a set of 77 features and systematically analyze their formal and structural properties in order to identify organizing principles that enable a categorization into meaningful groups. This leads to a novel taxonomy for audio features that assists the user in selecting adequate ones for a particular task. The taxonomy represents a novel perspective on audio features that associates techniques from different domains into one single structure.

The collection of features in this chapter gives an overview of existing techniques and may serve as reference for the reader to identify adequate features for her task. Furthermore, it may be the basis for the development of novel features and the improvement of existing techniques.

Additionally, we conclude that most of the surveyed publications perform retrieval tasks on their own audio databases and ground truths. Hence, the results are not comparable. We stress that the entire field of audio retrieval needs standardized benchmarking databases and ground truths specified by domain experts who have an unbiased view on the field. Although attempts of standardized benchmarking databases in the domains of speech and music retrieval have been made, more work has to be directed toward this task.

ACKNOWLEDGMENTS

We want to express our gratitude to Werner A. Deutsch (Austrian Academy of Sciences, Vienna) for his suggestions that lead to considerable improvements of this work. This work has received financial support from the Vienna Science and Technology Fund (WWTF) under grant no. CI06 024.

REFERENCES

[1] L. Rabiner, B. Juang, Fundamentals of Speech Recognition, Prentice-Hall, Upper Saddle River, NJ, 1993.
[2] J.S. Downie, Music information retrieval, Annu. Rev. Inform. Sci. Technol. 37 (2003) 295–340 (Chapter 7).
[3] A. Klapuri, M. Davy, Signal Processing Methods for Music Transcription, Springer, New York, NY, 2006.
[4] M. Cowling, R. Sitte, Comparison of techniques for environmental sound recognition, Pattern Recogn. Lett. 24 (15) (November 2003) 2895–2907.
[5] G. Salton, A. Wong, C.S. Yang, A vector space model for automatic indexing, Commun. ACM 18 (11) (1975) 613620.
[6] M.S. Lew, Principles of Visual Information Retrieval, Springer, London, January 2001.

[7] ISMIR, International Conference on Music Information Retrieval, 2004, http://ismir2004.ismir.net, last visited: September 2009.
[8] MIREX, Music Information Retrieval Evaluation Exchange, 2007, http://www.music-ir.org/mirexwiki, last visited: September 2009.
[9] ANSI, Bioacoustical Terminology, ANSI S3.20-1995 (R2003), American National Standards Institute, New York, NY, 1995.
[10] E. Terhardt, Zur Tonhöhenwahrnehmung von Klängen. I. Psychoakustische Grundlagen, Acustica 26 (1972) 173–186.
[11] E. Zwicker, H. Fastl, Psychoacoustics: Facts and Models, second ed., Springer, Berlin, 1999.
[12] G. Agostini, M. Longari, E. Pollastri, Musical instrument timbres classification with spectral features, in: Proceedings of the IEEE Workshop on Multimedia Signal Processing, Cannes, France, IEEE, Piscataway, NJ, October 2001, pp. 97–102.
[13] H. Terasawa, M. Slaney, J. Berger, Perceptual distance in timbre space, in: Proceedings of 11th Meeting of the International Conference on Auditory Display, Limerick, Ireland, July 2005, pp. 61–68.
[14] S.S. Stevens, The relation of pitch to intensity, J. Acoust. Soc. Am. 6 (3) (1935) 150–154.
[15] D. Byrd, T. Crawford, Problems of music information retrieval in the real world, Inform. Process. Manage. 38 (2) (March 2002) 249–272.
[16] A. Wang, A. Divakaran, A. Vetro, S.F. Chang, H. Sun, Survey of compressed-domain features used in audio–visual indexing and analysis, J. Vis. Commun. Image Represent. 14 (2) (June 2003) 150–183.
[17] Y. Wang, Z. Liu, J.C. Huang, Multimedia content analysis using both audio and visual clues, IEEE Signal Process. Mag. 17 (6) (November 2000) 12–36.
[18] G. Tzanetakis, Manipulation, Analysis and Retrieval Systems for Audio Signals, Ph.D. Thesis. Computer Science Department, Princeton University, 2002.
[19] M. Liu, C. Wan, Feature selection for automatic classification of musical instrument sounds, in: JCDL'01: Proceedings of the 1st ACM/IEEE-CS Joint Conference on Digital Libraries, ACM Press, New York, NY, 2001, pp. 247–248.
[20] T. Zhang, C.C.J. Kuo, Content-Based Audio Classification and Retrieval for Audiovisual Data Parsing, Kluwer Academic Publishers, Boston, MA, 2001.
[21] S. Ravindran, K. Schlemmer, D. Anderson, A physiologically inspired method for audio classification, EURASIP J. Appl. Signal Process. 2005 (9) (2005) 1374–1381.
[22] B. Gajic, K.K. Paliwal, Robust speech recognition using features based on zero crossings with peak amplitudes, in: Proceedings of the IEEE International Conference on Acoustics, Speech, and Signal Processing, vol. 1, Hong Kong, China, IEEE, Piscataway, NJ, April 2003, pp. 64–67.
[23] H. Hermansky, N. Morgan, Rasta processing of speech, IEEE Trans. Speech Audio Process. 2 (1994) 578–589.
[24] S. Sukittanon, L.E. Atlas, W.J. Pitton, Modulation-scale analysis for content identification, IEEE Trans. Signal Process. 52 (10) (2004) 3023–3035.
[25] I. Mierswa, K. Morik, Automatic feature extraction for classifying audio data, Mach. Learn. J. 58 (2–3) (February 2005) 127–149.
[26] B. Bogert, M. Healy, J. Tukey, The quefrency alanysis of time series for echoes: cepstrum, pseudo-autocovariance, cross-cepstrum, and saphe-cracking, in: M. Rosenblatt (Ed.), Proceedings of the Symposium on Time Series Analysis, Wiley, New York, NY, 1963, pp. 209–243.
[27] S. Greenberg, B.E.D. Kingsbury, The modulation spectrogram: in pursuit of an invariant representation of speech, in: Proceedings of the IEEE International Conference on Acoustics, Speech, and Signal Processing, vol. 3, IEEE, Piscataway, NJ, April 1997, pp. 1647–1650.

[28] E. Pampalk, A. Rauber, D. Merkl, Content-based organization and visualization of music archives, in: Proceedings of the 10th ACM International Conference on Multimedia, ACM Press, New York, NY, 2002, pp. 570–579.
[29] B. Kingsbury, N. Morgan, S. Greenberg, Robust speech recognition using the modulation spectrogram, Speech Commun. 25 (1998) 117–132.
[30] H. Abarbanel, Analysis of Observed Chaotic Data, Springer, New York, NY, 1996.
[31] A.C. Lindgren, M.T. Johnson, R.J. Povinelli, Speech recognition using reconstructed phase space features, in: Proceedings of the IEEE International Conference on Acoustics, Speech, and Signal Processing, vol. 1, Hong Kong, China, IEEE, Piscataway, NJ, April 2003, pp. 60–63.
[32] E. Zwicker, Subdivision of the audible frequency range into critical bands (Frequenzgruppen), J. Acoust. Soc. Am. 33 (1961) 248.
[33] C.J. Moore, R.W. Peters, B.R. Glasberg, Auditory filter shapes at low center frequencies, J. Acoust. Soc. Am. 88 (1) (1990) 132–140.
[34] S.S. Stevens, J. Volkmann, E.B. Newman, A scale for the measurement of the psychological magnitude pitch, J. Acoust. Soc. Am. 8 (3) (January 1937) 185–190.
[35] B.C.J. Moore, An Introduction to the Psychology of Hearing, fifth ed., Academic Press, Amsterdam, 2004.
[36] B.C.J. Moore, B.R. Glasberg, Suggested formulae for calculating auditory-filter bandwidths and excitation patterns, J. Acoust. Soc. Am. 74 (3) (September 1983) 750–753.
[37] R.L. Wegel, C.E. Lane, The auditory masking of one pure tone by another and its probable relation to the dynamics of the inner ear, Phys. Rev. 23 (February 1924) 266–285.
[38] H. Fletcher, W.A. Munson, Loudness, its definition, measurement and calculation, J. Acoust. Soc. Am. 5 (2) (October 1933) 82–108.
[39] International Organization for Standardization (ISO), International Standard 226, Acoustics—Normal Equal-Loudness Level Contours, 1987.
[40] S. Pfeiffer, The importance of perceptive adaptation of sound features for audio content processing, in: Proceedings SPIE Conferences, Electronic Imaging 1999, Storage and Retrieval for Image and Video Databases VII, San Jose, CA, January 1999, pp. 328–337.
[41] S.S. Stevens, On the psychophysical law, Psychol. Rev. 64 (3) (May 1957) 153–181.
[42] E. Allamanche, J. Herre, O. Helmuth, B. Frba, T. Kasten, M. Cremer, Content-based identification of audio material using mpeg-7 low level description, in: Proceedings of the International Symposium of Music Information Retrieval, 2001.
[43] C.J.C. Burges, J.C. Platt, S. Jana, Distortion discriminant analysis for audio fingerprinting, IEEE Trans. Speech Audio Process. 11 (3) (May 2003) 165–174.
[44] B.J. Shannon, K.K. Paliwal, MFCC computation from magnitude spectrum of higher lag autocorrelation coefficients for robust speech recognition, in: Proceedings of the International Conference on Spoken Language Processing, October 2004, pp. 129–132.
[45] K.H. Yuo, T.H. Hwang, H.C. Wang, Combination of autocorrelation-based features and projection measure technique for speaker identification, IEEE Trans. Speech Audio Process. 13 (4) (July 2005) 565–574.
[46] M. Müller, F. Kurth, M. Clausen, Audio matching via chroma-based statistical features, in: Proceedings of the 6th International Conference on Music Information Retrieval, London, September 2005, pp. 288–295.
[47] T. Li, M. Ogihara, Q. Li, A comparative study on content-based music genre classification, in: SIGIR'03: Proceedings of the 26th Annual International ACM SIGIR Conference on Research and Development in Information Retrieval, Toronto, ON, Canada, ACM Press, New York, NY, 2003, pp. 282–289.

[48] G. Peeters, A large set of audio features for sound description (similarity and classification) in the CUIDADO project, Technical Report, 2004.
[49] L. Lu, H.J. Zhang, S.Z. Li, Content-based audio classification and segmentation by using support vector machines, Multimedia Syst. 8 (6) (April 2003) 482–492.
[50] S. Essid, G. Richard, B. David, Inferring efficient hierarchical taxonomies for MIR tasks, application to musical instruments, in: Proceedings of the International Conference on Music Information Retrieval, September 2005.
[51] J.S. Bridle, M.D. Brown, An experimental automatic word recognition system, JSRU Report No. 1003, Joint Speech Research Unit, Ruislip, England, 1974.
[52] K. El-Maleh, M. Klein, G. Petrucci, P. Kabal, Speech/music discrimination for multimedia applications, in: Proceedings of the IEEE International Conference on Acoustics, Speech, and Signal Processing, vol. 6, Istanbul, Turkey, IEEE, Piscataway, NJ, June 2000, pp. 2445–2448.
[53] B. Kedem, Spectral analysis and discrimination by zero-crossings, IEEE Proc. 74 (1986) 1477–1493.
[54] C. Panagiotakis, G. Tziritas, A speech/music discriminator based on RMS and zero-crossings, IEEE Trans. Multimedia 7 (1) (February 2005) 155–166.
[55] J. Saunders, Real-time discrimination of broadcast speech/music, in: Proceedings of the IEEE International Conference on Acoustics, Speech, and Signal Processing, vol. 2, Atlanta, GA, IEEE, Piscataway, NJ, May 1996, pp. 993–996.
[56] M.F. McKinney, J. Breebaart, Features for audio and music classification, in: Proceedings of the International Conference on Music Information Retrieval, October 2003.
[57] C.C. Cheng, C.T. Hsu, Fusion of audio and motion information on hmm-based highlight extraction for baseball games, IEEE Trans. Multimedia 8 (3) (June 2006) 585–599.
[58] Z.J. Chuang, C.H. Wu, Emotion recognition using acoustic features and textual content, in: Proceedings of the IEEE International Conference on Multimedia and Expo, vol. 1, Taipei, Taiwan, IEEE, Piscataway, NJ, June 2004, pp. 53–56.
[59] T. Zhang, Automatic singer identification, in: Proceedings of the IEEE International Conference on Multimedia and Expo, vol. 1, IEEE, Piscataway, NJ, July 2003, pp. 33–36.
[60] R. Cai, L. Lu, A. Hanjalic, H.J. Zhang, L.H. Cai, A flexible framework for key audio effects detection and auditory context inference, IEEE Trans. Speech Audio Process. 14 (May 2006) 1026–1039.
[61] D.-S. Kim, J.-H. Jeong, J.-W. Kim, S.-Y. Lee, Feature extraction based on zero-crossings with peak amplitudes for robust speech recognition in noisy environments, in: Proceedings of the International Conference on Acoustics, Speech, and Signal Processing, vol. 1, IEEE, Piscataway, NJ, October 1996, pp. 61–64.
[62] D.S. Kim, S.Y. Lee, R.M. Kil, Auditory processing of speech signals for robust speech recognition in real-world noisy environments, IEEE Trans. Speech Audio Process. 7 (1) (January 1999) 55–69.
[63] M. Ghulam, T. Fukuda, J. Horikawa, T. Nitta, A noise-robust feature extraction method based on pitch-synchronous ZCPA for ASR, in: Proceedings of the International Conference on Spoken Language Processing, Jeju Island, Korea, October 2004, pp. 133–136.
[64] M. Ghulam, T. Fukuda, J. Horikawa, T. Nitta, Pitch-synchronous ZCPA (PS-ZCPA)-based feature extraction with auditory masking, in: Proceedings of the IEEE International Conference on Acoustics, Speech, and Signal Processing, vol. 1, Philadelphia, PA, IEEE, Piscataway, NJ, March 2005, pp. 517–520.
[65] ISO-IEC, Information Technology—Multimedia Content Description Interface—Part 4: Audio (Number 15938), ISO/IEC, Moving Pictures Expert Group, first ed., 2002.

[66] D. Mitrovic, M. Zeppelzauer, C. Breiteneder, Discrimination and retrieval of animal sounds, in: Proceedings of IEEE Multimedia Modelling Conference, Beijing, China, IEEE, Piscataway, NJ, January 2006, pp. 339–343.
[67] W.T. Chu, W.H. Cheng, J.Y.J. Hsu, J.L. Wu, Toward semantic indexing and retrieval using hierarchical audio models, Multimedia Syst. 10 (6) (May 2005) 570–583.
[68] H. Jiang, J. Bai, S. Zhang, B. Xu, SVM-based audio scene classification, in: Proceedings of the IEEE International Conference on Natural Language Processing and Knowledge Engineering, Wuhan, China, IEEE, Piscataway, NJ, October 2005, pp. 131–136.
[69] T. Wold, D. Blum, J. Wheaton, Content-based classification, search, and retrieval of audio, IEEE Multimedia 3 (3) (1996) 2736.
[70] Z. Liu, Y. Wang, T. Chen, Audio feature extraction and analysis for scene segmentation and classification, J. VLSI Signal Process. 20 (1–2) (October 1998) 61–79.
[71] L. Rabiner, R. Schafer, Digital Processing of Speech Signals, Prentice-Hall, Englewood Cliffs, NJ, 1978.
[72] T. Tremain, The government standard linear predictive coding algorithm: LPC-10, Speech Technol. Mag. 1 (April 1982) 40–49.
[73] M.K.S. Khan, W.G. Al-Khatib, M. Moinuddin, Automatic classification of speech and music using neural networks, in: MMDB'04: Proceedings of the 2nd ACM International Workshop on Multimedia Databases, ACM Press, New York, NY, 2004, pp. 94–99.
[74] M.K.S. Khan, W.G. Al-Khatib, Machine-learning based classification of speech and music, Multimedia Syst. 12 (1) (August 2006) 55–67.
[75] M. Liu, C. Wan, A study on content-based classification and retrieval of audio database, in: Proceedings of the International Symposium on Database Engineering and Applications, Grenoble, France, IEEE Computer Society, Washington, DC, July 2001, pp. 339–345.
[76] C. Xu, N.C. Maddage, X. Shao, Automatic music classification and summarization, IEEE Trans. Speech Audio Process. 13 (3) (May 2005) 441–450.
[77] J.P. Campbell, Speaker recognition: A tutorial, Proc. IEEE 85 (9) (September 1997) 1437–1462.
[78] A.G. Krishna, T.V. Sreenivas, Music instrument recognition: from isolated notes to solo phrases, in: Proceedings of the IEEE International Conference on Acoustics, Speech, and Signal Processing, vol. 4, Montreal, QC, Canada, IEEE, Piscataway, NJ, May 2004, pp. 265–268.
[79] J.Y. Tourneret, Statistical properties of line spectrum pairs, Signal Process. 65 (2) (March 1998) 239–255.
[80] L. Lu, H.J. Zhang, Unsupervised speaker segmentation and tracking in real-time audio content analysis, Multimedia Syst. 10 (4) (April 2005) 332–343.
[81] T. Li, M. Ogihara, Music artist style identification by semi-supervised learning from both lyrics and content, in: Proceedings of the 12th Annual ACM International Conference on Multimedia, ACM Press, New York, NY, 2004, pp. 364–367.
[82] T. Li, M. Ogihara, Toward intelligent music information retrieval, IEEE Trans. Multimedia 8 (3) (June 2006) 564–574.
[83] K. Umapathy, S. Krishnan, S. Jimaa, Multigroup classification of audio signals using time-frequency parameters, IEEE Trans. Multimedia 7 (2) (April 2005) 308–315.
[84] S. Srinivasan, D. Petkovic, D. Ponceleon, Towards robust features for classifying audio in the CueVideo system, in: Proceedings of the 7th ACM International Conference on Multimedia (Part 1), ACM Press, New York, NY, 1999, pp. 393–400.
[85] F. Mörchen, A. Ultsch, M. Thies, I. Löhken, Modeling timbre distance with temporal statistics from polyphonic music, IEEE Trans. Audio Speech Lang. Process. 14 (1) (January 2006) 81–90.

[86] E. Scheirer, M. Slaney, Construction and evaluation of a robust multi-feature speech/music discriminator, in: Proceedings of the IEEE International Conference on Acoustics, Speech, and Signal Processing, vol. 2, Munich, Germany, April 1997, pp. 1331–1334.

[87] L. Lu, H. Jiang, H.J. Zhang, A robust audio classification and segmentation method, in: Proceedings of the 9th ACM International Conference on Multimedia, Ottawa, ON, Canada, ACM Press, New York, NY, 2001, pp. 203–211.

[88] G. Tzanetakis, Musical genre classification of audio signals, IEEE Trans. Speech Audio Process. 10 (5) (July 2002) 293–302.

[89] G. Tzanetakis, Audio-based gender identification using bootstrapping, in: Proceedings of the IEEE Pacific Rim Conference on Communications, Computers and Signal Processing, Victoria, BC, Canada, IEEE, Piscataway, NJ, August 2005, pp. 432–433.

[90] A. Wang, An industrial strength audio search algorithm, in: Proceedings of the International Conference on Music Information Retrieval, Baltimore, MD, October 2003, pp. 7–13.

[91] A. Wang, The Shazam music recognition service, Commun. ACM 49 (8) (August 2006) 44–48.

[92] B. Yegnanarayan, H.A. Murthy, Significance of group delay functions in spectrum estimation, IEEE Trans. Signal Process. 40 (9) (September 1992) 2281–2289.

[93] R. Smits, B. Yegnanarayana, Determination of instants of significant excitation in speech using group delay function, IEEE Trans. Speech Audio Process. 3 (5) (September 1995) 325–333.

[94] W.A. Sethares, R.D. Morris, J.C. Sethares, Beat tracking of musical performances using low-level audio features, IEEE Trans. Speech Audio Process. 13 (2) (March 2005) 275–285.

[95] L.D. Alsteris, K.K. Paliwal, Evaluation of the modified group delay feature for isolated word recognition, in: Proceedings of the International Symposium on Signal Processing and Its Applications, vol. 2, Sydney, Australia, IEEE, Piscataway, NJ, August 2005, pp. 715–718.

[96] M. Hegde, H.A. Murthy, V.R. Gadde, Significance of joint features derived from the modified group delay function in speech processing, EURASIP J. Appl. Signal Process. 15 (1) (January 2007) 190–202, doi:10.1155/2007/79032.

[97] R.M. Hegde, H.A. Murthy, G.V.R. Rao, Application of the modified group delay function to speaker identification and discrimination, in: Proceedings of the IEEE International Conference on Acoustics, Speech, and Signal Processing, vol. 1, Montreal, QC, Canada, IEEE, Piscataway, NJ, May 2004, pp. 517–520.

[98] H.A. Murthy, V. Gadde, The modified group delay function and its application to phoneme recognition, in: Proceedings of the IEEE International Conference on Acoustics, Speech, and Signal Processing, vol. 1, Hong Kong, China, IEEE, Piscataway, NJ, April 2003, pp. 68–71.

[99] T. Nagarajan, H.A. Murthy, Subband-based group delay segmentation of spontaneous speech into syllable-like units, EURASIP J. Appl. Signal Process. 2004 (17) (2004) 2614–2625.

[100] H.A. Murthy, K.V.M. Murthy, B. Yegnarayana, Formant extraction from Fourier transform phase, in: International Conference on Acoustics, Speech, and Signal Processing vol. 1, May 1989, pp. 484–487.

[101] T. Li, G. Tzanetakis, Factors in automatic musical genre classification of audio signals, in: Proceedings of the IEEE Workshop on Applications of Signal Processing to Audio and Acoustics, New Paltz, NY, IEEE, Piscataway, NJ, October 2003, pp. 143–146.

[102] H. Srinivasan, M. Kankanhalli, Harmonicity and dynamics-based features for audio, in: Proceedings of the IEEE International Conference on Acoustics, Speech, and Signal Processing, vol. 4, Montreal, QC, Canada, IEEE, Piscataway, NJ, May 2004, pp. 321–324.

[103] A. Ramalingam, S. Krishnan, Gaussian mixture modeling using short time Fourier transform features for audio fingerprinting, in: Proceedings of the IEEE International Conference on Multimedia and Expo, Amsterdam, The Netherlands, IEEE, Piscataway, NJ, July 2005, pp. 1146–1149.

[104] H. Kim, N. Moreau, T. Sikora, MPEG-7 Audio and Beyond, Wiley, West Sussex, England, 2005.
[105] J. Herre, E. Allamanche, C. Ertel, How similar do songs sound? Towards modeling human perception of musical similarity, in: Proceedings of the IEEE Workshop on Applications of Signal Processing to Audio and Acoustics, New Paltz, NY, IEEE, Piscataway, NJ, October 2003, pp. 83–86.
[106] Z. Liu, J. Huang, Y. Wang, T. Chen, Audio feature extraction and analysis for scene classification, in: Proceedings of the IEEE Workshop on Multimedia Signal Processing, Princeton, NJ, IEEE, Piscataway, NJ, June 1997, pp. 343–348.
[107] G. Agostini, M. Longari, E. Pollastri, Musical instrument timbres classification with spectral features, EURASIP J. Appl. Signal Process. 2003 (1) (2003) 5–14.
[108] T. Li, M. Ogihara, Music genre classification with taxonomy, in: Proceedings of the IEEE International Conference on Acoustics, Speech, and Signal Processing, vol. 5, IEEE, Piscataway, NJ, March 2005, pp. 197–200.
[109] N.S. Jayant, P. Noll, Digital Coding of Waveforms: Principles and Applications to Speech and Video, Prentice-Hall Signal Processing Series, Prentice-Hall, Englewood Cliffs, NJ, 1984.
[110] E. Guaus, E. Batlle, Visualization of metre and other rhythm features, in: Proceedings of the IEEE International Symposium on Signal Processing and Information Technology, Darmstadt, Germany, IEEE, Piscataway, NJ, December 2003, pp. 282–285.
[111] R. Lancini, F. Mapelli, R. Pezzano, Audio content identification by using perceptual hashing, in: Proceedings of the IEEE International Conference on Multimedia and Expo, vol. 1, Taipei, Taiwan, IEEE, Piscataway, NJ, June 2004, pp. 739–742.
[112] J. Herre, E. Allamanche, O. Hellmuth, Robust matching of audio signals using spectral flatness features, in: Proceedings of the IEEE Workshop on Applications of Signal Processing to Audio and Acoustics, New Paltz, NY, IEEE, Piscataway, NJ, October 2001, pp. 127–130.
[113] H. Misra, S. Ikbal, H. Bourlard, H. Hermansky, Spectral entropy based feature for robust ASR, in: Proceedings of the IEEE International Conference on Acoustics, Speech, and Signal Processing, vol. 1, Montreal, QC, Canada, IEEE, Piscataway, NJ, May 2004, pp. 193–196.
[114] H. Misra, S. Ikbal, S. Sivadas, H. Bourlard, Multi-resolution spectral entropy feature for robust ASR, in: Proceedings of the IEEE International Conference on Acoustics, Speech, and Signal Processing, vol. 1, Philadelphia, PA, IEEE, Piscataway, NJ, March 2005, pp. 253–256.
[115] F. Mörchen, A. Ultsch, M. Thies, I. Löhken, M. Nöcker, C. Stamm, N. Efthymiou, M. Kümmerer, Musicminer: Visualizing timbre distances of music as topographical maps, Technical Report, 2005.
[116] S. Pfeiffer, R. Lienhart, W. Effelsberg, Scene determination based on video and audio features, Multimedia Tools Appl. 15 (1) (September 2001) 59–81.
[117] W. Hess, Pitch Determination of Speech Signals: Algorithms and Devices, Springer, Berlin, 1983.
[118] Y.D. Cho, M.Y. Kim, S.R. Kim, A spectrally mixed excitation (SMX) vocoder with robust parameter determination, in: Proceedings of the International Conference on Acoustics, Speech and Signal Processing, vol. 2, May 1998, pp. 601–604.
[119] R. Meddis, L. O'Mard, A unitary model of pitch perception, J. Acoust. Soc. Am. 102 (3) (September 1997) 1811–1820.
[120] R.N. Shepard, Circularity in judgements of relative pitch, J. Acoust. Soc. Am. 36 (1964) 2346–2353.
[121] M.A. Bartsch, G.H. Wakefield, Audio thumbnailing of popular music using chroma-based representations, IEEE Trans. Multimedia 7 (1) (February 2005) 96–104.
[122] M. Goto, A chorus-section detecting method for musical audio signals, in: Proceedings of the IEEE International Conference on Acoustics, Speech, and Signal Processing, vol. 5, Hong Kong, China, IEEE, Piscataway, NJ, April 2003, pp. 437–440.

[123] M. Müller, Information Retrieval for Music and Motion, Springer, Berlin, 2007.
[124] Y. Zhu, M.S. Kankanhalli, Precise pitch profile feature extraction from musical audio for key detection, IEEE Trans. Multimedia 8 (3) (June 2006) 575–584.
[125] W. Chou, L. Gu, Robust singing detection in speech/music discriminator design, in: Proceedings of the IEEE International Conference on Acoustics, Speech, and Signal Processing, vol. 2, Salt Lake City, UT, IEEE, Piscataway, NJ, May 2001, pp. 865–868.
[126] G. Peeters, S. McAdams, P. Herrera, Instrument description in the context of MPEG-7, in: Proceedings of International Computer Music Conference, Berlin, Germany, August, 2000.
[127] S. Davis, P. Mermelstein, Comparison of parametric representations for monosyllabic word recognition in continuously spoken sentences, IEEE Trans. Acoust. Speech Signal Process. 28 (4) (August 1980) 357–366.
[128] A.M. Noll, Short-time spectrum and "cepstrum" techniques for vocal-pitch detection, J. Acoust. Soc. Am. 36 (2) (1964) 296–302.
[129] M. Xu, L. Duan, L. Chia, C. Xu, Audio keyword generation for sports video analysis, in: Proceedings of the ACM International Conference on Multimedia, 2004, pp. 758–759.
[130] X. Wang, Y. Dong, J. Hakkinen, O. Viikki, Noise robust Chinese speech recognition using feature vector normalization and higher-order cepstral coefficients, in: Proceedings of the 5th International Conference on Signal Processing, vol. 2, August 2000, pp. 738–741.
[131] N. Maddage, C. Xu, M. Kankanhalli, X. Shao, Content-based music structure analysis with applications to music semantics understanding, in: Proceedings of the ACM International Conference on Multimedia, ACM Press, New York, NY, 2004, pp. 112–119.
[132] E.H.C. Choi, On compensating the Mel-frequency cepstral coefficients for noisy speech recognition, in: Proceedings of the Australasian Computer Science Conference, Hobart, Australia, Australian Computer Society, Darlinghurst, NSW, 2006, pp. 49–54.
[133] Q. Li, F.K. Soong, O. Siohan, An auditory system-based feature for robust speech recognition, in: Proceedings of the European Conference on Speech Communication and Technology, Aalborg, Denmark, International Speech Communication Association, Geneva, September 2001, pp. 619–622.
[134] X. Yang, K. Wang, S. Shamma, Auditory representations of acoustic signals, IEEE Trans. Inform. Theory 38 (2) (March 1992) 824–839.
[135] H. Hermansky, Perceptual linear predictive (PLP) analysis of speech, J. Acoust. Soc. Am. 87 (4) (April 1990) 1738–1752.
[136] B.S. Atal, Effectiveness of linear prediction characteristics of the speech wave for automatic speaker identification and verification, J. Acoust. Soc. Am. 55 (6) (June 1974) 1304–1312.
[137] A. Adami, D. Barone, A speaker identification system using a model of artificial neural networks for an elevator application, Inform. Sci. 138 (1–4) (October 2001) 1–5.
[138] H. Fastl, Fluctuation strength and temporal masking patterns of amplitude-modulated broadband noise, Hear. Res. 8 (1) (September 1982) 59–69.
[139] T. Houtgast, H.J. Steeneken, A review of the MTF concept in room acoustics and its use for estimating speech intelligibility in auditoria, J. Acoust. Soc. Am. 77 (3) (March 1985) 1069–1077.
[140] S. Sukittanon, L.E. Atlas, Modulation frequency features for audio fingerprinting, in: Proceedings of the IEEE International Conference on Acoustics, Speech, and Signal Processing, vol. 2, Orlando, FL, IEEE, Piscataway, NJ, May 2002, pp. 1773–1776.
[141] J. Foote, Automatic audio segmentation using a measure of audio novelty, in: Proceedings of the IEEE International Conference on Multimedia and Expo, vol. 1, New York, NY, IEEE, Piscataway, NJ, August 2000, pp. 452–455.

[142] J. Foote, S. Uchihashi, The beat spectrum: a new approach to rhythm analysis, in: Proceedings of the IEEE International Conference on Multimedia and Expo, IEEE, Piscataway, NJ, 2001, pp. 881–884.
[143] F. Kurth, T. Gehrmann, M. Müller, The cyclic beat spectrum: tempo-related audio features for time-scale invariant audio identification, in: Proceedings of the 7th International Conference on Music Information Retrieval, Victoria, BC, Canada, October 2006, pp. 35–40.
[144] E. Scheirer, Tempo and beat analysis of acoustic musical signals, J. Acoust. Soc. Am. 103 (1) (January 1998) 588–601.
[145] E. Scheirer, Music-Listening Systems. Program in Media Arts and Sciences, Ph.D. Thesis, MIT, Cambridge, MA, 2000.
[146] G. Tzanetakis, G. Essl, P. Cook, Audio analysis using the discrete wavelet transform, in: Proceedings of the International Conference on Acoustics and Music: Theory and Applications, Malta, September 2001.
[147] G. Tzanetakis, G. Essl, P. Cook, Human perception and computer extraction of musical beat strength, in: Proceedings of the International Conference on Digital Audio Effects, Hamburg, Germany, September 2002, pp. 257–261.
[148] M. Grimaldi, P. Cunningham, A. Kokaram, A wavelet packet representation of audio signals for music genre classification using different ensemble and feature selection techniques, in: Proceedings of the ACM SIGMM International Workshop on Multimedia Information Retrieval, Berkeley, CA, ACM Press, New York, NY, 2003, pp. 102–108.
[149] S. Mallat, A Wavelet Tour of Signal Processing, Academic Press, San Diego, CA, 1999.
[150] A. Rauber, E. Pampalk, D. Merkl, Using psycho-acoustic models and self-organizing maps to create a hierarchical structuring of music by sound similarity, in: Proceedings of the International Conference on Music Information Retrieval, Paris, France, IRCAM-Centre Pompidou, Paris, October 2002.
[151] H. Kim, N. Moreau, T. Sikora, Audio classification based on MPEG-7 spectral basis representations, IEEE Trans. Circuits Syst. Video Technol. 14 (2004) 716–725.
[152] C.J.C. Burges, J.C. Platt, S. Jana, Extracting noise-robust features from audio data, in: Proceedings of the IEEE International Conference on Acoustics, Speech, and Signal Processing, vol. 1, Orlando, FL, IEEE, Piscataway, NJ, May 2002, pp. 1021–1024.
[153] H. Malvar, A modulated complex lapped transform and its applications to audio processing, in: Proceedings of the IEEE International Conference on Acoustics, Speech, and Signal Processing, vol. 3, Phoenix, AZ, IEEE, Piscataway, NJ, March 1999, pp. 1421–1424.
[154] I. Kokkinos, P. Maragos, Nonlinear speech analysis using models for chaotic systems, IEEE Trans. Speech Audio Process. 13 (6) (November 2005) 1098–1109.
[155] V. Pitsikalis, P. Maragos, Speech analysis and feature extraction using chaotic models, in: Proceedings of the IEEE International Conference on Acoustics, Speech, and Signal Processing, vol. 1, Orlando, FL, IEEE, Piscataway, NJ, May 2002, pp. 533–536.
[156] L. Bai, Y. Hu, S. Lao, J. Chen, L. Wu, Feature analysis and extraction for audio automatic classification, in: Proceedings of the IEEE International Conference on Systems, Man and Cybernetics, Big Island, HI, IEEE, Piscataway, NJ, vol. 1, October 2005, pp. 767–772.
[157] R. Becker, G. Corsetti, J. Guedes Silveira, R. Balbinot, F. Castello, A silence detection and suppression technique design for voice over IP systems, in: Proceedings of the IEEE Pacific Rim Conference on Communications, Computers, and Signal Processing, Victoria, BC, Canada, IEEE, Piscataway, NJ, August 2005, pp. 173–176.

[158] S. Pfeiffer, Pause concepts for audio segmentation at different semantic levels, in: Proceedings of the ACM International Conference on Multimedia, Ottawa, ON, Canada, ACM Press, New York, NY, 2001, pp. 187–193.
[159] R. Huang, J.H.L. Hansen, High-level feature weighted GMM network for audio stream classification, in: Proceedings of the International Conference on Spoken Language Processing, Jeju Island, Korea, October 2004, pp. 1061–1064.
[160] K.Y. Lee, Local fuzzy PCA based GMM with dimension reduction on speaker identification, Pattern Recogn. Lett. 25 (16) (2004) 1811–1817.
[161] J. Farinas, F.C. Pellegrino, J.-L. Rouas, F. Andre-Obrech, Merging segmental and rhythmic features for automatic language identification, in: Proceedings of the IEEE International Conference on Acoustics, Speech, and Signal Processing, vol. 1, Orlando, FL, IEEE, Piscataway, NJ, May 2002, pp. 753–756.
[162] Q.R. Gu, T. Shibata, Speaker and text independent language identification using predictive error histogram vectors, in: Proceedings of the IEEE International Conference on Acoustics, Speech, and Signal Processing, vol. 1, Hong Kong, China, IEEE, Piscataway, NJ, April 2003, pp. 36–39.
[163] J. Navratil, Spoken language recognition—a step toward multilinguality in speech processing, IEEE Trans. Speech Audio Process. 9 (6) (September 2001) 678–685.
[164] P.A. Torres-Carrasquillo, E. Singer, M.A. Kohler, R.J. Greene, D.A. Reynolds, J.R. Deller Jr., Approaches to language identification using Gaussian mixture models and shifted delta cepstral features, in: Proceedings of the International Conference on Spoken Language Processing, Denver, CO, September 2002, pp. 89–92.
[165] Z. Inanoglu, R. Caneel, Emotive alert: HMM-based emotion detection in voicemail messages, in: Proceedings of the International Conference on Intelligent User Interfaces, San Diego, CA, ACM Press, New York, NY, 2005, pp. 251–253.
[166] T.L. Nwe, S.W. Foo, L.C. De Silva, Classification of stress in speech using linear and nonlinear features, in: Proceedings of the IEEE International Conference on Acoustics, Speech, and Signal Processing, vol. 2, Hong Kong, China, IEEE, Piscataway, NJ, April 2003, pp. 9–12.
[167] A.A. Razak, M.H.M. Yusof, R. Komiya, Towards automatic recognition of emotion in speech, in: Proceedings of the IEEE International Symposium on Signal Processing and Information Technology, Darmstadt, Germany, IEEE, Piscataway, NJ, December 2003, pp. 548–551.
[168] N. Minematsu, M. Sekiguchi, K. Hirose, Automatic estimation of one's age with his/her speech based upon acoustic modeling techniques of speakers, in: Proceedings of the IEEE International Conference on Acoustics, Speech, and Signal Processing, vol. 1, Orlando, FL, IEEE, Piscataway, NJ, May 2002, pp. 137–140.
[169] R. Behroozmand, F. Almasganj, Comparison of neural networks and support vector machines applied to optimized features extracted from patients' speech signal for classification of vocal fold inflammation, in: Proceedings of the IEEE International Symposium on Signal Processing and Information Technology, Athens, Greece, IEEE, Piscataway, NJ, December 2005, pp. 844–849.
[170] Y.C. Cho, S. Choi, S.Y. Bang, Non-negative component parts of sound for classification, in: Proceedings of the IEEE International Symposium on Signal Processing and Information Technology, Darmstadt, Germany, IEEE, Piscataway, NJ, December 2003, pp. 633–636.
[171] L. Owsley, L. Atlas, C. Heinemann, Use of modulation spectra for representation and classification of acoustic transients from sniper fire, in: Proceedings of the IEEE International Conference on Acoustics, Speech, and Signal Processing, vol. 4, Philadelphia, PA, IEEE, Piscataway, NJ, March 2005, pp. 1129–1132.

[172] R. Radhakrishnan, A. Divakaran, P. Smaragdis, Audio analysis for surveillance applications, in: Proceedings of the IEEE Workshop on Applications of Signal Processing to Audio and Acoustics, New Paltz, NY, IEEE, Piscataway, NJ, October 2003, pp. 158–161.
[173] S. Pfeiffer, S. Fischer, E. Effelsberg, Automatic audio content analysis, in: Proceedings of the ACM International Conference on Multimedia, Boston, MA, ACM Press, New York, NY, 1996, pp. 21–30.
[174] K. Wang, C. Xu, Robust soccer highlight generation with a novel dominant-speech feature extractor, in: Proceedings of the IEEE International Conference on Multimedia and Expo, Taipei, Taiwan, IEEE, Piscataway, NJ, vol. 1, June 2004, pp. 591–594.
[175] C.G. Chan, G.J.F. Jones, Affect-based indexing and retrieval of films, in: Proceedings of the Annual ACM International Conference on Multimedia, Singapore, ACM Press, Berkeley, 2005, pp. 427–430.
[176] S. Esmaili, S. Krishnan, K. Raahemifar, Content based audio classification and retrieval using joint time-frequency analysis, in: Proceedings of the IEEE International Conference on Acoustics, Speech, and Signal Processing, vol. 5, Montreal, QC, Canada, IEEE, Piscataway, NJ, May 2004, pp. 665–668.
[177] A.M. Fanelli, L. Caponetti, G. Castellano, C.A. Buscicchio, Content-based recognition of musical instruments, in: Proceedings of the IEEE International Symposium on Signal Processing and Information Technology, Rome, Italy, IEEE, Piscataway, NJ, December 2004, pp. 361–364.
[178] M. Grimaldi, P. Cunningham, A. Kokaram, Discrete wavelet packet transform and ensembles of lazy and eager learners for music genre classification, Multimedia Syst. 11 (5) (April 2006) 422–437.
[179] A. Mesaros, E. Lupu, C. Rusu, Singing voice features by time-frequency representations, in: Proceedings of the International Symposium on Image and Signal Processing and Analysis, vol. 1, Rome, Italy, IEEE, Piscataway, NJ, September 2003, pp. 471–475.
[180] J.-J. Aucouturier, F. Pachet, M. Sandler, The way it sounds: timbre models for analysis and retrieval of music signals, IEEE Trans. Multimedia 7 (6) (December 2005) 1028–1035.
[181] T. Jehan, Hierarchical multi-class self similarities, in: Proceedings of the IEEE Workshop on Applications of Signal Processing to Audio and Acoustics, New Paltz, NY, IEEE, Piscataway, NJ, October 2005, pp. 311–314.
[182] T. Li, M. Ogihara, Content-based music similarity search and emotion detection, in: Proceedings of the IEEE International Conference on Acoustics, Speech, and Signal Processing, vol. 5, Montreal, QC, Canada, IEEE, Piscataway, NJ, May 2004, pp. 705–708.
[183] M. Clausen, F. Kurth, A unified approach to content-based and fault-tolerant music recognition, IEEE Trans. Multimedia 6 (5) (October 2004) 717–731.
[184] L. Lu, M. Wang, H.J. Zhang, Repeating pattern discovery and structure analysis from acoustic music data, in: Proceedings of the ACM SIGMM International Workshop on Multimedia Information Retrieval, New York, NY, ACM Press, New York, NY, 2004, pp. 275–282.
[185] S.W. Foo, W.T. Leem, Recognition of piano notes with the aid of FRM filters, in: Proceedings of the International Symposium on Control, Communications and Signal Processing, Hammamet, Tunisia, IEEE, Piscataway, NJ, March 2004, pp. 409–413.
[186] M. Cooper, J. Foote, Summarizing popular music via structural similarity analysis, in: Proceedings of the IEEE Workshop on Applications of Signal Processing to Audio and Acoustics, New Paltz, NY, IEEE, Piscataway, NJ, October 2003, pp. 127–130.
[187] S. Pauws, Cubyhum: a fully operational query by humming system, in: Proceedings of the International Conference on Music Information Retrieval, Paris, France, IRCAM-Centre Pompidou, Paris, October 2002.

[188] E. Brazil, M. Fernström, G. Tzanetakis, P. Cook, Enhancing sonic browsing using audio information retrieval, in: Proceedings of the International Conference on Auditory Display, Kyoto, Japan, July 2002.
[189] A. Meng, P. Ahrendt, J. Larsen, Improving music genre classification by short time feature integration, in: Proceedings of the IEEE International Conference on Acoustics, Speech, and Signal Processing, vol. 5, Philadelphia, PA, IEEE, Piscataway, NJ, March 2005, pp. 497–500.
[190] X. Changsheng, N.C. Maddage, S. Xi, C. Fang, T. Qi, Musical genre classification using support vector machines, in: Proceedings of the IEEE International Conference on Acoustics, Speech, and Signal Processing, vol. 5, Hong Kong, China, IEEE, Piscataway, NJ, April 2003, pp. 429–432.
[191] M.A. Bartsch, G.H. Wakefield, To catch a chorus: using chroma-based representations for audio thumbnailing, in: Proceedings of the IEEE Workshop on the Applications of Signal Processing to Audio and Acoustics, New Paltz, NY, IEEE, Piscataway, NJ, October 2001, pp. 15–18.
[192] C. Lvy, G. Linars, P. Nocera, Comparison of several acoustic modeling techniques and decoding algorithms for embedded speech recognition systems, in: Proceedings of the Workshop on DSP in Mobile and Vehicular Systems, Nagoya, Japan, April 2003.
[193] R. Muralishankar, A.G. Ramakrishnan, Pseudo complex cepstrum using discrete cosine transform, Int. J. Speech Technol. 8 (2) (June 2005) 181–191.
[194] L. Baojie, K. Hirose, Speaker adaptive speech recognition using phone pair model, in: Proceedings of the 5th International Conference on Signal Processing, vol. 2, Beijing, China, August 2000, pp. 714–717.
[195] X. Wang, Y. Dong, J. Häkkinen, O. Viikki, Noise robust Chinese speech recognition using feature vector normalization and higher-order cepstral coefficients, in: Proceedings of the 5th International Conference on Signal Processing, vol. 2, Beijing, China, August 2000, pp. 738–741.
[196] D. Dimitriadis, P. Maragos, A. Potamianos, Modulation features for speech recognition, in: Proceedings of the IEEE International Conference on Acoustics, Speech, and Signal Processing, vol. 1, Orlando, FL, IEEE, Piscataway, NJ, May 2002, pp. 377–380.
[197] T. Kinnunen, Joint acoustic-modulation frequency for speaker recognition, in: Proceedings of the IEEE International Conference on Acoustics, Speech, and Signal Processing, vol. 1, Toulouse, France, IEEE, Piscataway, NJ, May 2006, pp. 665–668.
[198] T. Pohle, E. Pampalk, G. Widmer, Evaluation of frequently used audio features for classification of music into perceptual categories, in: Proceedings of the 4th International Workshop Content-Based Multimedia Indexing, Riga, Latvia, 2005.
[199] V. Pitsikalis, I. Kokkinos, P. Maragos, Nonlinear analysis of speech signals: generalized dimensions and Lyapunov exponents, in: Proceedings of the European Conference on Speech Communication and Technology, Geneva, Switzerland, September 2003, pp. 817–820.
[200] G. Tzanetakis, A. Ermolinskyi, P. Cook, Pitch histograms in audio and symbolic music information retrieval, J. New Music Res. 32 (2) (June 2003) 143–152.
[201] Y.C. Huang, S.K. Jenor, An audio recommendation system based on audio signature description scheme in MPEG-7 audio, in: Proceedings of the IEEE International Conference on Multimedia and Expo, Taipei, Taiwan, vol. 1, IEEE, Piscataway, NJ, June 2004, pp. 639–642.
[202] G. Lu, Indexing and retrieval of audio: a survey, Multimedia Tools Appl. 15 (3) (December 2001) 269–290.
[203] M. Davy, S.J. Godsill, Audio information retrieval a bibliographical study, Technical Report, February 2002.
[204] P. Cano, E. Batle, T. Kalker, J. Haitsma, A review of algorithms for audio fingerprinting, in: Proceedings of the IEEE Workshop on Multimedia Signal Processing, St. Thomas, VI, IEEE, Piscataway, NJ, December 2002, pp. 169–173.

Multimedia Services over Wireless Metropolitan Area Networks

KOSTAS PENTIKOUSIS

VTT Technical Research Centre of Finland, Oulu, Finland

JARNO PINOLA

VTT Technical Research Centre of Finland, Oulu, Finland

ESA PIRI

VTT Technical Research Centre of Finland, Oulu, Finland

PEDRO NEVES

Portugal Telecom Inovação, R. Eng. José Ferreira Pinto Basto, Aveiro, Portugal

SUSANA SARGENTO

Institute of Telecommunications, University of Aveiro, Campus Universitário de Santiago, Aveiro, Portugal

Abstract

The IEEE 802.16 family of wireless local and metropolitan area network (LAN/MAN) standards has received plenty of attention in recent years due to its potential to change the field of telecommunications operations. WiMAX, which is based on IEEE 802.16 and a network reference architecture defined by the WiMAX Forum, emerges as a potent proposal for building next-generation broadband wireless networks. This chapter is a comprehensive review of wireless MANs using WiMAX technologies, focusing in particular on multimedia service delivery over such networks. We start by summarizing recent trends in Internet traffic, which demonstrate the proliferation of multimedia services alongside the strong need for cost-efficient broadband access solutions, such as WiMAX. We then provide an overview of both fixed and mobile WiMAX, which includes a short introduction to the physical (PHY) and medium access control (MAC) layers of the IEEE 802.16 standard. We focus on the key architectural elements, including Quality of Service (QoS) support, and relate them with multimedia service delivery. To better illustrate the opportunities emerging in the broadband wireless access telecommunications segment, we also describe a set of reference WiMAX deployment scenarios. We then switch gears and present the advances in voice over Internet Protocol (VoIP) technology presenting ITU-T standard codecs such as G.711, G.723.1, and G.729.1, open-source codecs such as Speex, as well as codecs employed in popular VoIP applications. Finally, this chapter presents a thorough empirical evaluation of multimedia service delivery over WiMAX. We measure multimedia service delivery performance over standards-conforming WiMAX testbeds in Finland and Portugal, and evaluate the benefits of voice sample aggregation and robust header compression (ROHC) in practice.

1. Introduction 154
2. WiMAX Overview 157
    2.1. Physical Layer 158
    2.2. Medium Access Control Layer 164
    2.3. Mobile WiMAX Network Reference Model 167
3. Multimedia over WiMAX Reference Scenarios 170
    3.1. Fixed and Mobile WiMAX Generic Scenarios 171
    3.2. Telemedicine WiMAX Applications 174
    3.3. Environmental Monitoring WiMAX Applications 175
4. Advances in Telephony and the Emergence of Voice over IP 178
    4.1. ITU-T Codecs: From G.711 to G.723.1 to G.729.1 184
    4.2. Speex: An Open-Source Option for Voice Coding 188
    4.3. Voice Codecs in Popular VoIP Applications 189
    4.4. Summary 191
5. VoIP over WiMAX 191
    5.1. Testbed Description 193
    5.2. VoIP Aggregation 194
    5.3. VoIP over WiMAX with Robust Header Compression 203
    5.4. WiMAX with MIMO 206
    5.5. Quality of Service Support for VoIP over WiMAX 208

6. Remote Surveillance and IPTV over WiMAX 209
7. Summary and Outlook 216
   Acknowledgments 219
   References 219

Acronyms

3GPP, Third-Generation Partnership Project; A/V, Audio/video; AAS, Adaptive antenna system; ACELP, Algebraic code-excited linear prediction; ADSL, Asynchronous Digital Subscriber Line; AMC, Adaptive modulation and coding; AMR-WB, Adaptive multirate wideband; ARQ, Automatic repeat request; ASN, Access service network; ASN-GW, Access service network gateway; BE, Best effort; BS, Base station; BWA, Broadband wireless access; CBR, Constant bitrate; CELP, Code-excited linear prediction; CID, Connection identifier; CN, Core network; CPE, Customer-premises equipment; CRC, Cyclic redundancy check; CS, Convergence sublayer; CSIT, Channel state information at the transmitter; CSN, Connectivity service network; DSL, Digital Subscriber Line; DSLAM, Digital subscriber line access multiplexer; DTX, Discontinuous transmission; ertPS, Extended real-time polling service; ESP, Encapsulating security protocol; FDD, Frequency division duplexing; FEC, Forward error correction; FFT, Fast Fourier transform; FMIP, Fast Mobile Internet Protocol; GIPS, Global IP Solutions; GPRS, General Packet Radio Service; GSM, Global system for mobile communications; HARQ, Hybrid automatic repeat request; HSPA, High-speed packet access; IETF, Internet Engineering Task Force; iLBC, Internet low bitrate codec; IMT, International Mobile Telecommunications; IMT-A, International Mobile Telecommunications—Advanced; IP, Internet Protocol; iPCM-wb, Internet pulse-code modulation wideband; IPR, Intellectual property rights; IPTV, Internet Protocol television; iSAC, Internet speech audio codec; ISI, Intersymbol interference; ITU-T, International Telecommunication Union—Telecommunication Standardization Sector; LOS, Line-of-sight; LTE, Long-term evolution; MAC, Medium access control; MIB, Management information base; MIMO, Multiple-input multiple-output; MIP, Mobile IP; MOS, Mean opinion score; MPDU, Medium access control protocol data unit; MP-MLQ, Multipulse maximum-likelihood quantization; MS, Mobile station; MSDU, Medium access control service data unit; MTU, Maximum Transmission Unit; NB, Narrowband; NLOS, Non-line-of-sight; NWG, Network Working Group; OFDM, Orthogonal frequency division multiplexing; OFDMA, Orthogonal frequency division multiple access; PBX, Private branch exchange; PCM, Pulse-code modulation; PEP, Performance enhancing proxy;

PHY, Physical layer; POTS, Plain old telephone service; PSTN, Public switched telephone network; QAM, Quadrature amplitude modulation; QoS, Quality of service; RAN, Radio access network; RFC, Request for comments; ROHC, Robust header compression; RTP, Real-time transport protocol; rtPS, Real-time polling service; SC, Single carrier; SDU, Service data unit; SFID, Service flow identifier; SID, Silence insertion descriptor (frame); SINR, Signal-to-interference-plus-noise ratio; SMS, Short message service; SNR, Signal-to-noise ratio; SS, Subscriber station; STBC, Space–time block coding; STC, Space–time coding; TCP, Transmission Control Protocol; TDAC, Time-domain aliasing cancellation; TDBWE, Time-domain bandwidth extension; TDD, Time division duplexing; UDP, User Datagram Protocol; UGS, Unsolicited Grant Service; UMTS, Universal Mobile Telecommunications System; UWB, Ultra-wideband; VAD, Voice activity detection; VBR, Variable bitrate; VoIP, Voice over Internet Protocol; WB, Wideband; WiMAX, Worldwide interoperability for microwave access; WMAN, Wireless metropolitan area network

1. Introduction

The current networking environment is increasingly dominated by two major factors. First, users are becoming accustomed with *broadband access* both at home and at work, which enables a dramatic increase in multimedia traffic. Multimedia traffic here denotes any type of audio/visual material, irrespective of whether it is distributed in real time or on demand. For example, a video watched on a Web-based video platform requires a high-speed connection regardless of the transport protocol used. Early on, ordinary users acted mainly as consumers of digital content and information which was mainly created by major producers. Today, however, advances in hardware and the commoditization of digital content creation devices (from mobile phones with integrated high-resolution cameras to high-definition cameras with integrated Wi-Fi) have given rise to unprecedented quantities of *digital media* becoming available online by the users. In the process, users have become competent digital content producers as well as consumers.

These developments form a positive reinforcing cycle especially when combined with the viral distribution effects accompanying social networking. For instance, a decade ago, Web users searched for interesting cooking recipes in online forums and magazine Web sites. These recipes were typically described in text and were sometimes accompanied by digital photos. Today, however, there are countless videos made by everyday folks *showing* how to cook a meal or a delicacy from a country far away. These videos, often just home movies, are distributed widely over Web-based video-on-demand (VoD) services. Once online, other users can post their own videos in response to the original, thus putting even more audio/video (A/V) material online. As more users connect with broadband speeds and can therefore share their digital content online in an expedient manner, more content becomes available for consumption and more people become interested in joining the "broadband experience." At the same time, as the broadband access market becomes larger, and economies of scale and competition put pressure on prices, a larger proportion of the population can partake in this new digital society. The net effect is a large increase in Internet traffic, which according to the Minnesota Internet Traffic Studies Web site (MINTS: www.dtc.umn.edu/mints) is in the order of 50–60% per year. According to Cisco estimates [1], the average Western European mobile broadband subscriber recorded monthly traffic of 856 MB in 2007. Cisco expects that this average will grow almost fivefold to more than 4 GB by 2012. For comparison, the respective estimates for fixed-line usage are 3 GB in 2007 and 18 GB by 2012. Multimedia services, in general, and video, in particular, are expected to play a dominant role in future traffic increases.

That said, for nearly two decades there has been perennial anticipation that multimedia traffic over UDP, without congestion control checks will eventually dominate the traffic. In practice, this never materialized, and several measurement studies have shown that TCP transfers have been consistently dominating the traffic mix for years. Today, multimedia traffic broadly defined as audio, video, and gaming is the dominant contributor and is mostly carried over TCP. In part, this can be attributed to the changing nature of A/V and multimedia. While in the early 1990s a key network research issue was how to synchronously transfer A/V streams in real time, today users are more interested in VoD type of services. Further, multicast distribution although highly efficient, in and of itself is no longer sufficient for catering for the observed multimedia consumption patterns. Although it is possible to use multicast for distributing hundreds of channels to millions of receivers, interactive and asynchronous viewing requires different network technologies. As an aside, note that the term "video on demand" today has little to do with earlier operator plans for pay-per-view services. With respect to its contribution to the observed traffic mix, video on demand today is manifested in YouTube and scores of other Web-based video services and social networking platforms. In addition, peer-to-peer (p2p) networks also make available vast amounts of A/V material, albeit often illegally. Last, but certainly not least, there is a move toward catering for the long tail, that is, content that is addressing not the masses but niche audiences [2].

In the meantime, the Internet has emerged as a critical support infrastructure. With respect to broadband Internet adoption and bridging the so-called digital divide, governments around the world, not least in the European Union, have come up with a range of initiatives. In Finland, for example, the government plans to increase high-speed broadband coverage across the country. As reported in the

Akamai 2008 Q4 State of Internet quarterly report, "the investment is part of an effort to increase access to 100 Mbps connectivity in Finland to 100% of the population by 2016. The expectation is that by 2015, approximately 95% of the population will have access to the higher broadband speeds through commercial development, and by 2010 all broadband users are expected to be able to receive at least 1 Mbps." Further, the same report mentions that the French government is the first in the world to call on all telecommunications providers to offer broadband services at affordable rates across all its territory for a maximum of 35 Euros per month.

Finland is a very interesting case in point for broadband deployment. Mobile and wireless communications play a central role in research, development, and everyday life, and as a result, only about a third of the population has a fixed-line voice connection according to FICORA (Finnish Communications Regulatory Authority). Despite this, a FICORA study found that more than 68% of Finnish households have access to broadband services. The same study points out that four out of five respondents think that they need a connection speed of at least 1 Mb/s and that voice over Internet protocol (VoIP) calls are becoming more common: about one in four Internet users replied that they used VoIP in 2008.

Given the prohibitive costs of wired infrastructures and the ease of deploying wireless technologies, we argue that broadband wireless access (BWA) will play a key role in delivering Internet connectivity and access to multimedia content to the next billions of Internet users.

Worldwide interoperability for microwave access (WiMAX), often cited as a technology that could serve as the substrate for next-generation mobile broadband networks, is based on IEEE 802.16 standards [3–9]. WiMAX aims at providing a cost-effective and efficient platform for network operators. As we will see in Section 3, WiMAX is suitable for a variety of deployment scenarios, ranging from simple fixed ("DSL replacement") and mobile Internet access and content distribution, to more specialized but critical applications such as e-health and environmental monitoring. WiMAX networks can provide point-to-point and point-to-multipoint (PMP) broadband Internet protocol (IP) connectivity to both fixed and mobile hosts with Quality of Service (QoS) guarantees and robust security [10,11]. Moreover, WiMAX may become the mainstay of double and triple play offerings (VoIP, Internet, and IPTV) especially in developing countries where wired infrastructure is limited or nonexistent. WiMAX is ideal for such deployments due to lower infrastructure and construction costs, the ability to expand incrementally network coverage and the associated service offerings, and native support for QoS in the last wireless hop.

In theory, and according to vendor field trials and demonstrations, WiMAX can deliver cell bitrates greater than 100 Mb/s, covering large areas (up to 50 km radius from a single base station site using directional antennae), and serving tens of subscribers [12]. These are impressive figures. However, the currently available, commercial off-the-shelf (COTS) equipment deliver significantly less application-layer throughput. We empirically evaluated WiMAX, aiming to improve our understanding of what is realistically possible using COTS equipment. In previous work [13–15], we employed the VTT Converging Networks Laboratory infrastructure, which includes fixed and mobile WiMAX base stations (BSs) and subscriber stations (SSs), and studied voice over IP (VoIP) and live IPTV streaming uplink and downlink performance. WiMAX was used both as backhaul for voice and data services as well as a last-mile network access technology. In particular, we designed experiments with multiple competing traffic sources over a PMP WiMAX topology and measured capacity in terms of number of synthetic bidirectional VoIP "calls" between subscriber stations while concurrently delivering a variable number of video streams with negligible loss. We measured throughput, packet loss, and one-way delay for both line-of-sight (LOS) and non-line-of-sight (NLOS) conditions. We also calculated the corresponding mean opinion score (MOS) based on the International Telecommunication Union—Telecommunication Standardization Sector (ITU-T) E-model, for the experiments with emulated G.723.1-encoded conversations. We accurately measured one-way delay by employing a software-only implementation of the IEEE 1588 Precision Time Protocol (PTP). In this chapter, we summarize and discuss these results putting them in perspective with respect to key deployment scenarios for WiMAX, which are introduced in Section 3, and the general trends in broadband wireless access and multimedia service delivery. We also present an environmental monitoring WiMAX application which was deployed and tested in Portugal.

The rest of this chapter is organized as follows. Section 2 provides an overview of WiMAX and the underlying IEEE physical (PHY)- and medium access control (MAC)-layer protocols. After discussing reference scenarios for multimedia delivery over fixed and mobile WiMAX networks in Section 3, we summarize recent advances in VoIP codecs in Section 4. We then present our empirical evaluation results from VoIP over WiMAX in Section 5. Section 6 presents results from a fixed WiMAX testbed deployed in a mountainous region in Portugal as part of an environmental monitoring and fire prevention initiative. We conclude the chapter with a summary of the results presented and an outlook of future developments in the wireless metropolitan area networks area.

2. WiMAX Overview

Figure 1 illustrates WiMAX deployment around the world as of June 2009. According to WiMAX Forum, since February 2009 WiMAX operators provide services to areas inhabited by more than 430 million people. In the map of Fig. 1,

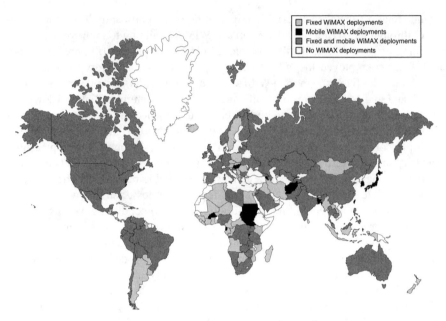

Fig. 1. WiMAX deployment map. (June 2009; source: WiMAX Forum)

countries marked with light gray color have operational fixed WiMAX networks whereas countries marked with black color have mobile WiMAX deployments only. Countries which have both fixed and mobile WiMAX networks deployed are marked with dark gray. Countries without any WiMAX deployment are shown in white. Up-to-date data can be retrieved from the WiMAX Forum Web site (see www.wimaxforum.org and www.wimaxmaps.org); however, it becomes clear that WiMAX deployments are on the rise and the technology is already in use across the world.

In the following sections, we summarize the salient aspects of the technologies underlying WiMAX networks, namely, the IEEE standardized PHY and MAC layers, along with the WiMAX Forum network reference model. Interested readers can find more details about the functionality of fixed and mobile WiMAX in Refs. [16–18].

2.1 Physical Layer

The physical layer of WiMAX is based on the multicarrier techniques called orthogonal frequency division multiplexing (OFDM) and orthogonal frequency division multiple access (OFDMA) [19]. The air interfaces parameters are specified in the IEEE 802.16-2009 [4] standard and are referred to as WirelessMAN-OFDM and WirelessMAN-OFDMA for OFDM and OFDMA, respectively (see also "The Role of IEEE in WiMAX" inset box for more information).

The Role of IEEE in WiMAX

The IEEE 802.16 working group (WG) has provided the series of standards for BWA systems upon which WiMAX networks can operate. Although the WiMAX Forum has system profiles defined with a large variety of configuration parameters, it only uses a subset of the vast selection of features specified in the IEEE 802.16 standards for fundamentally different PHY-layer implementations, MAC architectures, frequency bands, and duplexing methods. Current state-of-the-art WiMAX equipment is compliant with the WirelessMAN-OFDM and WirelessMAN-OFDMA air interfaces, which are specified in the IEEE standards, as these multicarrier air interfaces were chosen by the WiMAX Forum to be the ones used in WiMAX system profiles.

The IEEE 802.16 WG has been operational since 1999. It published its first standard (IEEE 802.16-2001) in the beginning of 2002. The first version of the standard included specifications for single carrier (SC) BWA systems operating in LOS conditions at the 10–66 GHz frequency band. During the first half of 2003 the standard was expanded with the IEEE 802.16c-2002 and IEEE 802.16a-2003 amendments. IEEE 802.16c-2002 introduced detailed system profiles for the original standard. IEEE 802.16a-2003 added specifications for OFDM- and OFDMA-based multicarrier air interfaces, NLOS support, and expanded the operational frequency band to include the frequencies between 2 and 11 GHz.

In the second half of 2004 all existing standards were collected into one document, namely IEEE 802.16-2004, which became the base standard for fixed WiMAX systems. In the end of 2005, IEEE 802.16f-2005 introduced the management information base (MIB) for IEEE 802.16-2004 compliant systems. Subsequently, the IEEE 802.16e-2005 amendment was published in the beginning of 2006. It introduced mobility support specifications for the IEEE 802.16 wireless multicarrier systems as well as corrections to fixed operations as the document also included a corrigendum for IEEE 802.16-2004.

In the end of 2007, two additional amendments were published. First, an amendment to IEEE 802.1D MAC bridging standard was published under the name IEEE 802.16k-2007. It added the bridging of IEEE 802.16 into IEEE 802.1D. The second amendment, IEEE 802.16g-2007, contained specifications for management plane procedures and services in IEEE 802.16 networks.

The latest version of the IEEE 802.16 standard is IEEE 802.16-2009 [4]. Published during the first half of 2009, it collected the previous standard versions and amendments into a single document and also introduced new functionalities, such as persistent air interface resource assignments for VoIP operation and MIB for mobile systems, by merging previously unpublished draft amendments into the standard. In addition, the first amendment to this standard was published almost simultaneously under the name IEEE 802.16j-2009 and it includes the multihop relay specifications for IEEE 802.16 networks.

Currently, the IEEE 802.16 working group has two documents under development. IEEE 802.16h will include specifications for operation in license exempt frequencies. In addition, IEEE 802.16m will specify a new enhanced air interface for IEEE 802.16 systems which will be compliant with the International Mobile Telecommunications—Advanced (IMT-A) system requirements. The IEEE 802.16 WG has also published recommendations for fixed BWA network deployments in order to control cross-interference between systems and facilitate their coexistence. The latest version of the coexistence standard IEEE 802.16.2-2004 was published in the beginning of 2004.

Table I summarizes all the active standard releases published by the IEEE 802.16 working group as of October 2009.

TABLE I
IEEE 802.16 STANDARDS (ACTIVE AS OF OCTOBER 2009)

Standard title	Date	IEEE standard version
Air Interface for Broadband Wireless Access Systems	May 2009	802.16-2009
Multihop Relay Specification	June 2009	802.16j-2009
Bridging of IEEE 802.16	August 2007	802.16k-2007
Protocol Implementation Conformance Statement (PICS) Proforma for Frequencies below 11 GHz	January 2007	802.16/Conformance 04-2006
Radio Conformance Tests (RCT) for 11–66 GHz WirelessMAN-SC Air Interface	June 2004	802.16/Conformance 03-2004

In OFDM, the available bandwidth of the communication channel is divided into parallel orthogonal subcarriers. The wideband channel is effectively transformed into a series of channels with lower bandwidth and can be defined as a group of adjacent narrowband channels. When a high bitrate data stream is divided into these subcarriers, multiple narrowband data streams are transmitted over the air as a result. OFDM has become the *de facto* PHY layer for state-of-the-art BWA systems due to its relatively easy hardware implementation with fast Fourier transform (FFT) algorithms, its suitability for adaptive modulation and coding (AMC) techniques, and its capability to reduce intersymbol interference (ISI) in high bitrate transmissions.

ISI is caused by multipath propagation in the wireless communication medium when multiple copies of different transmitted data symbols arrive at the receiver with large intervals due to reflections in the propagation path and, hence, overlap with each other in the time domain. As the data symbol duration is inversely proportional to the transmission bitrate, the transmitted symbol duration is increased when the high bitrate data stream is divided into several low bitrate streams for the narrowband channels and the level of ISI can be reduced. When guard intervals and cyclic prefixes are used between successive data symbols, the impact of ISI can be decreased even more.

OFDMA extends OFDM functionality and enables it to be used as a multiaccess scheme for the wireless medium. In OFDMA, the narrowband subcarriers are grouped into subchannels, which can be assigned to different users contending for the data link. Each subchannel can contain a different amount of subcarriers and by altering the subcarrier group sizes and observing the channel conditions, it is possible to use differentiation in the channel allocation for several users. This makes OFDMA a very flexible multiaccess scheme which enables adaptive resource allocation and is suitable for both fixed and mobile deployments. OFDMA also makes subchannelization possible for both uplink and downlink, which enhances efficiency of the uplink capacity usage. In plain OFDM, partial subchannelization is possible in the uplink direction only. Otherwise, all subcarriers are allocated to a single user at a time in each OFDM frame. The physical-layer transmission frame size in WiMAX systems can vary between 2.5 and 20 ms. In practice, however, current WiMAX configurations mainly use a frame size of 5 ms.

Depending on the deployment, OFDMA subchannelization provides two different methods for allocating the subchannels to users. In fixed and low mobility WiMAX systems, the formation of subchannels out of concurrently assigned subcarriers is an efficient way to increase the capacity of the wireless link. By choosing subchannels from different sections of the available bandwidth and allocating them to the users by intelligently adjusting the modulation scheme and coding rate, the received signal-to-interference-plus-noise ratio (SINR) of each user and, hence, the capacity of the wireless link, can be maximized. In WiMAX systems, the subchannelization scheme based on concurrent subchannels is called the band AMC. On the other hand, in mobile WiMAX deployments, the subcarriers assigned to one subchannel can be distributed over the available bandwidth and, hence, frequency diversity can be attained. In fully mobile WiMAX systems, this is advantageous because frequency diversity can be used to make the transmission link more robust against fast fading.

WiMAX systems support both time division duplexing (TDD) and frequency division duplexing (FDD) methods. By dividing the OFDM frames into uplink and downlink slots in time, TDD uses the same channel frequency for both directions. State-of-the-art WiMAX systems use primarily TDD due to its flexibility and lower bandwidth demands. FDD, in contrast, allows for a simpler implementation alternative, as the synchronization requirements in the network are not as strict as in TDD. Nonetheless, as FDD uses separate carrier frequencies for uplink and downlink, it requires more bandwidth than TDD systems in order to provide the same capacity to users. Originally, TDD was the only duplexing method defined for mobile WiMAX systems. However, due to the deployment complexity and frequency regulation issues in some market areas, the WiMAX Forum has specified an option for an FDD-based mobile WiMAX system in the most recent release of the mobile WiMAX system profile (see also "The Role of the WiMAX Forum in WiMAX" inset box for more information).

The Role of the WiMAX Forum in WiMAX

The WiMAX Forum is a nonprofit organization formed in 2001 and its aim is to enhance the compatibility and interoperability of equipment based on the IEEE 802.16 family of standards. Although the IEEE standards allow multiplicity for deployment in very diverse environments, they may also lead to either solely vertical, single vendor deployments or no deployment at all, as operators do not want to be locked into any particular implementation. Thus, a major motivation for establishing the WiMAX Forum was to develop *predefined system profiles* for equipment manufacturers, which include a subset of the features included in the IEEE 802.16 standards. WiMAX Forum certified products are guaranteed to be interoperable and to support wireless broadband services from fixed to fully mobile scenarios. The aim is to enable rapid market introduction of new standard-compliant WiMAX equipment and to promote the use of the technology in different sectors. The WiMAX Forum is also working on the international roaming specifications for WiMAX networks.

Unlike many other industry driven technology forums, the WiMAX Forum has also a strong technical role in WiMAX technology in addition to the marketing and certification activities. Hence, the WiMAX Forum produces several technical documents through different working groups. In addition to the WiMAX system profiles [20–22], which specify the PHY- and MAC-layer configurations for WiMAX systems, and the corresponding certification profiles, the WiMAX Forum Network Architecture documents [23, 24] include the end-to-end specifications for WiMAX network operations (see Section 2.3).

Another salient feature of the WiMAX physical layer is its support for adaptive antenna systems (AASs). AAS support in WiMAX includes both beamforming and multiple-input multiple-output (MIMO) techniques, which can be used separately or in parallel. Adaptive beamforming in the WiMAX BS concentrates the energy of the transceiver by narrowing down the antenna beam using antenna element adjustment. With a narrow and concentrated beam, the available antenna gain is higher for a certain power level and as the level of interference received from the surrounding cells is also lower, the signal-to-noise ratio (SNR) of the received signal is higher. In mobile systems, the antenna beam can also be made to follow its communication counterpart moving around in the base station transceiver's range. By using adaptive beamforming, both capacity and range of the WiMAX system can be increased considerably.

In MIMO, multiple antenna elements are used to transmit and receive either the same or individual data streams over the air interface simultaneously. When the same data content is transmitted and received through multiple antennae, the MIMO method is called space–time coding (STC). By using STC, multiple copies of the same data bits are transmitted over independent communication channels and the robustness of the WiMAX link against fast fading in mobility is increased. With a fixed transmission power, STC increases the SNR of the received data signal, which is a maximum ratio combined version of the signals received through each antenna element. On the other hand, if independent data streams are transmitted and received through multiple antenna elements, the MIMO technique is called spatial multiplexing. With spatial multiplexing, in optimal circumstances, the spectral efficiency and, thus, the capacity of the wireless link can be multiplied by the minimum number of used antennae either in the transmitting or receiving end of the wireless link. It is also possible to combine adaptive beamforming with MIMO techniques in order to achieve additional gains in the wireless link capacity. For more information on MIMO techniques in WiMAX systems, see Section 5.4.

WiMAX also supports hybrid automatic repeat-request (HARQ) retransmissions at the OFDMA physical layer as an optional feature which adds forward error correction (FEC) coding on top of the cyclic redundancy check (CRC) error detection bits of plain automatic repeat-request (ARQ) functionality. HARQ operation can be managed on a per-connection or per-terminal basis and the operational parameters are negotiated during the control message exchange upon network entry. Per-connection HARQ is used with mobile stations (MSs) in mobile WiMAX. Per-terminal HARQ is employed with fixed MSs or SSs. In practice, a mobile station can only support one of the two HARQ management methods at a time.

During HARQ operation, the receiver must acknowledge all physical-layer data bursts it receives through the wireless link either with a positive or a negative acknowledgment, depending on the outcome of the decoding process of the data burst in question. If the transmitter receives a negative acknowledgment for one of the data bursts it previously transmitted, that data burst must be retransmitted. When retransmitting, HARQ does not operate like a simple ARQ scheme. Instead, the receiver holds on to the negatively acknowledged data bursts and uses diversity combining between the original erroneously decoded data and the retransmitted data in order to increase the probability of successful decoding at the receiver. The process is repeated until the data burst is decoded successfully or until the predefined maximum number of retransmissions is reached.

In Type I HARQ, also known as chase combining, the redundancy bits and the user data are kept the same between successive retransmissions of a physical-layer data packet. In Type II HARQ, referred to as incremental redundancy, the coding and punctuation pattern can also be changed between successive retransmissions. In Type II HARQ, only parts of the original physical-layer data packet is required to be included in each retransmission which means that less bandwidth is wasted if retransmissions are needed. Type I HARQ is used with mobile WiMAX whereas Type II HARQ can be used in both fixed and mobile WiMAX systems.

2.2 Medium Access Control Layer

The MAC layer functions as an adaptation layer between the physical layer and the upper protocol layers. Its main task is to receive MAC service data units (MSDUs) from the layer above, encapsulate them into MAC protocol data units (MPDUs) and pass them down to the physical layer for transmission. At the receiving side, the MAC layer takes MPDUs from the physical layer, decapsulates them into MSDUs, and passes them on to the upper protocol layers. In addition, the MAC layer of a WiMAX base station manages bandwidth allocations for both uplink and downlink directions. The BS assigns bandwidth for the downlink according to incoming network traffic, whereas for the uplink, bandwidth is allocated based on the requests received from the user equipment.

The WiMAX MAC layer is organized into three separate functional entities as shown in Fig. 2. An additional layer between the generic MAC part and upper protocol layers, called the service-specific convergence sublayer (CS), functions as an adaptation interface to the actual MAC layer. Several different convergence sublayer configurations are presented in Ref. [4] for a variety of different protocols. However, the WiMAX Forum specifications currently include support only for IP- and Ethernet-specific convergence layers. Other upper-layer protocols can operate over the IP and Ethernet convergence sublayers via encapsulation. The service-specific convergence sublayer may also support header compression for the upper protocol-layers packets.

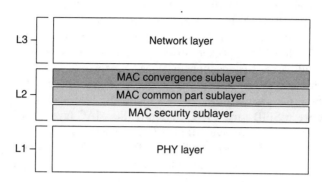

Fig. 2. The MAC-layer structure of WiMAX.

Whereas the convergence sublayer adapts the upper-layer protocols to the WiMAX MAC functionalities, the common part sublayer is responsible for all protocol-independent operations, such as, data fragmentation, packing, transmission, QoS control, and ARQ operations. The security sublayer between the MAC and PH layers handles encryption, authorization, and key exchange.

The MAC layer of WiMAX enables flexible allocation of transmission capacity among different users. A single data burst can comprise variably sized MPDUs from different user data flows before it is handed over to the physical layer for transmission. Multiple small MSDUs can be packed into one MPDU and, conversely, one big MSDU can be fragmented into multiple small ones at the MAC layer in order to further enhance the performance of the system. By bundling up several MPDUs or MSDUs, both the PH- and MAC-layer header overheads can be reduced, respectively. The MAC layer must also perform the converse operations to the data packets it receives from the physical layer. To do to that, it must unpack and possibly defragment the incoming MPDUs and reorganize the data into the original MSDUs before it can pass them up the protocol stack.

In addition to the HARQ functionality at the PH layer, the MAC layer can also use traditional ARQ as a selective retransmission method in order to enhance the performance of the WiMAX air interface. MAC-layer retransmissions function as a fall-back mechanism for the considerably faster physical-layer retransmissions if the HARQ process malfunctions due to, for example, mobility-inflicted signal quality fluctuations or dropped packets during handover execution.

A unique connection identifier (CID) is assigned to each uplink/downlink connection pair at the WiMAX MAC layer before any data transmission takes place over the air interface. A CID serves as a temporary address for the transmitted data packets over the WiMAX link. In order for the base station to manage the QoS requirements of individual data flows, the same service flow identifier (SFID) is assigned to unidirectional packet flows which have the same QoS parameters. As basically all wireless connections in WiMAX networks are centrally controlled by the serving base stations, it is reasonable to implement also the QoS control functionality at the base stations. After all, it is the base station that handles the mapping between SFIDs and CIDs. Currently, the MAC layer of a WiMAX base station includes support for five different QoS classes which can be used to adjust system behavior to be more favorable to different traffic types and applications. A summary of the data delivery services with individual QoS guarantees is given in Table II along with the respective scheduling service categories (see also Section 5.5).

To support battery powered user terminals, the WiMAX MAC layer also incorporates sophisticated power management features. The savings in power consumption are achieved by defining different operational modes for the user equipment. When there is no user data to be transmitted or received by the system, both WiMAX power saving operation modes, that is, the sleep and idle modes, switch off certain parts of the transceiver that are not currently required for communication purposes. However, the methods for controlling the switch off intervals differ between the sleep and idle modes. The idle mode allows for greater savings in power consumption by using longer and more extensive power switch off sequences than the sleep mode.

In sleep mode, the WiMAX user terminal switches itself off for a set period, after which it turns on again and checks for incoming data. In idle mode, the user terminal can switch off the transmitter completely and receive only the downlink broadcast traffic from the nearby base stations without registering with them. When the base station, which served the user terminal before it switched to the idle mode, receives incoming data for the user terminal from the network, that base station and a group of adjacent base stations (called the paging group) page for the user terminal through the broadcast channel. Upon receipt of the page, the user terminal powers up and prepares for data reception. By using paging groups instead of a single paging base station, a mobile terminal can be located by the network even if it has moved from one cell to another during the idle period. During long idle periods, the user terminal updates its paging group on a regular basis. Currently, support for the idle mode is optional in WiMAX systems. In addition to reducing energy consumption at the user equipment, both the sleep and idle operation modes save network resources by reducing the amount of control signaling in the network.

TABLE II
THE DATA DELIVERY AND SCHEDULING SERVICES OF WiMAX

Data delivery service/QoS class	Scheduling service	Features	Example applications
Unsolicited Grant Service (UGS)	Unsolicited Grant Service (UGS)	–Real-time data transmission with CBR and fixed or variable size PDUs –Periodic and fixed size uplink bandwidth allocations without continuous solicitations	–Constant bitrate VoIP
Real-Time Variable Rate Service (RT-VR)	Real-Time Polling Service (rtPS)	–Real-time data transmission with VBR, guaranteed data rate, and delay –Periodic and variable size uplink bandwidth allocations on demand	–Video streaming –Audio streaming
Non-Real-Time Variable Rate Service (NRT-VR)	Non-Real-Time Polling Service (nrtPS)	–Delay insensitive data transmission with guaranteed data rate –Slots reserved for uplink unicast transmission requests on a regular basis	–File transfer
Best-Effort Service (BE)	Best-Effort Service (BE)	–Data transmission without data rate and delay guarantees –Uplink transmission in vacant slots	–Web browsing –Instant messaging –Data transfer
Extended Real-Time Variable Rate Service (ERT-VR)	Extended Real-Time Polling Service (ertPS)	–Real-time data transmission with VBR, guaranteed data rate, and delay –Uplink scheduling extended from rtPS with dynamic variable size bandwidth allocations without continuous solicitations	–Variable bitrate VoIP

2.3 Mobile WiMAX Network Reference Model

The WiMAX Forum Network Working Group (NWG) has defined a network reference model [23, 24], which specifies the placement, role, and functionality of the operational entities in WiMAX networks. In short, the WiMAX Forum has defined a standardized end-to-end network model for IEEE 802.16-based wireless technologies, aiming to guarantee interoperability not only in the air interface but also at the system level.

In the WiMAX network reference model, three basic architectural entities are defined: (i) the user equipment, that is, a SS or a MS; (ii) the access service network

(ASN); and (iii) the connectivity service network (CSN). The role of the user equipment is to provide the wireless connection to the WiMAX network. The ASN is the radio access network (RAN) with which the user equipment establishes a connection to. An ASN can consist of one or more base stations (BSs) and access service network gateways (ASN-GWs), which are all managed by a single network access provider (NAP). The CSN, on the other hand, is the entity which provides all the IP core network (CN) functionalities to the WiMAX radio equipment residing in ASNs. CSN interconnects the WiMAX user with the rest of the Internet. CSNs are typically maintained by network service providers (NSPs).

Both ASNs and CSNs are further divided into smaller functional entities such as base stations, gateways, mobility agents, and so on. These network entities communicate with each other using standardized interfaces called reference points which guarantee that a certain set of communication protocols and procedures are always supported between particular network entities irrespective of the underlying hardware. The defined reference points are used for various control and management messaging purposes, as well as for user data bearing inside the network architecture. The WiMAX network reference model and the specified reference points are illustrated in Fig. 3.

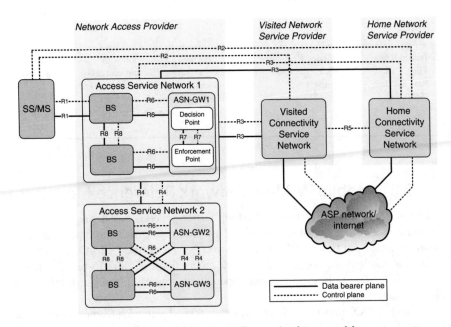

FIG. 3. The WiMAX Forum NWG network reference model.

Reference point R_1 resides between a SS/a MS and a WiMAX BS. It implements the specifications of the air interface consisting of both control and data-bearing protocols and procedures compliant with IEEE 802.16 standards. Reference point R_2 is a logical control interface between the user equipment and a CSN. R_2 is used solely for management purposes, such as authentication, authorization, and accounting (AAA), mobility management, and IP host configuration. No direct protocol interface connection between a SS/MS and a CSN exists in an actual network. Reference point R_3 serves a purpose similar to R_2 between an ASN and a CSN by supporting AAA and mobility management capabilities in addition to the user data-bearing services. Mobility management between two separate ASNs is handled via reference point R_4 which supports the required control and bearer plane protocols enabling continuous data transmission during inter-ASN link-layer handovers. The control and bearer plane protocols of reference point R_5 enable interworking between the home CSN and a visited CSN for network-layer mobility management.

As already mentioned above, an ASN can contain one or more BSs and ASN-GWs. In Fig. 3, two example ASN deployments are presented. Access service network 1 comprises two base stations and a single ASN gateway (ASN-GW1). Through ASN-GW1 all control and data traffic is centrally routed to and from the core network when the users inside a base station cell of access service network 1 are communicating with the outside world. When an ASN includes more than one ASN gateways, as in access service network 2 in Fig. 3, all ASN gateways must be interconnected via the R_4 reference point.

Other reference points used in the intra-ASN control and data message exchange are R_6, which consists of control and data-bearing protocols for the interconnection between base stations and ASN gateways; and R_8, which supports control and data-bearing protocols between separate base stations inside a single ASN. R_8 enables fast and seamless intra-ASN handovers.

An optional ASN-GW decomposition is also defined in the WiMAX network reference model and it is presented inside the ASN-GW1 block in Fig. 3. In this decomposed layout, the decision point part of an ASN gateway includes the control plane functions whereas the enforcement point handles the data bearer plane functions. If the aforementioned decomposition is adopted, the intra-ASN-GW reference point R_7, supports an optional set of control plane protocols, which are used, for example, for AAA and policy coordination purposes. R_7 interconnects the separate decision and enforcement point entities of an ASN gateway.

All in all, the WiMAX Forum network reference model includes definitions for the usage of mobile Internet protocol (MIP) in various forms for network level mobility management, network selection, IP addressing, and QoS, as well as considerations for security, radio resource management, and AAA (see also "The Role of IETF in WiMAX" inset box for more information). The WiMAX Forum is also investing heavily developing the specifications that will enable international roaming. WiMAX is taking the final steps in the evolution from a mere LOS fixed BWA technology to a fully standardized mobile cellular system.

The Role of IETF in WiMAX

As WiMAX networks are fully IP-based, the WiMAX Forum has recommended that Internet engineering task force (IETF)-standardized protocols should be used in WiMAX networks as extensively as possible. By employing standardized protocols for IP networking, creating networking products, applications, and services for WiMAX networks should be easier as proprietary protocols are not complicating interoperability and portability of the designed solutions. However, the salient features of WiMAX links, such as the connection-oriented, PMP architecture and native multicast support available only in the downlink direction, inflict problems on the operation of some of the IETF protocols, especially, as many of the salient features of IPv6 protocols rely on bidirectional multicasting. Hence, IETF has formed a working group called *IP over IEEE 802.16 Networks* (16ng) which has been working on the specifications for IPv4 and IPv6 operation over IEEE 802.16 and WiMAX networks.

The 16 ng working group has taken the leading role in the standardization and released already three request for comments (RFC) documents [25–27] on IP operation over IEEE 802.16 and has two specifications in an Internet Draft state under development. Another IETF working group has also contributed to specification work. The working group on *Mobility for IP: Performance, Signaling and Handoff Optimization* (MIPSHOP) has released an RFC concerning fast mobile IP (FMIP) operation in IEEE 802.16 compliant mobile networks [28].

---

3. Multimedia over WiMAX Reference Scenarios

Network operators and service providers consider WiMAX, the technological driver for the BWA market, as a great opportunity for a large number of application scenarios. In fact, the positive results in terms of performance combined with the increasing interest showed by many governments worldwide, as demonstrated by the hundreds of trials and the associated regulatory activities currently going on in the world, are boosting WiMAX acceptance within the wireless market. In this section, we briefly introduce some of the most relevant WiMAX deployment scenarios, covering both fixed and mobile environments.

3.1 Fixed and Mobile WiMAX Generic Scenarios

As a BWA technology, WiMAX is capable to deliver high data rates for stationary and mobile terminals, by capitalizing on IEEE 802.16 standards. A very interesting and promising scenario for fixed WiMAX is to provide broadband connectivity in rural and urban areas. Figure 4 illustrates the fixed WiMAX scenarios we describe next.

Rural broadband connectivity is, in fact, one of the most challenging scenarios for network operators in both developed and emerging countries. The difficulty to provide such type of connectivity in these environments is mostly because of the large distances between the remote rural areas and the network operator points of presence (PoP), which requires the installation and deployment of an extremely expensive wired infrastructure. Furthermore, these regions are usually home to a small number of potential broadband subscribers, hinting to rather low operator revenues, which are not enticing enough in order to undertake the enormous investments for building out a wired network infrastructure. Even in developed countries, it is not always cost-effective to deploy a wired solution in many remote areas. Wired solutions, such as digital subscriber line (DSL), coaxial cable, or optical fiber, are very expensive to install in environments without any preinstalled infrastructure. Consequently, these areas are left uncovered, and contribute significantly in increasing the digital divide. In such cases, WiMAX can play a central role since it provides

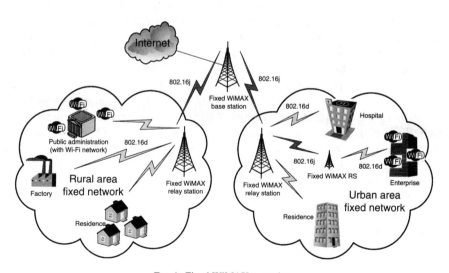

FIG. 4. Fixed WiMAX scenarios.

broadband wireless connectivity over large distances. As illustrated on the left-hand side of Fig. 4, different types of clients from the rural area can be served by the fixed WiMAX link, including residential and enterprise users, factories, and public administration buildings.

Besides being a cost-efficient fixed last-mile broadband wireless solution, WiMAX can also be used as a first-mile solution. That is, one can employ a WiMAX network to backhaul traffic from both wireless and wired local area network technologies. An example, also illustrated in Fig. 4, is to use fixed WiMAX as a backhaul link for a wireless local area network (WLAN) based, for example, on IEEE 802.11 (Wi-Fi), deployed in a remote rural area. In this case, one or more Wi-Fi access points (APs) are directly connected to the WiMAX SS, creating a concatenated BWA network, employing fixed WiMAX on the first hop and Wi-Fi on the second and last hop. In this scenario, the WiMAX network is transparent for the end-user terminal. That is, users do not need WiMAX-enabled devices, but only simple and inexpensive Wi-Fi compliant cards or devices with integrated Wi-Fi interfaces. Another possibility is to connect a wired technology, such as DSL, to the WiMAX link. In this case, the end users connect to the network infrastructure using DSL modems that are connected to the digital subscriber line access multiplexer (DSLAM) with copper lines.

Nomadic scenarios are very interesting from the user point of view. With WiMAX, the user can literary take his indoor WiMAX SS to another location and connect automatically with the WiMAX network without the need to contact the network operator. For example, during construction operations, a company can benefit from the nomadic capability provided by WiMAX and connect the construction head-office with different construction sites temporarily, without having to purchase several broadband access subscriptions from the network provider. This is simply not possible with wired broadband access alternatives.

Similar to rural areas, in certain cases, metropolitan environments can also be quite challenging for operators who want to provide cost-effective solutions for broadband connectivity. A characteristic of urban areas, especially in developed countries, which restricts the spread of cost-effective wireless broadband solutions, is the large number of high rises and even skyscrapers. This urban terrain makes the propagation conditions for the wireless signal very problematic. As in rural areas, fixed WiMAX can also be used in urban environments to overcome the aforesaid limitations. Moreover, a fixed WiMAX solution is very straightforward to mount and operate, requiring only the installation of an indoor or outdoor WiMAX receiver antenna, directly connected to the user local area network. WiMAX terminals can operate in line-of-sight (LOS), optical line-of-sight (OLOS), and non-line-of-sight (NLOS) environments.

NLOS signal propagation is the typical situation in a metropolitan area. A typical urban scenario using fixed WiMAX as the access technology is shown on the right-hand side of Fig. 4. Residential buildings, small and big enterprises, and hospitals are some of the potential customers presented in the schematic.

To overcome the propagation difficulties created by big buildings, the urban scenario presented in Fig. 4 includes the novel WiMAX relay system, based on the IEEE 802.16j standard [5], which provides mobile multihop relay (MMR) connectivity. The WiMAX relay system is composed of relay stations (RS), positioned between the WiMAX base stations and the WiMAX receiving nodes, also known as WiMAX subscriber stations. The main goal of the WiMAX relay system is to extend cell coverage, surpassing NLOS environments, as illustrated in Fig. 4.

In addition to the fixed BWA scenarios described so far, WiMAX can also cater for users requiring always-on, mobile connectivity. Therefore, one of the most promising broadband solutions for next-generation networks is mobile WiMAX, based on IEEE 802.16-2009 [4]. Combining the mobile and broadband characteristics provided by IEEE 802.16, mobile WiMAX could easily be used for ubiquitous connectivity in next-generation environments, as depicted in Fig. 5. With mobile WiMAX end users have the capability to connect to the Internet and have access to multimedia services independently of their location, and at anytime, using WiMAX-enabled laptops, phones or personal digital assistants (PDAs).

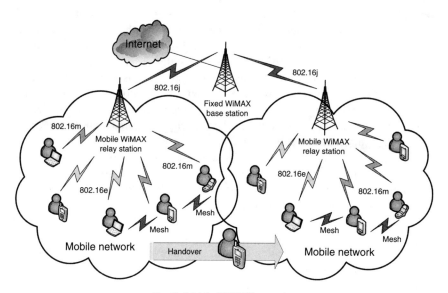

FIG. 5. Mobile WiMAX scenario.

Although IEEE 802.16-2009-compliant equipment provide broadband connectivity, ITU IMT—Advanced specifies that the next-generation wireless technologies must provide data rates up to 100 Mb/s for high mobility users and 1 Gb/s for low mobility or stationary terminals. To address this requirement, the IEEE 802.16 working group initiated the specification of IEEE 802.16m [9], also known as mobile WiMAX advanced air interface.

Another interesting aspect illustrated in Fig. 5 is the use of the WiMAX relay system in a mobile environment. In this scenario, the WiMAX MMR is essential for backhauling traffic for the mobile WiMAX BSs in the field, connecting them with the mobile operator network infrastructure. Furthermore, Fig. 5 also shows how to introduce and take advantage of the WiMAX network mesh capabilities. In this case, the mesh capability add-on enables operators to route traffic directly between WiMAX MSs, without having to send all their packets toward the WiMAX RS/BS.

3.2 Telemedicine WiMAX Applications

e-Health is meant to play a key role in making health care an accessible, high quality, cost-effective, economically viable, and safe service for citizens. e-Health is also one of the areas where WiMAX technologies can substantially contribute to improve the daily activities of people with special needs and thus enhance their quality of life. Given the inability of previous wireless access technologies to provide true broadband mobile connectivity [29], a significant number of medical-related procedures are still very complex and expensive for health care institutions.

For example, *remote follow-up* is a very appealing telemedicine application that can benefit from mobile WiMAX features. Typically, after therapies or surgical procedures, patients need to be constantly monitored and assisted by the medical staff. Up to now, for monitoring purposes, patients had to move to distant hospitals or specialized medical staff had to travel to the patient accommodation. By deploying mobile WiMAX networks, however, the capacity, QoS, and security guarantees are in place to allow remote follow-up and assistance of patients. Voice and video over IP connections can be established to support real-time communications with QoS assurances between the patient location and the e-health premises.

The remote follow-up telemedicine application is illustrated in Fig. 6. Two different variants of this application are presented: (1) a patient connected to a local/home Wi-Fi access network, backhauled by fixed WiMAX and (2) a patient directly monitored via a mobile WiMAX link to the e-health care center.

Another telemedicine application that can be improved through mobile WiMAX is always-on medical assistance. In this scenario, the medical staff is able to have a permanent wireless broadband communication link with the health care center. For example, in case of childbirth at home, street accidents or other emergencies, it is

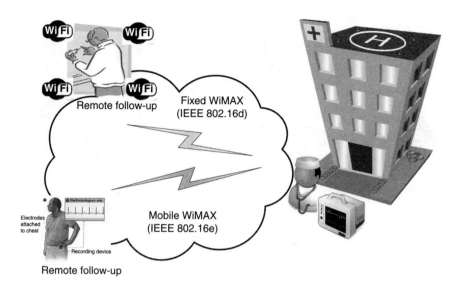

FIG. 6. Remote follow-up WiMAX application.

required to transmit critical data urgently in order to make an immediate diagnosis. In this case, mobile WiMAX can be used to provide connectivity to doctors in the field where broadband access is usually unavailable. Thus, voice and video over IP connections are established to transmit real-time information with an adequate level of QoS. Figure 7 illustrates the always-on medical assistance scenario. Two applications are presented: (1) a doctor providing assistance at the patient's house, communicating in real time with the medical staff located at the health care center and (2) a doctor giving first aid to a patient inside an ambulance, retrieving information about the patient from the hospital and simultaneously discussing with the hospital medical staff the surgical procedures to be undertaken upon arrival.

3.3 Environmental Monitoring WiMAX Applications

Due to the features offered by WiMAX technology to reach high data rates over long distances, it is very useful in scenarios of difficult access, or where human presence cannot be continuously granted. These application scenarios include monitoring of remote areas, such as a volcanic and seismic sites and fire prevention. Figure 8 is a schematic of the environmental monitoring application scenarios described below.

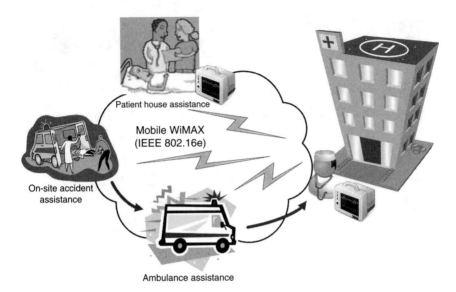

FIG. 7. Always-on medical assistance WiMAX application.

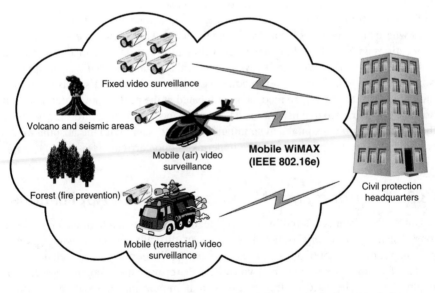

FIG. 8. Environmental monitoring (fire prevention, volcano, and seismic monitoring).

Active volcanoes represent one of the highest natural risks in the world. Volcano eruptions can cause heavy damages to private property and public infrastructure, as well as a large number of human casualties, in particular to communities living in close proximity to volcanic areas. Constant monitoring of volcanic activity can prove instrumental in foreseeing possible eruptions: With an early warning system in place, authorities can follow the best strategy in order to reduce the overall damage and the disruptive consequences of natural disasters.

Nowadays, seismic methods are used to monitor volcanic activity, as a large variety of different seismic signals occurs before an eruption. For instance, sensors and video cameras positioned close to a volcanic site can collect data continuously and feed them into a real-time seismic monitoring application. A WiMAX network can be used in this scenario to provide the broadband wireless connection necessary to transmit the data collected by the sensors as well as the video camera stream(s) to the Civil Protection Headquarters and from there to other research centers, where scientists can analyze the received information. This solution allows the real-time evaluation of data and avoids the need for operators to reach every station to download collected data. Moreover, WiMAX connectivity can also be used in critical situations to offer voice and video communication between the Civil Protection Headquarters and the teams of scientists involved in field activities.

Additionally, climate changes and land use practices have turned forest fires into a major disaster in extensive areas of the world. Fire prevention and firefighting efforts are still largely unsuccessful, in part due to the difficulty to guarantee early fire detection. Small fires, which go undetected can quickly develop into wildfires causing havoc to entire regions. This is especially true in areas that are scarcely populated or difficult to reach. Today, human spotters, stationed on fixed surveillance towers or, in some cases, flying in small airplanes, detect fires by searching for smoke indications and then relaying the collected field information to the Civil Protection Headquarters. These methods are not cost-efficient: each tower requires the permanent presence of human watchers, while airplane patrols become increasingly more expensive.

Communication between surveillance towers and control centers is another problem in fire prevention. Broadband technologies such as ADSL or 3G/UMTS are not available in remote forests. Alternatives like GPRS, GSM, or UHF radio links, besides their coverage limitations, cannot provide the bandwidth required for more automated fire detection. WiMAX appears to fit well in this scenario. Given its potential range, bandwidth, and adaptability to environmental conditions, WiMAX may provide connectivity to remote monitoring systems (fixed and mobile, airborne and terrestrial), capable of supporting early fire detection in a more efficient and cost-effective manner. Cameras mounted in surveillance towers can automatically scan an entire field at 360°, with 20–40 km radius, and send a variety of collected data (live surveillance video streams, infrared sensor readings, geolocation data, meteorological data) to control centers, where prospective fires are further investigated (for instance remotely pointing and zooming the surveillance cameras to the target area). This solution is both cost-effective, since the number of required human operators is drastically reduced, and efficient, as the data received by the control center is more precise.

In Section 6, we revisit this scenario and present an actual fire prevention testbed developed in Portugal based on a WiMAX network deployed in a remote mountainous region. As we will see, the testbed includes a remote control application which lets operators to monitor areas spanning dozens of square kilometers of mountainous and forested terrain.

4. Advances in Telephony and the Emergence of Voice over IP

Voice telephony has been traditionally carried over a separate, dedicated infrastructure known as the public switched telephone network (PSTN) which originally comprised of copper wired connections. Later, microwave, satellite, and optical fiber connections have also been added to the infrastructure. In PSTN, voice and all associated call establishment and termination signaling go through dedicated telephone switching centers which route calls to their destinations. Making a phone call requires the establishment of an end-to-end circuit, which originally was simply the concatenation of physical cables. Starting in the late 1870s, human operators would plug and unplug connecting cables at switching centers in order to establish and teardown these physical circuits (see Fig. 9).

Automation was gradually introduced in several stages, but the essential aspects of the plain old telephone service (POTS) remained the same: voice is transmitted in analog form as a physical signal over an (automated) circuit-switched network using, for example crossbar switches (Fig. 10). However, due to the analog nature of the communication and the supported bandwidth of the twisted pair copper wire in POTS, fundamental restrictions exist with respect to the voice quality it can support without sophisticated higher order modulation schemes and performance enhancing source and channel coding methods. That is, as twisted pair copper wire in frequency division multiplexed POTS can only deliver a bandwidth range of 300–3400 Hz for voice and audio signaling, the Nyquist rate dictates the fundamental limits of the maximum capacity available for voice transmission over the communication medium. If the symbol rate of the transmission is increased higher than the Nyquist rate, ISI will occur in the receiver and distort the signal [30]. Effectively,

Fig. 9. A human operator physically connects a phone call using patch cables and a telephone switchboard. (photo by Joseph A. Carr, 1975)

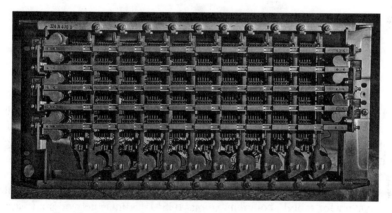

Fig. 10. Crossbar switch manufactured by Western Electric (1970).

due to the aforementioned physical and technological restrictions, analog voice transmission has hard bounds on the quality of sound it can deliver and allows only for narrowband communication with small user densities.

In the 1980s the major telecommunication vendors and operators pushed for the gradual replacement of PSTN by the integrated services digital network (ISDN). ISDN featured several new services and was no longer a voice-only telecommunications network. Alongside the integration and joined management of voice and data,

Fig. 11. ISDN telephone.

ISDN also assisted in moving faster toward digitized voice calls by introducing modern digital voice coding and better speech quality, as well as hosting of additional services, such as caller identification, call forwarding, and so on (see Fig. 11). Despite the ambitious goals, however, in most of the world, ISDN was largely a failure. Although it was deployed in many countries, it quickly became obsolete. Nonetheless, nearly all ideas and services introduced in ISDN were incorporated in global system for mobile communications (GSM) cellular networks and are now standard for the evolved mobile/fixed telephone networks. In fact, most users have come to expect the full spectrum of ISDN-introduced services in all modern voice communication environments, including VoIP.

As the age of IP-based computer networks dawned, operators started looking into gradually moving away from a dedicated circuit-switched infrastructure and toward a converged network that can deliver a variety of services. As such, we have seen significant efforts during the last two decades to move voice transmissions to the constantly growing packet-switched domain. The transmission of voice data and signaling over IP-based computer networks is referred to as VoIP. Open-source software to run a fully functional private branch exchange (PBX) on commodity hardware, such as Asterisk [31] are already very popular, and are perceived by many as the future of telecommunications. Figure 12 illustrates the source code from the Asterisk VoIP PBX telephony software. Moreover, Fig. 13 illustrates the Asterisk Voicemail for iPhone application, which allows users to check their voice mail via a Web-based interface using an iPhone. Figures 9–13 point to the evolution of the telecommunication infrastructure from an all-hardware system to a computerized one, where software plays the major role.

The move to VoIP enabled several innovations both in terms of services and in terms of speech quality. By taking the voice samples and putting them into packetized format, new digital coding methods can be used to enhance the quality of the transmitted voice. When compressed, packetized, and transmitted over IP networks, more voice data can now be included in the same amount of bandwidth the old PSTN offered for the analog voice transmissions. Hence, voice codecs are an essential part of VoIP functionality as they are used to decrease the bandwidth requirements for a certain quality of voice through coding and data compression.

Figure 14 illustrates how voice packet generation works in most VoIP codecs. First, the analog voice signal from the microphone is quantized so that speech can be presented in digital format. In the quantization process, the voice signal is sampled with the sampling rate specified by the used codec and each sample is assigned a specific code word which is the bit representation of the voice signal level at the sampling time instant. These digitized voice samples are then collected into a voice frame at the output of the voice codec. The number of voice samples per frame is also an encoder-specific parameter; however, normally the larger the number of voice samples per frame the better the voice quality. When the codec voice frames are collected from the encoder output they can be transmitted over the network. Usually a single voice frame is treated as payload data which is encapsulated with Real-Time Protocol (RTP) [32], User Datagram Protocol (UDP), and Internet Protocol (IP) headers in the packetization phase. However, both the number of voice frames per network protocol data (NPD) payload and used protocols can be changed depending on the application in use. For example, as the voice frames produced by VoIP encoders are normally just in the order of tens of bytes in length, by simply adding more than one voice sample into the payload of the transmitted VoIP packet, network overhead can be reduced and the overall performance of the system can be increased significantly, as shown in Section 5.

The VoIP codecs currently in use have evolved over the years from narrowband constant bitrate (CBR) encoders to adaptive codecs which support variable bitrates (VBRs) and wideband encoding. Today, it is also common to have the ability to adjust to different network environments depending on available capacity, the fluctuation of end-to-end delay, or varying packet loss rate. The design basis for

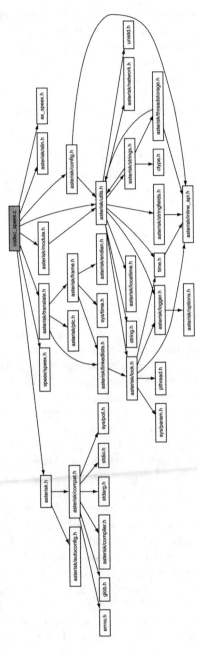

FIG. 12. A schematic of the source code dependencies in Asterisk's `codec_speex.c` file.

FIG. 13. The Asterisk Voicemail for iPhone application by Chris Carey.

the codecs has also changed as voice traffic is moving from circuit-switched delivery paths to IP-based packet-switched networks. That is, a voice codec designer can make different decisions about the employed algorithms depending on whether the expected underlying infrastructure is predominantly a circuit-switched network with minimal guarantees (e.g., PSTN or ISDN) or a best-effort packet-switched all-IP network. Consequently, codecs optimized for VoIP transmissions over packet-switched networks have also emerged in recent years. These codecs are designed to take into consideration the larger packet loss rates and varying delay jitter encountered in both packet-switched networks and wireless access links and with different methods in the encoding/decoding process, it is possible to conceal imperfections in the received voice quality resulting from, for example, burst errors in the delivery path [33, p. 309].

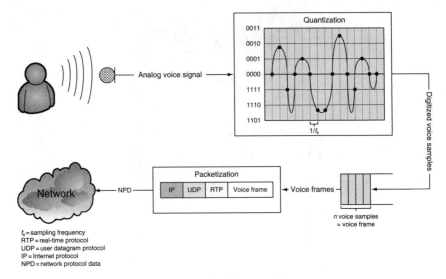

Fig. 14. VoIP packet generation.

This section summarizes recent developments in voice codecs commonly used in VoIP applications. Both standardized and open-source codecs are presented in addition to VoIP applications which use these codecs in their operation. Of course, we do not claim that the list of codecs presented in this section is exhaustive. Instead, our aim is to introduce codecs of reference and those that are prevailing in the current state of the art. The section concludes with a summary and comparative evaluation of the respective codec features.

4.1 ITU-T Codecs: From G.711 to G.723.1 to G.729.1

ITU-T has been one of the leading actors in the standardization efforts of voice codecs. Its widely cited recommendation documents serve as the main point of reference in this field. In this section, we will introduce some of the ITU-T standardized voice codecs and highlight the evolution of the codecs from the early days of digital telephony to current state-of-the-art VoIP applications. In particular, we focus on three key codecs here, namely, the one based on the ITU-T G.711 [34–37], G.723.1 [38], and G.729.1 [39, 40] recommendations. The ITU-T G.711 recommendation was published in 1988 and was originally designed for digital circuit-switched telephony. The ITU-T G.723.1 recommendation was published in 1996 aiming in particular video telephony applications including teleconferencing.

Finally, G.729.1 is the latest major addition to the list of ITU-T codec recommendations. It was finalized in 2006 and it is designed for packet-switched networks targeting carrier-grade VoIP service with high-quality audio.

4.1.1 Digital Telephony (ITU-T G.711)

One of the first and most used voice codecs for digital telephony is ITU-T G.711 [34]. It was firstly introduced in 1972 and was formally standardized by ITU-T in 1988. Due to its simplicity, G.711 is also widely used in a variety of VoIP applications. As the codec does not use data compression algorithms for the voice frames it encodes and packetizes, fast voice packet generation is enabled and excessive data processing is not required in the encoder. As a downside, in order to provide toll-quality voice, G.711 requires more bandwidth than most of the codecs that use data compression resulting in an output data rate of 64 kb/s. The codec uses a sampling frequency of 8 kHz and produces 8-bit voice samples using pulse-code modulation (PCM). The encoded VoIP frames are 160 bytes in size and are transmitted to the network in 20 ms intervals, that is, 50 packets per second. As compression is not used, each voice packet contains a 20 ms sample of speech.

The G.711 codec has two different versions of which are referred to as A-law and μ-law encodings. They conform to different digital trunk standards deployed around the world, that is, E1 and T1 carrier system standards, respectively. Currently, the T1 carrier system is used in North America, Japan, and Korea, whereas the E1 carrier system, which is an improved version of the original T1 standard, is used in the rest of the world. The A-law encoding converts a linear 13-bit PCM sample into a logarithmic 8-bit sample. The μ-law encoding does the same conversion to a linear 14-bit PCM sample. In general, by using the A-law encoding, it is possible to achieve better voice quality due to the difference in the amount of encoding quantization levels.

Two appendices have also been published to the original G.711 standard. These appendices introduced packet loss concealment [35] and comfort noise payload definitions [36] on top of the original codec functionality. Packet loss concealment is used to improve the user-perceived quality of the received voice signal. Concealment mainly addresses packet loss due to burst errors, which can be caused by congested delivery paths in wired networks or fading wireless access links in either end of an otherwise well-provisioned delivery path. The small comfort noise payloads, on the other hand, are used during the silent phases of a conversation in order to decrease the amount of bandwidth required by the VoIP application when there is no meaningful voice data to be sent.

In addition to these appendices, an extension standard G.711.1 [37] was approved in 2008. G.711.1 enhances the original codec functionality by adding narrowband

and wideband coding capabilities over the well-established G.711 algorithms. The G.711.1 codec can use the original 8 kHz sampling rate when operating in narrowband mode or at a 16 kHz sampling rate when operating in wideband mode. The functionalities are structured as additional coding layers on top of the original G.711 encoding so that the first additional layer provides enhanced narrowband coding capabilities and increases the transmission data rate by 16 kb/s. That is, the G.711.1 codec operating with an additional narrowband coding layer has a transmission data rate of 80 kb/s. The wideband enhancement increases the data rate by an additional 16 kb/s, resulting to a maximum encoder output data rate of 96 kb/s. The adjustment between these encoding rates can be done "on the fly," that is, a G.711.1-compatible VoIP application can accommodate all three transmission rates and change between them in an adaptive manner as a response to varying network conditions.

4.1.2 Video Telephony and Teleconferencing (ITU-T G.723.1)

ITU-T G.723.1 [38] is another widely used narrowband voice codec, standardized in 1996. Even though G.723.1 has been designed for video telephony and teleconferencing applications, it is currently also used in a number of VoIP applications. The codec is very flexible and is more suitable for bandwidth-limited environments, unlike G.711, as it uses data compression in its operation. Through data compression for voice, the transmission data rate of the codec is dramatically decreased and more bandwidth can be reserved, for example, for video streaming in video telephony. However, as data compression decreases the required transmission data rate of the VoIP application, it also degrades the quality of G.723.1-encoded data which does not achieve toll-quality voice.

G.723.1-compatible VoIP applications can operate in two different data rates, namely 5.3 and 6.3 kb/s. Both data rates are mandatory in codec implementations and they correspond to 20 and 24 byte voice frame sizes, respectively. An additional payload size of 4 bytes can be used as silence insertion descriptor (SID) frames. SID frames can be instrumental in reducing the bandwidth used by VoIP streams. They are inserted in the voice sample packet flow when conversational breaks are detected during a VoIP session. G.723.1 can also be used as a VBR codec when discontinuous transmission (DTX) is used in conjunction with the SID frames. For example, the transmission data rate can be changed at any frame boundary during the codec's operation.

G.723.1 compresses 30 ms of voice into each VoIP frame which corresponds to a transmission rate of approximately 34 packets per second. The codec also has an additional look-ahead delay of 7.5 ms per packet due to the data processing required in the codec operation. This yields a total algorithmic delay of 37.5 ms per transmitted packet. G.723.1 uses an 8 kHz sampling rate and 16-bit voice samples. In other words, the 30 ms voice frames packetized for transmission in the VoIP application contain 240 voice samples each. The two operational data rates of the codec use different algorithms to compress the voice samples into the packetized voice frames. The higher data rate operation mode with higher voice quality results in a transmitted data rate of 6.3 kb/s and uses a multipulse maximum-likelihood quantization (MP-MLQ) algorithm in its encoding process. In the lower data rate operation mode of 5.3 kb/s, an algebraic code-excited linear prediction (ACELP) algorithm is used.

4.1.3 High-Quality Voice over Packet Data Networks (ITU-T G729.1)

ITU-T G729.1 [40] is a more recent state-of-the-art voice and audio codec designed for packet data networks and supports wideband audio coding. Standardized in 2006, G.729.1 is based on the popular G.729 [39] narrowband codec and extends its functionality by introducing wideband voice/audio sampling capabilities and a sophisticated layered and adaptive voice frame structure. Through wideband coding, the codec is capable of delivering high-quality audio and still operate at relatively low data transmission rates.

The operational data rate of the G.729.1 can vary between 8 and 32 kb/s. The core layer, which is fully bitstream compatible with the G.729 narrowband codec, is sent at 8 kb/s. The first additional layer improves the narrowband quality of the codec and increases the bitrate to 12 kb/s, and after that, each additional coding layer increases the data rate with 2 kb/s. These extra layers are providing the wideband extension on top of the narrowband layers and, thus, improve the audio quality. Consequently, the maximum number of coding layers in G.729.1 is 12, which corresponds to total of 12 different operational data rates. However, only the core layer is required to be successfully decoded in order for the received VoIP packets to be transformed back to understandable speech.

The G.729.1 codec uses a sampling rate of 16 kHz by default, but a sampling rate of 8 kHz is also supported. The voice frame duration of the codec is 20 ms and the frame size varies between 20 and 80 bytes. The bandwidth of the codec can be changed at the transmitting side at any time during the codec operation and the receiver will be able to decode the VoIP packets properly. This enables the codec to adjust its parameters to changing link or network conditions without outband control signaling or additional codec parameter negotiation between the transmitter and receiver.

The bitstream interoperability of the codec's core layer and G.729 also makes the G.729.1 codec suitable for seamless transition from narrowband to wideband VoIP. Three different algorithms are used to code the voice samples in G.729.1 depending on the audio frequency band in question. At the lower band (50–4000 Hz), code-excited linear prediction (CELP) is used and at the higher band (4000–7000 Hz), time-domain bandwidth extension (TDBWE) coding is employed. Optionally, time-domain aliasing cancellation (TDAC) coding can be used for the full audio band (50–7000 Hz). The algorithmic delay of G.729.1 is approximately 48.9 ms.

4.2 Speex: An Open-Source Option for Voice Coding

In addition to ITU-T standardized voice codecs, open-source options have also emerged during the last decade. One of the most popular and widely used open-source codecs is Speex [41]. It is a VBR audio codec specifically designed for voice data compression and VoIP applications delivering voice over packet-switched data networks. The Speex project has been ongoing since 2002 and the first version of the software was released in 2003. Since then, the codec has been enhanced with new features incrementally and, as of today, it is one of the most versatile and advanced codecs available for voice transmission over packet data networks.

Speex supports three different voice sampling rates ranging from narrowband at 8 kHz, to wideband at 16 kHz, and finally, to ultra-wideband at 32 kHz. Interestingly, all sampling rates can be used within the same bitstream. Due to its flexibility, Speex features a wide range of operational data rates varying from 2.15 to 44 kb/s. Data rates between 2.15 and 24.6 kb/s are considered "narrowband," while data rates between 3.95 and 44 kb/s are referred to as "wideband." The operational data rate can be changed dynamically and at any point of time during a Speex-encoded VoIP session. The voice frame duration in Speex is 20 ms and the frame size varies between 5.375 and 110 bytes. The codec has an additional 10 and 14 ms look-ahead delay for narrowband and wideband operation modes, respectively. Speex uses CELP for encoding voice samples with a total algorithmic delay of 30 ms in the narrowband operation mode, and 34 ms in the wideband operation mode, which include the 20 ms frame delay and the corresponding look-ahead delays.

As Speex is designed from the beginning for packet-switched, best-effort networks, it has several features which make it robust against packet loss and network state changes. Features that make the codec operation resilient in the presence of dropped packets include packet loss concealment and, as the codec supports voice activity detection (VAD) and DTX functionalities, it can adapt to varying network conditions using its multirate features. The codec also supports intensity stereo encoding, acoustic echo cancellation, and noise suppression. With its wide range of features and good voice quality, Speex has gained ground as a viable open-source option for encoding audio and speech in several applications. A number of diverse voice applications use it, ranging from VoIP and teleconferencing applications to Microsoft's Xbox Live, Adobe Flash Player version 10, and even the U.S. Army Land Warrior system.

4.3 Voice Codecs in Popular VoIP Applications

This section presents a short overview of the features of today's most popular VoIP applications. In particular we survey the voice codecs used in popular applications such Skype, Google Talk, Windows Live Messenger, and Yahoo! Messenger. These applications are all freely downloadable and support VoIP, instant messaging, file transfer, and video delivery services. We also shortly summarize the codecs supported by the Asterisk open-source PBX telephone switching software.

All VoIP client applications presented in this section have a few things in common. They all provide both the software and basic PC-to-PC VoIP services for free, if both end clients are using the application. Interworking between Yahoo! Messenger and Windows Live Messenger VoIP applications is available without extra charge as well. As of December 2009, with the exception of Google's GTalk, all applications enable PC-to-phone VoIP calls to landline and cellular telephone networks for a fee. Lastly, all of these applications rely primarily on proprietary voice encoders in their VoIP communications, such as those developed by Global IP Solutions (GIPS) Corporation.

Skype is probably the most advanced VoIP client application on the market today. First released in 2003, Skype has since been at the forefront of VoIP communication. In addition to the services mentioned above, Skype has several advanced features such as call forwarding, virtual online phone numbers, conferencing, videoconferencing, high-quality video calls, voicemail, and even a short message service (SMS) compatible will all major cellular networks. The Skype client is a standalone application, based on sophisticated p2p software. Currently, the default codec used in Skype-to-Skype VoIP calls is SILK. SILK is a proprietary codec which Skype makes available to third-party users through a royalty-free license. In addition, SILK is currently also being standardized in IETF [42].

SILK is a scalable superwideband audio codec with sampling rates at 8, 12, 16, and 24 kHz, and operational data rates ranging from 6 to 40 kb/s. With an algorithmic delay of 25 ms, comprising 20 ms voice frame size and 5 ms look-ahead delay, SILK is both high performance and lightweight. SILK was introduced in early 2009, replacing the sinusoidal voice over packet coder (SVOPC) which was previously used as Skype's default codec. Other codecs supported by Skype include the GIPS' royalty-free narrowband codec called Internet low bitrate codec (iLBC) and the proprietary wideband codec called Internet speech audio codec (iSAC). Furthermore,

Skype also supports the ITU-T standardized G.711 and G.729 codecs for narrowband VoIP, as well as G.722.2 [43] for wideband VoIP. ITU-T G.722.2 is also known as adaptive multirate wideband (AMR-WB) voice codec.

Both Microsoft's Windows Live Messenger and Yahoo! Messenger were originally instant messaging software which have grown to include VoIP functionality as well. Both were released more than a decade ago and, similarly with Skype, provide both free PC-to-PC and chargeable PC-to-phone VoIP communication. Although originally competitive, the two applications have become interoperable so that free VoIP calls between client software of both providers are now supported. However, both applications are primarily meant for instant messaging, and do not offer the large variety of features mentioned above. Instead, emphasis is placed on ease of use and the main market target is personal usage. With their already established and large combined user base spread out all over the world, these two applications are serving a large portion of today's VoIP users in their everyday communication needs. Both Windows Live Messenger and Yahoo! Messenger use standard ITU-T codecs such as G.711, proprietary voice codecs from GIPS, as well as Speex.

Google's GTalk, first released in 2005, takes a different approach on VoIP communications than the previously described standalone VoIP software. GTalk does not require any dedicated VoIP client software to be installed on the user's computer. Instead, the user can use Google Talk simply as a Web-based application with an Internet browser plug-in though Google's Web mail service (Gmail). GTalk supports advanced VoIP features such as conference calling, videoconferencing, and can send SMS messages (in the United States only). GTalk is mainly targeting private rather than business users, a sort of extension to Google's online communications toolbox. Nonetheless, by supporting both the open-source Speex codec and the ITU-T standardized G.711 and G.726 [44] codecs, as well as the widely used proprietary codecs from GIPS (such as, iLBC, iSAC, Internet pulse-code modulation wideband—iPCM-wb, and an enhanced G.711 version), a variety of options is available for voice communications in different network conditions and usage scenarios. In addition, by using open-source protocols also in other parts of its communication framework, the aim of GTalk is to provide a highly interoperable and flexible platform for VoIP communications.

In addition to the high availability of free end-user VoIP software, call switching software has also become available. One of the most used applications in this area is the open-source telephony engine Asterisk [31], which can be used, for instance as a VoIP call switch, network gateway, or a media server. Asterisk has a wide set of features typically available from traditional PBX equipment and switching centers. Asterisk supports a variety of different VoIP codecs ranging from ITU-T codecs (including G.711, G.722 [45], G.723.1, G.726, and G.729), to GSM and GIPS' iLBC codecs. The list of supported codecs also naturally includes the open-source Speex codec.

4.4 Summary

Table III summarizes the features of the presented codecs and lists additional information not included in Sections 4.1 and 4.2. The presented figures are collected and derived mainly from Refs. [34, 38, 40, 41].

The discussion up to now, as well as Table III, documents the evolution of digital telephony since the 1980s. Clearly, codec features shift from real-time narrowband encodings toward compressed wideband audio. At the same time, the target deployment environment changes from a dedicated circuit-switched telephone network toward general-purpose packet-switched networks.

5. VoIP over WiMAX

As discussed earlier in this chapter, the evolution of voice and video telephony has been steadily pointing to packet-switched networks and, currently, all major telecommunication standardization forums put forward all-IP architectures. Besides the WiMAX Forum reference (all-IP) architecture described earlier in Section 2.3, the 3GPP systems architecture evolution/long-term evolution (LTE) standards adopt for the time an all-IP network design and operation. VoIP will be an essential application in tomorrow's wireless networks. As a result, network operators of next-generation wireless technologies should be able to provide high capacity for VoIP services alongside other data traffic. A particular characteristic of VoIP applications is the generation of large quantities of small data packets. Typically, a VoIP packet carries a single audio sample, which contains voice fragments lasting 10–60 ms [33]. In Section 4, we saw that most currently used codecs produce voice frames with sizes ranging between 20 and 100 bytes. This imposes significant challenges for many current wireless technologies as they are often optimized for data packet sizes in the order of the Ethernet maximum segment size (1500 bytes). A flood of small packets generated by VoIP applications could lead to underutilization of radio access resources. Moreover, VoIP applications have strict one-way delay and jitter requirements if a good conversational quality is to be maintained. In short, wireless network technologies have to strike a balance between being able to handle a large number of good quality VoIP calls while delivering high rates to elastic data traffic.

The starting point of the IEEE 802.16 family of standards has been to specify a data-oriented BWA technology. This often implies technology for data transmission without strict delay and bandwidth requirements. Multimedia applications and their QoS requirements are attended to in IEEE 802.16d and IEEE 802.16e-2005 standards, and were recently consolidated in the IEEE 802.16-2009 standard

TABLE III
COMPARISON OF VoIP CODEC FEATURES

	ITU-T G.711	ITU-T G.723.1	ITU-T G.729.1	Speex
Publication year	1972 (standardized in 1988)	1996	2006	2003
Sampling rate (kHz)	8	8	8 (NB) 16 (WB)	8 (NB) 16 (WB) 32 (UWB)
Acoustic frequency range (Hz)	300–3400	300–3400	50–7000	50–7000
Data rate (kb/s)	64	5.3 or 6.3	8–32	2.15–44
Frame length (ms)	20	30	20	20
Frame size (bytes)	160	20 or 24	20–80	5.375–110
Sample size (bits)	8	16	16	16
Samples/frame	160	240	160 (NB) 320 (WB)	160 (NB) 320 (WB) 640 (UWB)
Algorithmic delay (ms)	Real time	37.5	48.9375	30–34
Original purpose	Digital telephony	Video telephony and teleconferencing	Voice over IP	Voice over IP
Licensing	Patent expired, public domain	Patented	Patented	Open source, revised GNU, BSD license
Prominent applications	A large variety of proprietary and free VoIP applications, including Skype, Windows Live Messenger, Yahoo! Messenger, GTalk	Several proprietary VoIP applications	Several proprietary wideband VoIP applications	Several VoIP applications; Microsoft Xbox Live, Adobe Flash Player; GTalk Windows Live Messenger, Yahoo! Messenger, Asterisk

NB, narrowband; WB, wideband; UWB, ultra-wideband.

(see Section 2). The standards specify several QoS classes for different types of traffic, from real-time VoIP and video streaming to non-real-time data transmission (see Table II). These QoS classes assure the quality of real-time VoIP and video over the wireless WiMAX link.

The IEEE 802.16 Task Group m is currently working on a backward compatible amendment standard (IEEE 802.16m), which will further improve VoIP and other multimedia quality over the WiMAX. Besides higher data rates, IEEE 802.16m air interface provides improved mobility, more extensive set of MIMO schemes, lower power consumption, and lower latencies compared to the current standards. Latency should be less than 10 ms. The aim of the standard is to fulfill the requirements of the IMT—Advanced systems [46], specified by the International Telecommunication Union's Radio communication Sector (ITU-R). The standard is expected to be finalized in late 2009–early 2010 [12].

In this section, we present our empirical performance evaluation of VoIP over fixed and mobile WiMAX. We discuss efficient ways to improve VoIP quality over the wireless link, taking into account the distinctive characteristics of VoIP traffic. When inspecting the content of a VoIP packet, one can notice the significant network protocol header overhead introduced, most commonly, by RTP/UDP/IP. For example, if a voice sample frame is 20 bytes long, the total packet size after adding the standard RTP (12 bytes), UDP (8 bytes), and IP (20 bytes) headers adds up to 60 bytes. Here, the proportion of actual voice data per packet is only 33%. If one considers IPv6 networks [47], where the IP header (at least 40 bytes) is significantly larger than in IPv4, the overhead is even larger. Many of the techniques to increase VoIP capacity in a WiMAX cell aim at decreasing protocol header overhead, namely through packet aggregation and the use of robust header compression (ROHC) [48]. In addition, we also present how the use of a multiantenna system improves performance over the wireless WiMAX link.

5.1 Testbed Description

The measurements presented in this section are performed using the testbed illustrated in Fig. 15. The testbed is part of the VTT Converging Networks Laboratory and includes two Alcatel-Lucent mobile WiMAX BSs and one Airspan fixed WiMAX BS. Both mobile WiMAX and fixed WiMAX equipment operate in the 3.5-GHz frequency band, but otherwise have very different parameters. The main differences are on the physical layer (fixed WiMAX uses OFDM while mobile WiMAX uses OFDMA; see Section 2.1).

In a fixed WiMAX system profile, the WirelessMAN-OFDM air interface uses either TDD or FDD method. In our fixed WiMAX testbed, FDD is employed. The mobile WiMAX system profile, based on the WirelessMAN-OFDMA air

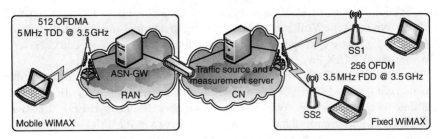

FIG. 15. Schematic of the mobile WiMAX and fixed WiMAX testbed.

interface, is designed to use TDD. The duplexing method affects the total bandwidth employed. While our mobile WiMAX BSs use the same 5-MHz channel bandwidth for both uplink and downlink, the fixed WiMAX testbed uses two separate 3.5-MHz channels for uplink and downlink, and thus, overall, uses 7 MHz of bandwidth.

It is important to keep in mind that in the results presented in this section, we never use the mobile and fixed WiMAX BSs simultaneously. We use several different terminals, commonly referred to as customer-premises equipment (CPE) in the case of mobile WiMAX and SSs in the case of fixed WiMAX. In addition to the base stations, the mobile WiMAX RAN, also known as ASN, includes the ASN-GW as specified in the WiMAX Forum Network Architecture by WiMAX Forum NWG. As we discussed in Section 2.3, ASN-GW handles, among others, data path creations between the core network and the terminals. The core network is also referred to as CSN. The RAN includes network elements enabling both RAN and CN anchored mobility based on mobile IP (MIP) [49,50]. As per the WiMAX Forum Network Architecture, the core network includes essential network elements for the mobile WiMAX operations such as, for instance, terminal and end-user AAA management functions. The fixed WiMAX testbed consists only of the BS connected to our CN using gigabit Ethernet.

We used different traffic sources and measurement tools on the wired and wireless sides for generating and evaluating various types of traffic over the WiMAX link. Further details, specific to each type of measurement (number of measurement entities, traffic types, directionality, and bitrate), as well as the employed modulation schemes are given in the following sections.

5.2 VoIP Aggregation

The small packet size and real-time aspects of VoIP traffic streams have a distinctive flavor when compared to other types of traffic. When a single voice sample is the only payload of an IP packet, the fixed cost incurred due to the protocol packet headers is proportionally very large, leading to low effective resource utilization. Packet aggregation, that is, the bundling of multiple voice frames into one packet, is an efficient way to improve the effective wireless link capacity when carrying VoIP conversations. However, this is done at the expense of delay increases, due to the holding time that the first voice sample frame needs to wait in order to complete the bundling operation. This delay increase depends on the codec used. Typically, a voice frame includes data corresponding to 10–60 ms of voice (see also Table III). Given that the total end-to-end one-way delay (also referred to as "mouth-to-ear" delay) should be limited to less than 150 ms, the sample frame rate and end-to-end transmission and queuing delays restrict the maximum aggregation level to two or three voice samples per IP packet for the majority of VoIP codecs.

VoIP aggregation can be performed at three layers of the protocol stack: application, network, and MAC. Aggregation at the MAC layer (see also Section 2.2) requires vendor support and changes to the access technology specific layer. MAC-layer aggregation is not supported in our testbed, which employs off-the-shelf equipment. In practice, however, capacity gains achieved through network- and application-layer aggregation can often be as high as those achieved by MAC-layer aggregation. In addition, upper-layer aggregation techniques are access technology-independent solutions. One could argue that best results can be attained if aggregation is supported at all three layers, thereby allowing for more flexible system operation.

Figures 16 and 17 illustrate application- and network-layer aggregation, respectively. In application-layer aggregation (Fig. 16), multiple voice sample frames are combined in a single VoIP packet before it is encapsulated in the RTP header [32]. In network-layer aggregation (Fig. 17), multiple IP packets are bundled into a new one. Network-layer aggregation requires adding a performance enhancing proxy (PEP) header to the generated packet in order to indicate that the received IP packet consists of multiple IP packets. A PEP can be inserted, for example, at the two ends

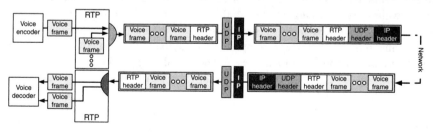

FIG. 16. VoIP application-layer aggregation.

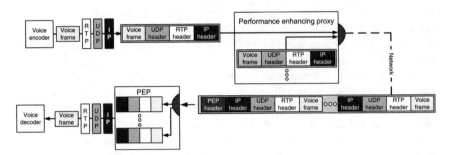

FIG. 17. VoIP network-layer aggregation.

of a WiMAX link. Application-layer VoIP aggregation would typically be applied to a single flow due to the end-to-end nature of the communication. Aggregated voice samples are deaggregated at the receiving application. On the other hand, performance enhancing proxies can be located only at the two ends of a WiMAX link: packets will be aggregated, possibly from different flows, solely for traversing the wireless link, and then will be deaggregated in order to continue their path toward their respective destinations. Network-layer aggregation does not restrict aggregation solely to one VoIP flow, but allows bundling voice sample frames from multiple VoIP flows into one packet.

5.2.1 VoIP Aggregation over WirelessMAN-OFDM

We have empirically evaluated VoIP aggregation over WiMAX testbeds; see Refs. [13, 51] for further details. In fixed WiMAX measurements [13], we connected one subscriber station to the base station and injected G.723.1 emulated VoIP flows into the downlink and uplink separately. This setup corresponds to using the WirelessMAN-OFDM (fixed WiMAX) link as backhaul (refer to Section 3.1). We use the 64 QAM FEC: 3/4 modulation scheme for both the uplink and the downlink. Without aggregation, 400 G.723.1-encoded VoIP flows can saturate the WirelessMAN-OFDM downlink, attaining a cumulative goodput of 3.3 Mb/s. We define *goodput* as the ratio of application payload to the time necessary for its transmission from one peer to another. When calculating goodput we only consider the application-layer payload and exclude all transport, network, and MAC-layer headers. This goodput level is only one-third of the maximum goodput measured using UDP, with an MTU of 1500 bytes and application payload of 1472 bytes, while experiencing negligible packet loss. The uplink was saturated by 160 VoIP flows, with a cumulative goodput to 1.5 Mb/s. This is only 27% of the maximum measured uplink goodput (5.5 Mb/s). The G.723.1 [38] codec produces, with the employed mode, 24 bytes long voice sample frames, which translates into an operational goodput of 6.3 kb/s. When adding RTP/UDP/IPv4 headers, the whole VoIP packet is 64 bytes all in all. Thus, the total protocol header overhead is 167%.

After these baseline measurements, we experimented with application-layer aggregation, bundling two voice sample frames into one packet. We will refer to this as "application Level-1 aggregation." In this case, the cumulative goodput increases to 5 Mb/s in the downlink. Now, the header overhead is almost half of the baseline one (80%). Using Level-2 aggregation (i.e., aggregating three voice sample frames in a single packet) saturates the downlink goodput at 6 Mb/s, reflecting the lower header overhead (55%). In the uplink, the goodput saturation points were 2.5 and 3.5 Mb/s with Level-1 and Level-2 aggregation, respectively.

Although the goodput saturation point can be increased by bundling even more voice sample frames into one IP packet, the excess delay caused by buffering the samples before transmitting them over the wireless link is a key restriction. To keep the end-to-end one-way delay at modest levels (preferably below 150 ms), Level-2 aggregation is the maximum aggregation level in our experimental measurements when employing the G.723.1 codec. Recall that in G.723.1 each voice sample frame is produced every 30 ms.

Network-layer aggregation does not deliver any gains with respect to the cumulative goodput saturation point. This is because the header overhead does not decrease, which is the case with application-layer aggregation. In fact, as more headers are introduced (the PEP header in Fig. 17, in addition to the individual RTP/UDP/IP headers) the overhead is actually increased. Nevertheless, network-layer aggregation delivers larger VoIP capacities in terms of number of emulated VoIP calls. The performance gain with network-layer aggregation is due to handling much more efficiently a smaller number of (larger) packets.

Determining the saturation points of the WiMAX links is an important network characterization exercise, but does not, in and of itself, paint the complete picture. VoIP quality is typically evaluated using three main metrics: packet loss, one-way end-to-end delay, and jitter. To objectively measure VoIP quality, the MOS gauge is used as well. MOS captures the essential performance characteristics for this traffic class, expressing the quality of VoIP call in the range of 1–5, where 1 indicates very bad quality and 5 excellent quality. Value 3 is considered as the threshold between tolerable and bad quality. We employed the ITU-T E-model [52] to calculate the MOS values and used delay and packet loss as the main quality metrics. The codec-specific values for G.723.1 are available from Ref. [53]; interested readers can find the full details of our evaluation in Ref. [13].

Figure 18 presents the G.723.1 VoIP MOS values in the downlink with and without aggregation. Without aggregation, the downlink can handle only 200 good

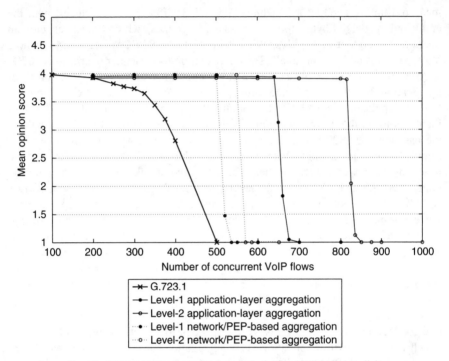

FIG. 18. G.723.1 VoIP mean opinion score in the fixed WiMAX downlink.

quality VoIP flows and 375 moderate quality VoIP flows. The one-way end-to-end delay with 400 VoIP flows remains below 30 ms, and is not a performance-limiting factor. Network-layer aggregation more than doubles the amount of good quality VoIP flows in the downlink. Application-layer aggregation is the most efficient way for increasing VoIP call capacity and scored the most gains with respect to number of good quality VoIP flows. Level-2 application-layer aggregation more than triples the number of lossless VoIP flows in the downlink. Note, however, that aggregated VoIP quality drops very rapidly, once congestion rises and the one-way end-to-end delay becomes too large.

As can be seen from Fig. 19, when no aggregation is employed, the uplink sustained no more than 175 concurrent VoIP flows with moderate quality. Level-1 and Level-2 network-layer aggregation and Level-1 application-layer aggregation performed similarly, allowing for approximately 320–330 VoIP flows to be sustained at an acceptable MOS level. Level-2 application-layer aggregation is once again the clear winner, allowing for more than 450 concurrent VoIP flows at very

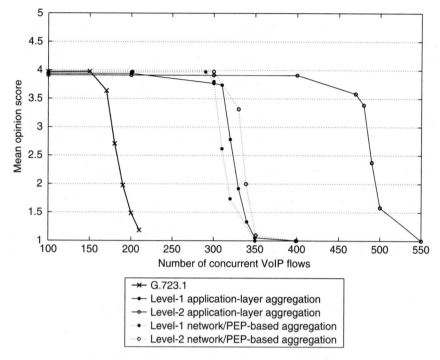

FIG. 19. G.723.1 VoIP mean opinion score in the fixed WiMAX uplink.

high quality. We also note that the one-way end-to-end delay remained under 100 ms throughout all experiments, another indication that Level-2 aggregation is a viable option in real-world deployments.

5.2.2 VoIP Aggregation over WirelessMAN-OFDMA

The experimental results presented in the previous section indicate that application-layer VoIP aggregation can deliver large performance gains for VoIP over WiMAX. To complete our study of the benefits of VoIP aggregation, we proceeded with an empirical evaluation of application-layer aggregation over the mobile WiMAX testbed in stationary link conditions. In contrast to the fixed WiMAX experiments presented above, in these measurements we opted to use bidirectional, concurrent VoIP flows, emulating more realistically VoIP conversations. We also experimented with the more modern layered wideband codec G.729.1 [40], instead of G.723.1. We opted to use the four-layer coding (base layer and three additional layers) which produces 40 byte long voice sample frames, reflecting an operational bitrate of 16 kb/s [54]. The employed modulation schemes in the WiMAX links were 16 QAM FEC: 1/2 and 64 QAM FEC: 1/2 in the uplink and downlink, respectively. The limiting factor in these measurements was the capacity of the uplink. That is, although the uplink was measured to sustain up to 1.5 Mb/s in terms of UDP goodput, the downlink can sustain slightly over 5 Mb/s. In theory, in order to measure concurrent bidirectional traffic, the uplink sets the threshold for packet loss. In practice, all significant losses were recorded in the downlink. Nevertheless, the mobile WiMAX testbed can sustain the equivalent of a T1 carrier in both directions simultaneously.

Figure 20 presents the goodput saturation points for nonaggregated, Level-1 and Level-2 application-layer aggregation. Without aggregation, the WirelessMAN-OFDMA link sustained 19 concurrent VoIP conversations, which yields a cumulative goodput of 280 kb/s in each direction of the WiMAX link. This is far from the

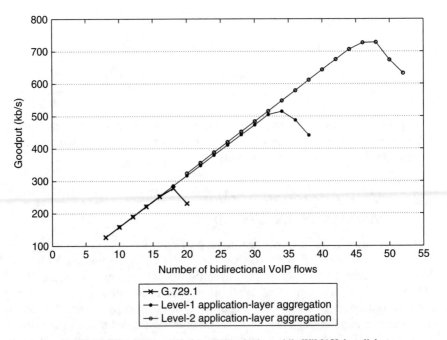

FIG. 20. G.729.1 VoIP cumulative goodput in the mobile WiMAX downlink.

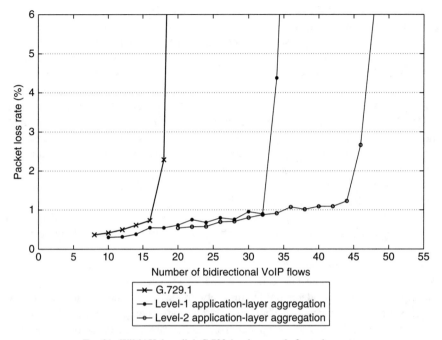

FIG. 21. WiMAX downlink G.729.1 voice sample frame loss rate.

maximum uplink capacity, which is 1.5 Mb/s with the employed modulation scheme and with 1400 bytes MTU. In fact, it is approximately only 18.6% of the maximum UDP goodput with insignificant packet loss—an abysmal performance. By bundling two voice sample frames into a single VoIP packet, we found that we can increase the cumulative goodput by 200 kb/s in each direction. Even further, Level-2 application-layer aggregation more than doubles the saturation point of the cumulative goodput, clocking in 720 kb/s. Effectively, this is half of the maximum capacity of the uplink with the used modulation scheme.

Figure 21 illustrates the downlink voice sample frame loss rate, which restricted the number of sustained VoIP conversations. When no voice sample frame aggregation is used, the downlink can sustain up to 18 simultaneous VoIP conversations before the packet loss rate exceeds the 5% threshold recommended in Ref. [33]. Average packet loss rate remained under 5% with 34 simultaneous VoIP conversations when two voice sample frames were aggregated. By aggregating three voice samples into a single VoIP packet, the amount of sustained simultaneous VoIP conversations increased to 47. In short, Level-1 and Level-2 application-layer aggregation increase VoIP call capacity in our mobile WiMAX testbed by 88.8% and 161% over nonaggregated VoIP, respectively.

When no aggregation is used, the one-way end-to-end delay ranged between 20 and 30 ms in the downlink. One-way delays increase slightly as we injected more VoIP flows in the WirelessMAN-OFDMA link. However, one-way delays increased rapidly as we approached the downlink saturation points, which are tightly correlated with the packet loss rates. One-way delays were measured in the uplink as well and their range stayed at 30–90 ms across all measurement configurations. When no aggregation is employed, the uplink one-way delays range 30–35 ms. Of course, the interpacket one-way delays are higher when voice sample frame aggregation is employed, and lower when no aggregation takes place. For example, when aggregating two G.729.1 voice sample frames in a VoIP packet, every second sample needs to be buffered at the very least an additional 20 ms (the encoding delay of a G.729.1 codec). In general, voice sample aggregation at the application layer introduces buffering delays, which need to be carefully considered by software developers and network operators.

With respect to deployment, we argue that adaptive application-layer aggregation is not hard to implement and can yield significant performance gains. Regarding intelligent aggregation decisions at the network layer, we note that packet flow monitoring may be necessary. The network monitor can classify packet flows based on their characteristics, such as packet size, packet rate, and service type. Based on the collected data, a PEP can take decisions about aggregating packets or not and, if so, which packets and from which particular flows. For example, if interpacket delays become large, it may not be preferable to employ aggregation. This way, aggregation bottlenecks can be avoided and bypassed by larger packets.

Finally, with respect to balancing the aggregation processing load among a set of PEPs, IPv6 [47] anycasting may come handy. An anycast IP address is an identifier of a set of network interfaces. When sending a packet to an anycast address, the packet is delivered to one of the entities this anycast address is assigned to, possibly the nearest host in network terms. IPv6 anycasting can be employed to carry out VoIP aggregation at the network layer. For instance, the PEP at either end of the WiMAX link can send aggregated packets to an anycast destination, that is, the PEP header (see Fig. 17) may be addressed to an anycast group. At the other end of the WiMAX link, there can be multiple PEP servers which can extract the encapsulated packets from the aggregated one and forward the unbundled packets to their Internet destination. These PEP servers will have the same anycast address assigned to their network interface. The PEP header added to the aggregated packet uses this anycast address as the destination IP and the aggregated packet is therefore forwarded to the nearest PEP server.

5.3 VoIP over WiMAX with Robust Header Compression

ROHC [48, 55–57] is one of the most commonly used and powerful ways to compress network headers. ROHC is one of two optional ways to reduce the length of network protocol headers in WiMAX systems. The schemes specified in Refs. [48, 57] define header compression profiles for the following protocols: RTP over UDP and IP (RTP/UDP/IP), UDP/IP, Encapsulating Security Protocol (ESP) over IP, and UDP Lite over IP. Further, a ROHC profile for TCP/IP is defined in Ref. [55].

ROHC takes note and exploits the redundancy and predictability of network and upper-layer protocol headers. For instance, typically the source and destination addresses in an IP header remain unchanged for the duration of a VoIP conversation. By capitalizing on this redundancy, the source and destination addresses can be replaced with proxy values in most of the packets in a VoIP flow. Effectively, ROHC can compress the RTP/UDP/IP headers to a few bytes only. The actual length of the compressed header varies depending on the profile and mode used [48, 55–57].

To decompress the headers, the receiving side must maintain context state for each ROHC-compressed packet flow. The context state includes previously received headers and other data describing each packet flow. This information is used to decompress the ROHC header and regenerate the original (uncompressed) headers as they were originally transmitted. Such context state information is created in the initialization phase, during which packets are sent uncompressed. In other words, ROHC cannot be used end-to-end across the entire path without back-to-back compression and decompression operations on each hop. This is because the IP addresses, among other fields, are compressed thus making Layer-3 routing impossible without maintaining ROHC context state on a per-flow basis at each router in the path. In practice, ROHC can be used over single hops, such as a WiMAX link, where resources are limited. For example, a city bus can be fitted with a mobile WiMAX CPE in order to provide Internet access to passengers. The WiMAX CPE can also incorporate a Wi-Fi access point as we saw in Section 3. Then, bus passengers can connect to the Internet using the IEEE 802.11 WLAN. In this kind of topology, bus passengers can use VoIP over Wi-Fi, which is relayed over mobile WiMAX to the rest of the Internet. The WiMAX CPE can use network-layer aggregation and/or ROHC to increase the effective capacity of the WiMAX link.

RFC 3095 [48] specifies three different modes for compressors and decompressors: unidirectional (U), optimistic (O), and reliable (R). In the optimistic and reliable modes, communication between the compressor and the decompressor is bidirectional. A feedback channel is used for error recovery requests and acknowledgments. In the U-mode communication is unidirectional. In this case, an optimistic approach is taken which relies on higher layer protocols to ensure data transmission reliability. Due to the lack of feedback channel in the unidirectional mode, the compressor sends periodically a packet with an uncompressed header in order to refresh the context state.

Recall from Section 2.2 that the WiMAX MAC comprises three parts: the service-specific convergence sublayer, the common part sublayer, and the security sublayer. CS functions as an intermediate layer categorizing and classifying service data units (SDUs) received from upper layers and handing them over to the appropriate MAC connection, and *vice versa*. Moreover, the convergence sublayer ensures that QoS requirements are met, taking care of proper bandwidth allocation. The most commonly supported packet convergence sublayers are defined for IP and Ethernet (IEEE 802.3). The IP-based convergence sublayer does not handle nor forward Ethernet packets. The convergence sublayer supports SDUs in two formats: ROHC and ECRTP [58]. This facilitates header compression of IP and upper-layer headers over WiMAX. However, the support for header compression or for header-compressed packets is left optional and vendors do yet support it in practice.

We empirically evaluated the performance of ROHC over fixed [15, 59] and mobile WiMAX [51] using synthetically generated VoIP traffic and emulating ROHC. The emulated VoIP codec was G.729.1 with four-layer encoding generating 40 byte long voice sample frames at a bitrate of 16 kb/s as per Ref. [54].

5.3.1 WirelessMAN-OFDM

In the fixed WiMAX measurement topology, two subscriber stations were symmetrically connected to one base station (recall Fig. 15). Thus, the simultaneous bidirectional VoIP conversations traversed through two WiMAX links, where 64 QAM FEC 3/4 modulation was employed in the uplink and downlink. Figure 22 illustrates the downlink voice sample frame loss rate over fixed WiMAX. We can inject 65 simultaneous G.729.1-encoded VoIP conversations before packet loss exceeds 5% [15]. This corresponds to cumulative goodput of slightly over 1.35 Mb/s. This level of goodput is only 24.5% of the uplink capacity of the employed WiMAX link. ROHC moderately decreases the overhead and our WiMAX equipment can handle 69 VoIP conversations with voice sample frame loss rate less than 5%. Thus, ROHC allows us to carry 6.2% more VoIP conversations over the WiMAX link.

As shown in Fig. 22, our WiMAX testbed can sustain 97 bidirectional VoIP flows with an average voice frame loss rate below 5% when we use Level-1 application-layer aggregation. However, when using voice sample frame aggregation at the application layer and ROHC in unison, we observe even further gains in terms of the number of sustained bidirectional VoIP conversations. By using ROHC and aggregation in unison the testbed can carry G.729.1-encoded 121 bidirectional VoIP

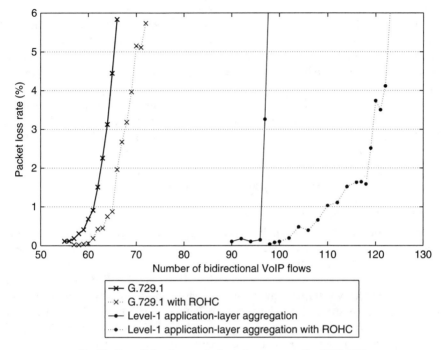

FIG. 22. Downlink voice sample frame loss rate over fixed WiMAX.

conversations with an overall packet loss rate below 5%. The recorded one-way end-to-end delay, including the aggregation-related buffering delay, remained below 150 ms throughout all experiments. To sum up, ROHC and application-layer aggregation outperformed the nonaggregated standard header VoIP performance by 86.1%.

5.3.2 WirelessMAN-OFDMA

We repeated the measurements over our mobile WiMAX testbed using one stationary mobile WiMAX CPE connected to the BS [51]. We found that ROHC does not provide for significant gains. The baseline experiment (no ROHC, no aggregation) showed that the testbed can carry 17 bidirectional VoIP flows. ROHC allowed us to inject one additional bidirectional VoIP call with moderate quality (sample loss rate below 5%). Level-1 application aggregation is far more effective, allowing the testbed to carry twice as many flows (34) as the baseline scenario.

When ROHC and application-layer aggregation are used in unison, 35 flows can be sustained. Level-2 application-layer aggregation increases the amount of sustained simultaneous VoIP conversations to 47. If ROHC and Level-2 aggregation is employed, 48 simultaneous VoIP conversations can be sustained without quality degradation. With respect to one-way end-to-end delays, we recorded 20–30 ms in the downlink and 30–35 ms in the uplink during the baseline experiments.

Overall, ROHC contributed significantly less gains than VoIP aggregation. Nevertheless, when the link capacity is higher, as it was the case with the fixed WiMAX testbed, ROHC does increase the effective WiMAX capacity as it makes better use of the underlying resources.

5.4 WiMAX with MIMO

IEEE 802.16-2009 [4] specifies multiple advanced antenna system technologies to improve data propagation reliability, capacity, and coverage area. These technologies combat interference and deal with multipath phenomena typical of wireless communication. In particular, the MIMO technique is adopted by many novel and upcoming wireless access technologies such as IEEE 802.11n, HSPA+, LTE, and mobile WiMAX. MIMO uses multiple antennas at the transmitting and receiving sides to improve spectral efficiency by capitalizing on transmission and spatial diversities along with multipath propagation. Mobile WiMAX supports the use of one, two, and four antennas at the base station and at most two antennas at the mobile station.

IEEE 802.16-2009 defines various ways for MIMO processing, not all of which are included in the WiMAX Forum Mobile System Profile. The Mobile System Profile Release 1 introduced in 2006 as the main downlink MIMO techniques open-loop transmit diversity, open-loop spatial multiplexing, and beamforming. Open-loop MIMO schemes refer to techniques that do not require channel state information at the transmitter (CSIT) and are most suitable for high mobility environments. The Mobile System Profile Release 1.5 [20] provides a more extensive set of advanced MIMO techniques, such as, for example, expanding the transmit diversity and spatial multiplexing for the uplink as well.

IEEE 802.16-2009 adopts STC as the main transmit diversity scheme. STC capitalizes on spatial diversity, that is, transmission diversity and possibly receiving diversity. In transmission diversity, multiple redundant copies (according to the number of transmission antennas) of a packet flow are sent over the WiMAX link using orthogonal coding, defining individual fading channels for each data stream. When receiving diversity is employed, the receiver uses multiple appropriately spaced antennas to receive signals from independent channels established between the received and transmitter. This allows the receiver to use simple combining of signals in order to regenerate the transmitted signal. STC improves reliability significantly (in terms of bit error rates) without the need for an increased transmission power.

In spatial multiplexing, a packet flow is split into multiple streams and each of stream is transmitted using different antennas. In the ideal case, spatial multiplexing increases the capacity by the number of transmitting antennas. A transmitter can switch between spatial multiplexing and diversity in order to find the optimal reliability/throughput combination.

The basic idea of beamforming is to direct transmission power toward the receiver by utilizing interference. By enhancing the transmit signal toward the desired direction one can extend the coverage area per BS and increase the received signal strength at the receiver. Beamforming requires channel state information at the transmitter and, as such, is not the most suitable MIMO scheme for high mobility environments. Beamforming and other methods that exploit CSIT are commonly referred to as closed-loop MIMO techniques.

Our empirical experiments with a 2×2 MIMO (i.e., using two transmitting antennas and two receiving antennas) in mobile WiMAX verified the advantage of MIMO in practical mobility scenarios. Our mobile WiMAX testbed uses space–time block coding (STBC), an STC technique, in the downlink, which is based on Alamouti coding [60]. The Alamouti STBC used is a rate one transmit diversity code, which means that it does not increase or decrease data throughput, as spatial multiplexing or conventional error correction codes would do, but increases the data propagation reliability. In other words, it improves the bit error rate without increasing the transmission power.

Figure 23 illustrates TCP goodput, when we use the same mobile WiMAX testbed with and without MIMO. Our experiment has the mobile node downloading 64 byte-long files in a vehicular mobility scenario. The measurement is repeated three times while our car is moving at 25–40 km/h going through the same circuit. In each measurement batch we downloaded 1000 files when using the single-antenna mode and 1200 files when MIMO was enabled. The file size reflects the typical voice sample frame size. In these measurements ARQ and HARQ (see Section 2.1) are not used to recover fast from lost packets. Figure 22 shows that the median goodput with and without MIMO is the same (~ 3.2 kb/s). However, there is a clear difference in the spread of the distribution, which is a manifestation of significantly lower number of TCP retransmissions.

It is important to note that these measurements are conducted with the first MIMO implementation released outside the vendor's testing facilities. We observed that the MIMO implementation at the time of the experiments was not as complete as we would have desired. For example, in low signal strength areas and MIMO antenna mode, data transmission stagnated. Nevertheless this study is one of the few that have been performed by third parties. Interested readers can find more details in Ref. [61].

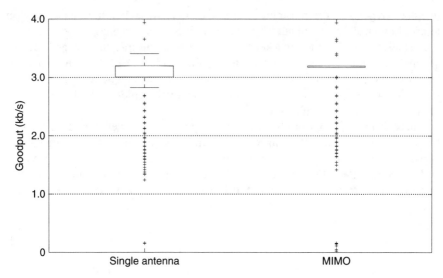

FIG. 23. TCP goodput with a single antenna and MIMO.

5.5 Quality of Service Support for VoIP over WiMAX

As discussed earlier in Section 2.2, the WiMAX MAC ensures that the QoS requirements of packet flows belonging to different service classes, such as, for example, VoIP, video streaming, and data transfers are met. Currently, the WiMAX MAC supports five different QoS classes (see Table II), which allow the system behavior to be quite customizable with respect to different traffic profiles and the corresponding applications. The use of QoS classes aims to serve constant and variable bitrate real-time applications, non-real-time data transmission applications, and applications that do not require any service level. In practice, the best-effort service class is still commonly used for all types of packet flows, irrespective of their traffic profile characteristics.

For VoIP, IEEE 802.16-2009 [4] defines three attractive QoS classes, the use of which depends on the codec. For CBR voice codecs, the first choice should be the *Unsolicited Grant Service* (UGS). UGS is designed for services that generate CBR traffic. This is the case with simple VoIP codecs that do not support silence suppression and do not employ a layered structure to dynamically scale VoIP quality. UGS assures that the fixed-size grants offered to the traffic flow based on its real-time needs are met. The mandatory QoS parameters involved in UGS are minimum reserved traffic rate and maximum latency.

The *Real-Time Polling Service* (rtPS) class provides QoS assurance to real-time network services generating variable size packets on a periodic basis, while requiring strict data rate and delay levels. In rtPS, the WiMAX BS can use unicast polling so that mobile hosts can request bandwidth. Latency requirements are met when the provided unicast polling opportunities are frequent enough. The rtPS service class is more demanding in terms of request overhead when compared to UGS, but is more efficient for variable size packet flows. The *Extended Real-Time Polling Service* (ertPS) combines the advantages of UGS and rtPS. This QoS service class enables the accommodation of packet flows whose bandwidth requirements vary with time. The ertPS QoS class parameters include maximum latency, tolerated jitter, and minimum and maximum reserved traffic rate.

It is important to keep in mind that these QoS classes can assure the required QoS levels only over the WiMAX link, not the end-to-end delay. For example, maximum latency here refers to the period between the time that a packet is received by the convergence sublayer (WiMAX MAC) and until the packet is handed over to the PHY layer for transmission.

6. Remote Surveillance and IPTV over WiMAX

To assess the suitability of WiMAX for the fire prevention scenario introduced in Section 3.3, a simple testbed was set up in the mountainous area near Coimbra, Portugal. The testbed encompasses two fire surveillance towers, located in *Lousã Mountain* and *Carvalho Mountain*. The surveillance towers were previously in use by regional fire prevention services and were equipped with surveillance cameras, digital compasses for geographical location of detected fires, and weather stations (wind, temperature, humidity).

For the testbed mount preparation, several places were considered in order to identify the most suitable place for the WiMAX antennas installation. Some of the requirements for the WiMAX system installation were a high-altitude location with clear view over a large area, availability of electrical power supply and the existence of infrastructure to protect the equipment from adverse weather conditions or from vandalism. After different locations were evaluated, three candidate sites were identified: *Lousã Mountain* (LM), *Carvalho Mountain* (CM), and *University of Coimbra* (UC), all of which satisfied the set constraints. Using simulation results and geographical data, the radio links between UC–LM, LM–CM, and UC–CM were evaluated. Figures 24–26 illustrate the terrain profiles and the Fresnel ellipsoids considering 66% and 90% of clearance.

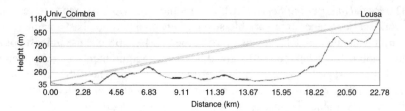

FIG. 24. University of Coimbra—Lousã Mountain terrain profile.

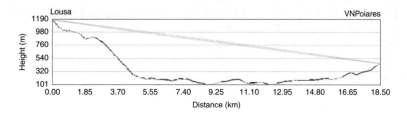

FIG. 25. Lousã Mountain—Carvalho Mountain terrain profile.

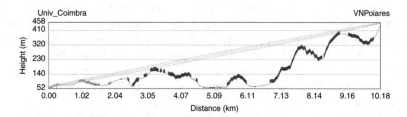

FIG. 26. University of Coimbra—Carvalho Mountain terrain profile.

Due to very difficult propagation condition (NLOS in a forest environment without a terrain topological place, like a mountain, to be used as reflection surface), it became clear that a radio link between the University of Coimbra and Carvalho (CM) could be established. Therefore, the PMP network topology with the base station at the University of Coimbra and the subscriber stations at the Lousã and Carvalho mountains was ruled out as a viable option. The final decision was to deploy a network topology with two links, one between the University of Coimbra and Lousã (also known as *Aggregation_Link*) and the other one between Lousã and

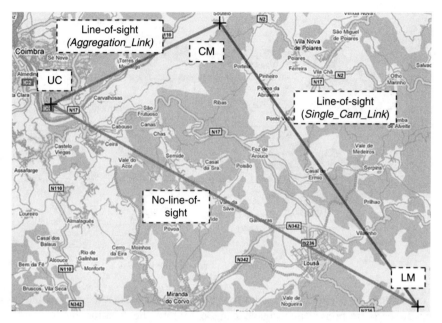

FIG. 27. WiMAX testbed antennas location.

Carvalho mountains (also known as *Single_Cam_Link*). To minimize interference, different frequencies on the radio link were assigned. Figure 27 maps the three antennas location and the link propagation conditions.

A schematic of the WiMAX testbed is shown in Fig. 28. A 24-km WiMAX link connects University of Coimbra with the first surveillance unit at the Lousã site. A second WiMAX link connects Lousã with Carvalho (19 km) further stretching WiMAX coverage into the remote forest. Both links are PMP (making it possible to add further surveillance units in the future, if deemed necessary). Neither of the surveillance towers has alternative broadband coverage, such as ADSL or 3G/UMTS.

The remote surveillance application comprises remote surveillance units, a central application server, and Web clients that can be installed in PCs or PDAs. Each remote surveillance unit consists of one or several outdoor cameras controlled by an IP video server, IP-enabled compasses, and an IP weather station, which can record temperature, wind, and humidity. The cameras automatically scan predefined sections of the forest and stream the video to the remote client. When the user detects a potential fire, she points the camera to the suspected location, performing remote rotation, tilt, and zoom, in order to conduct a more detailed inspection. If the alarm

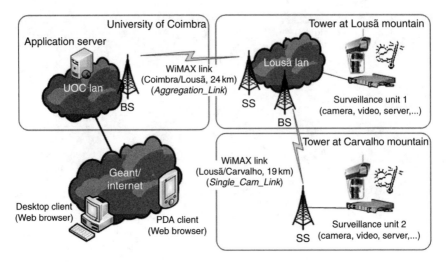

FIG. 28. Schematic of the fire prevention testbed.

is confirmed, the digital compass is used to determine the exact location of the fire and an early response fire brigade can be sent to the location by helicopter. Figures 29 and 30 illustrate the remote video surveillance application.

Each one of the remote surveillance units provides effective coverage in a radius of approximately 10 km from the Lousã and Carvalho monitoring sites. The central server is located at University of Coimbra, and client PCs can be located anywhere as long as there is Internet connectivity. The surveillance units were installed in watchtowers where the Portuguese Civil Protection already posts human watchers, equipped with binoculars, during the summer. It will be possible, therefore, to compare directly the performance of classic fire detection mechanisms and remote video surveillance. The key metric is the ability to detect early a fire, since the success of fire fighters greatly depends on the ability to reach the fire location within minutes from ignition, before it becomes uncontrollable. If the remote surveillance proves to be at least as effective as classic surveillance, it will be possible to reduce labor costs (a single operator can control several surveillance units) and to increase the number of surveillance posts, thus further increasing the probability of successful fire detection. Figures 31 and 32 display photos from the Lousã and Carvalho Mountain sites of the WiMAX testbed.

The WiMAX system is operating in the 3443–3471 MHz frequency range for the *Aggregation_Link* and on the 3543–3571 MHz band for the *Single_Cam_Link*. Detailed information about the WiMAX equipment configuration is given in Table IV.

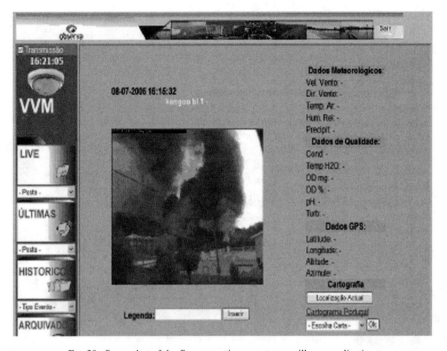

FIG. 29. Screenshot of the fire prevention remote surveillance application.

To study the performance of the installed testbed, a set of measurements were made for the two WiMAX links: (1) *Aggregation_Link* between University of Coimbra and Lousã and (2) the *Single_Cam_Link* between Lousã and Carvalho. The aim of these measurements was to verify that the installed WiMAX links are able to meet the fire prevention video application bandwidth requirements given that the surveillance cameras will have to stream video at a rate of 1.5 Mb/s.

It is important to differentiate between the two radio links. Although both of them have LOS propagation conditions, the distance for the *Aggregation_Link* is greater than the *Single_Cam_Link* (see Table V). Moreover, the *Aggregation_Link* must transport the video packets from both surveillance cameras: the streaming camera installed in Lousã and the one installed in Carvalho. For the *Single_Cam_Link*, since it only has to transport video streamed by the camera installed at Carvalho to Lousã, the bandwidth requirements are lower. Interested readers can find more details about the fire prevention WiMAX testbed in Ref. [62] and the references therein.

The *iperf* tool (see iperf.sourceforge.net) was used to measure the maximum available capacity on all links of the WiMAX testbed. The maximum throughput

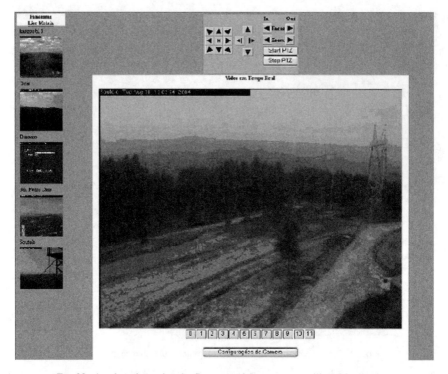

FIG. 30. Another view using the fire prevention remote surveillance application.

recorded for the *Aggregation_Link* and the *Single_Cam_Link* for each link configuration is given in Table V. When a 7-MHz channel bandwidth is allocated, the *Aggregation_Link* can provide up to 2.2 Mb/s of capacity in the uplink channel. In this case, it is possible to transport the video from both cameras, but not at the maximum bitrate (1.5 Mb/s per camera). Therefore, it is best to overprovision the link between the University of Coimbra and Lousã. By allocating 14 MHz to the *Aggregation_Link*, we can provide a capacity of 4.5 Mb/s, dedicated for the uplink direction, which enables both cameras to stream video at the maximum video encoding rate.

Concerning the *Single_Cam_Link*, since the distance is lower than the *Aggregation_Link*, the measured capacity is, as expected, higher. Since only the video from one camera is streamed over this link, a 3.5-MHz channel bandwidth is sufficient to satisfy the fire prevention application bandwidth requirements (1.5 Mb/s).

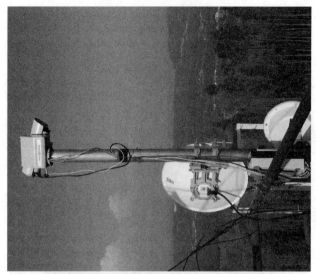

Fig. 31. Surveillance cameras at Lousã (left) and Carvalho mountains (right).

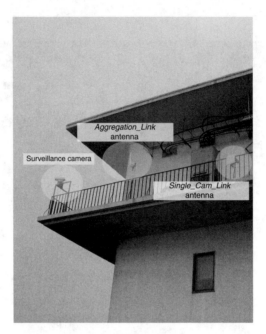

FIG. 32. Lousã Mountain surveillance tower: surveillance camera, subscriber station antenna (*Aggregation_Link* SS), and base station antenna (*Single_Cam_Link*).

TABLE IV
WiMAX FIRE PREVENTION TESTBED EQUIPMENT CONFIGURATION

Location of base station or subscriber station	BW (MHz)	Power (dbm)	Antenna
University of Coimbra (BS)	3.5, 7, and 14	23	90°
Lousã Mountain (SS)		16	15°
Lousã Mountain (BS)		23	90°
Carvalho Mountain (SS)		16	15°

7. Summary and Outlook

We presented a comprehensive review of IEEE 802.16 and WiMAX and their capacity to deliver multimedia services. First, we went through the salient aspects of the physical and MAC layers of IEEE 802.16 and reviewed the WiMAX Forum network reference model. We used a set of reference scenarios to motivate the

TABLE V
PROPAGATION CONDITIONS AND MEASURED THROUGHPUT

Link direction	BW (MHz)	Distance (km)	Propagation	Uplink (Mb/s)	Downlink (Mb/s)	Total (Mb/s)
Aggregation_Link	3.5	22.8	Line-of-sight	1.1	1.3	2.4
	7			2.2	2.7	4.9
	14			4.5	5.6	10.1
Single_Cam_Link	3.5	18.8	Line-of-sight	2.0	2.4	4.4
	7			4.1	4.8	8.9
	14			7.8	9.6	17.4

deployment of both fixed and mobile WiMAX in urban and rural areas, as well as for serving vertical applications. We discussed the emergence of broadband services, which include voice and video streaming over IP, and presented a large set of empirical measurements using COTS equipment. Based on our measurements, we concluded that state-of-the-art WiMAX equipment is suitable for deploying BWA services, including VoIP. For example, based on the measurements presented in Section 5, in a deployment such the one depicted in Fig. 33, we found that our fixed WiMAX base station featuring a WirelessMAN-OFDM interface could deliver cell rates in the order of 90 Mb/s (in a six sector layout with directional antennas), with a 2:1 downlink/uplink ratio. Besides serving business and residential customers, this capacity could be used to backhaul traffic from a Wi-Fi network in a small town, for example. As we saw, even without aggregation, each WirelessMAN-OFDM sector can carry hundreds of VoIP calls simultaneously. The results reported in Section 6 also show that WiMAX is suitable for backhauling high-quality video streams over links that span more than 20 km. Of course, we would like to emphasize that our results are based on testbeds; we expect that real-world deployments will be able to achieve even better rates.

When do we expect to see widespread deployment of WiMAX? According to studies carried out by the WiMAX Forum, as of August 2009, more than 500 WiMAX deployments around the world in 145 separate regional market areas. In North America, there are more than 1 million subscribers already. Although the majority of countries have deployed both fixed and mobile WiMAX networks as Fig. 1 illustrated, the number of fixed WiMAX networks still outnumbers the deployments with mobility support roughly by 2:1. Recent market data indicate that the large telecommunication operators opt for protecting their past investments in Third-Generation Partnership Project (3GPP) technologies and line behind 3GPP's LTE access technology. Fixed WiMAX deployments serving as last-mile links instead of wired connections in scarcely populated areas might be the main target for WiMAX

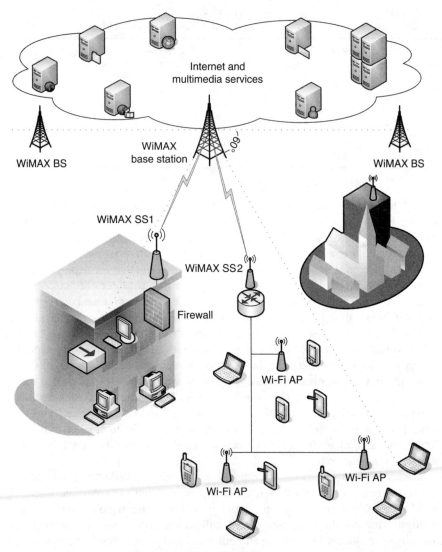

FIG. 33. WiMAX deployment schematic.

in the highly developed markets of North America and Western Europe. That is, key telecommunications players are considering WiMAX mainly as a fixed broadband replacement technology, for certain deployment areas, as it can provide broadband access at a fraction of the cost of an equivalent wired broadband alternative.

On the other hand, as there is still great room for developing a modern telecommunication infrastructure in Asia, Eastern Europe, South America, and Africa, the so-called "greenfield operators" may be able to take on the market opportunities by deploying fixed and mobile WiMAX systems. With an IP-based network core and low intellectual property rights (IPR) and patent portfolio considerations, WiMAX offers an alternative way to deploy wireless broadband networks with mobility support which is directly comparable to the latest 3GPP-standardized technologies that dominate current cellular network deployments. In particular, Russia and India have recently been active in planning and deploying new networks which will employ WiMAX to address the lack of sufficient wired infrastructure which has been the main obstacle in providing affordable broadband access.

As of late 2009, WiMAX still has a competitive advantage over other broadband access technologies, as it is a working solution for BWA. In contrast, LTE is not expected to be widely deployed before 2012. Nonetheless, past predictions about widespread deployment of WiMAX did not materialize so far. Earlier predictions claimed that by the end of 2008, there would be up to 10 million WiMAX subscribers worldwide. According to Maravedis (a market research firm), however, there were only 4 million WiMAX subscribers across the globe by mid-2009. We expect that several techno-economic factors, as well as government regulation and policy, will shape the future of next-generation BWA technologies. We can argue that WiMAX is well positioned in technical terms to play a major role in future developments, but dominance is far from guaranteed as of this writing.

ACKNOWLEDGMENTS

Part of this work was conducted within the framework of the IST Sixth Framework Programme Integrated Project WEIRD (IST-034622) [62], which was partially funded by the Commission of the European Union. Study sponsors had no role in study design, data collection and analysis, interpretation, or writing this chapter. The views expressed do not necessarily represent the views of the authors' employers, the WEIRD project, or the Commission of the European Union.

REFERENCES

[1] Cisco, Approaching the Zettabyte Era, June 2008. White paper.
[2] C. Anderson, The long tail, Wired 12 (10) October 2004.
[3] IEEE 802.16 WG, IEEE Recommended Practice for Local and Metropolitan Area Networks—Coexistence of Fixed Broadband Wireless Access Systems, March 2004. IEEE Std 802.16.2-2004.
[4] IEEE 802.16 WG, IEEE Standard for Local and Metropolitan Area Networks—Part 16: Air Interface for Broadband Wireless Access Systems, May 2009. IEEE Std 802.16-2009.
[5] IEEE 802.16 WG, IEEE Standard for Local and Metropolitan Area Networks—Part 16: Air Interface for Broadband Wireless Access Systems, Amendment 1: Multihop Relay Specification, June 2009. IEEE Std 802.16j-2009.

[6] IEEE 802.16 WG, IEEE Standard for Local and Metropolitan Area Networks: Media Access Control (MAC) Bridges—Amendment 2: Bridging of IEEE 802.16, August 2007. IEEE Std 802.16k-2007.
[7] IEEE 802.16 WG, IEEE Standard for Conformance to IEEE 802.16—Part 3: Radio Conformance Tests (RCT) for 10–66 GHz WirelessMAN-SC Air Interface, June 2004. IEEE Std 802.16/Conformance03-2004.
[8] IEEE 802.16 WG, IEEE Standard for Conformance to IEEE 802.16—Part 4: Protocol Implementation Conformance Statement (PICS) Proforma for Frequencies below 11 GHz, September 2006. IEEE Std 802.16/Conformance04-2006.
[9] IEEE 802.16 WG, IEEE 802.16m System Description Document [Draft] (work in progress), February 2009. IEEE 802.16m-08/003r7.
[10] WiMAX Forum Network Working Group, WiMAX Forum Network Architecture—Stage 2: Architecture Tenets, Reference Model and Reference Points, Release 1, Version 1.2, WiMAX Forum, January 2008.
[11] M. Katz, F.H.P. Fitzek (Eds.), WiMAX Evolution: Emerging Technologies and Applications, Wiley, Chichester, 2009.
[12] R.B. Marks, J.M. Costa, B.G. Kiernan, The evolution of WirelessMAN, IEEE Microwave Mag. 9 (4) (2008) 72–79.
[13] K. Pentikousis, E. Piri, J. Pinola, F. Fitzek, T. Nissilä, I. Harjula, Empirical evaluation of VoIP aggregation over a fixed WiMAX testbed, in: Proceedings of the 4th International Conference on Testbeds and Research Infrastructures for the Development of Networks and Communities (TRIDENTCOM), Innsbruck, Austria, March 2008. Article No. 19.
[14] K. Pentikousis, E. Piri, J. Pinola, F. Fitzek, An experimental investigation of VoIP and video streaming over fixed WiMAX, in: Proceedings of the 6th International Symposium on Modeling and Optimization in Mobile, Ad Hoc, and Wireless Networks and Workshops (WiOPT), Berlin, Germany, March 2008, pp. 8–15 10.1109/WIOPT.2008.4586026.
[15] E. Piri, J. Pinola, F. Fitzek, K. Pentikousis, ROHC and aggregated VoIP over fixed WiMAX: An empirical evaluation, in: Proceedings of the 13th IEEE Symposium on Computers and Communications (ISCC), Marrakech, Morocco, July 2008, pp. 1141–1146 10.1109/ISCC.2008.4625650.
[16] J. Pinola, K. Pentikousis, Mobile WiMAX, Internet Protocol J. 11 (2) (June 2008) 19–35.
[17] J.G. Andrews, R. Muhamed, Fundamentals of WiMAX: Understanding Broadband Wireless Networking, Prentice Hall, Upper Saddle River, NJ, 2007.
[18] K.-C. Chen, J.R.B. De Marca (Eds.), Mobile WiMAX, Wiley, Chichester, 2008.
[19] R. Van Nee, R. Prasad, OFDM for Wireless Multimedia Communications, Artech House Publishers, Norwood, MA, 2000.
[20] WiMAX Forum, WiMAX Forum® Mobile System Profile Specification, Release 1.5 Common Part, WMF-T23-001-R015v01, August 2009.
[21] WiMAX Forum, WiMAX Forum® Mobile System Profile Specification, Release 1.5 TDD Specific Part, WMF-T23-002-R015v01, August 2009.
[22] WiMAX Forum, WiMAX Forum® Mobile System Profile Specification, Release 1.5 FDD Specific Part, WMF-T23-003-R015v01, August 2009.
[23] WiMAX Forum, WiMAX Forum® Network Architecture—Stage 2: Architecture Tenets, Reference Model and Reference Points, Release 1.0, Version 4, WMF-T32-001-R010v04, February 2009.
[24] WiMAX Forum, WiMAX Forum® Network Architecture—Stage 3: Detailed Protocols and Procedures, Release 1.0, Version 4, WMF-T33-001-R010v04, February 2009.
[25] S. Madanapalli (Ed.), Analysis of IPv6 Link Models for IEEE 802.16 Based Networks, August 2007. IETF Request for Comments: 4968.

[26] B. Patil, F. Xia, B. Sarikaya, J. Choi, S. Madanapalli, Transmission of IPv6 via the IPv6 Convergence Sublayer over IEEE 802.16 Networks, February 2008. IETF Request for Comments: 5121.
[27] J. Jee, S. Madanapalli, J. Mandin, IP over IEEE 802.16 Problem Statement and Goals, April 2008. IETF Request for Comments: 5154.
[28] H. Jang, J. Jee, Y.-H. Han, S.D. Park, J. Cha, Mobile IPv6 Fast Handovers over IEEE 802.16e Networks, June 2008. IETF Request for Comments: 5270.
[29] K. Pentikousis, Wireless data networks, Internet Protocol J. 8 (1) (March 2005) 6–14.
[30] J. Proakis, Digital Communications, fourth ed., McGraw-Hill, Boston, MA, 2001.
[31] J. Van Meggelen, J. Smith, L. Madsen, Asterisk: The Future of Telephony, second ed., O'Reilly Media, Sebastopol, CA, 2007.
[32] H. Schulzrinne, S. Casner, R. Frederick, V. Jacobson, RTP: A Transport Protocol for Real-Time Applications, July 2003. IETF Request for Comments: 3550.
[33] A. Raake, Speech Quality of VoIP: Assessment and Prediction, Wiley, Chichester, 2006.
[34] ITU-T Recommendation G.711, Pulse Code Modulation (PCM) of Voice Frequencies, International Telecommunication Union, Geneva, November 1988.
[35] ITU-T Recommendation G.711—Appendix I, A High Quality Low-Complexity Algorithm for Packet Loss Concealment with G.711, International Telecommunication Union, Geneva, September 1999.
[36] ITU-T Recommendation G.711—Appendix II, A Comfort Noise Payload Definition for ITU-T G.711 Use in Packet-Based Multimedia Communication Systems, International Telecommunication Union, Geneva, February 2000.
[37] ITU-T Recommendation G.711.1, Wideband Embedded Extension for G.711 Pulse Code Modulation, International Telecommunication Union, Geneva, March 2008.
[38] ITU-T Recommendation G.723.1, Dual Rate Speech Coder for Multimedia Communications Transmitting at 5.3 and 6.3 kbit/s, International Telecommunication Union, Geneva, March 1996.
[39] ITU-T Recommendation G.729, Coding of Speech at 8 kbit/s Using Conjugate-Structure Algebraic-Code-Excited Linear Prediction (CS-ACELP), International Telecommunication Union, Geneva, January 2007.
[40] ITU-T Recommendation G.729.1, G.729 Based Embedded Variable Bit-Rate Coder: An 8–32 kbit/s Scalable Wideband Coder Bitstream Interoperable with G.729, International Telecommunication Union, Geneva, May 2006.
[41] J.-M. Valin, The Speex Codec Manual, Version 1.2 Beta 3, Xiph.org Foundation, December 2007. Available at http://www.speex.org/docs/manual/speex-manual.pdf.
[42] K. Vos, S. Jensen, K. Soerensen, SILK Speech Codec, IETF Internet-Draft: draft-vos-silk (work in progress), July 2009.
[43] ITU-T Recommendation G.722.2, Wideband Coding of Speech at around 16 kbit/s Using Adaptive Multi-Rate Wideband (AMR-WB), International Telecommunication Union, Geneva, July 2003.
[44] ITU-T Recommendation G.726, 40, 32, 24, 16 kbit/s Adaptive Differential Pulse Code Modulation (ADPCM), International Telecommunication Union, Geneva, December 1990.
[45] ITU-T Recommendation G.722, 7 kHZ Audio-Coding within 64 kbit/s, International Telecommunication Union, Geneva, November 1988.
[46] ITU-R, Requirements Related to Technical Performance for IMT Advanced Radio Interface(s), 2008. ITU-R Report M.2134.
[47] S. Deering, R. Hinden, Internet Protocol, Version 6 (IPv6) Specification, December 1998. IETF Request for Comments: 2460.
[48] C. Bormann, C. Burmeister, M. Degermark, H. Fukushima, H. Hannu, L. Jonsson, R. Hakenberg, T. Koren, K. Le, Z. Liu, A. Martensson, A. Miyazaki, et al., RObust Header Compression (ROHC):

Framework and Four Profiles: RTP, UDP, ESP, and Uncompressed, June 2001. IETF Request for Comments: 3095.
[49] C. Perkins (Ed.), IP Mobility Support for IPv4, August 2002. IETF Request for Comments: 3344.
[50] D. Johnson, C. Perkins, J. Arkko, Mobility Support in IPv6, June 2004. IETF Request for Comments: 3775.
[51] J. Pinola, E. Piri, K. Pentikousis, On the Performance Gains of VoIP Aggregation and ROHC over a WirelessMAN-OFDMA Air Interface, in: Proceedings of the IEEE Global Telecommunications Conference (GLOBECOM), Hawaii, November 2009.
[52] ITU-T Recommendation G.107, The E-Model, A Computational Model for Use in Transmission Planning, International Telecommunication Union, Geneva, December 1998.
[53] S. Sengupta, M. Chatterjee, S. Ganguly, R. Izmailov, Improving R-score of VoIP streams over WiMAX, Proc. IEEE ICC 2 (June 2006) 866–871.
[54] A. Sollaud, RTP Payload Format for the G.729.1 Audio Codec, October 2006. IETF Request for Comments: 4749.
[55] G. Pelletier, K. Sandlund, L.-E. Jonsson, M. West, RObust Header Compression (ROHC): A Profile for TCP/IP (ROHC-TCP), July 2007. IETF Request for Comments: 4996.
[56] L.-E. Jonsson, G. Pelletier, K. Sandlund, The RObust Header Compression (ROHC) Framework, July 2007. IETF Request for Comments: 4995.
[57] G. Pelletier, K. Sandlund, RObust Header Compression Version 2 (ROHCv2): Profiles for RTP, UDP, IP, ESP and UDP Lite, April 2008. IETF Request for Comments: 5225.
[58] T. Koren, S. Casner, J. Geevarghese, B. Thompson, P. Ruddyand, Enhanced Compressed RTP (CRTP) for Links with High Delay, Packet Loss and Reordering, July 2003. IETF Request for Comments: 3545.
[59] F.H.P. Fitzek, G. Schulte, E. Piri, J. Pinola, M.D. Katz, J. Huusko, K. Pentikousis, P. Seeling, Robust header compression for WiMAX femto cells, in: M. Katz, F.H.P. Fitzek (Eds.), WiMAX Evolution: Emerging Technologies and Applications, Wiley, Chichester, 2009, pp. 185–197.
[60] S.M. Alamouti, A simple transmit diversity technique for wireless communications, IEEE J. Sel. Areas Commun. 16 (October 1998) 1451–1458.
[61] E. Piri, J. Pinola, I. Harjula, K. Pentikousis, Empirical evaluation of mobile WiMAX with MIMO, in: Proceedings of the 2009 IEEE (GLOBECOM Workshops), Honolulu, Hawaii, USA, November 2009, pp. 1-6. 10.1109/GLOCOMW.2009.5360769.
[62] G. Martufi, M. Katz, P. Neves, M. Curado, M. Castrucci, P. Simoes, E. Piri, K. Pentikousis, Extending WiMAX to new scenarios: key results on system architecture and testbeds of the WEIRD project, in: Proceedings of the 2nd European Symposium on Mobile Media Delivery (EUMOB), Oulu, Finland, July 2008. 10.4108/ICST.MOBIMEDIA2008.3833.

An Overview of Web Effort Estimation

EMILIA MENDES

Computer Science Department, The University of Auckland, Auckland, New Zealand

Abstract

A cornerstone of Web project management is sound effort estimation, the process by which effort is predicted and used to determine costs and allocate resources effectively, thus enabling projects to be delivered on time and within budget. Effort estimation is a complex domain where the causal relationship among factors is nondeterministic with an inherently uncertain nature. For example, assuming there is a relationship between development effort and developers' experience using the development environment, it is not necessarily true that higher experience will lead to decreased effort. However, as experience increases so does the *probability* of decreased effort. The objective of this chapter is to provide an introduction to the process of estimating effort, discuss existing techniques used for effort estimation, and explain how a Web company can take into account the uncertainty inherent to effort estimation when preparing a quote. Therefore, this chapter is aimed to provide Web companies, researchers, and students with an introduction to the topic of Web effort estimation.

1. Introduction 224
   1.1. An Overview of Effort Estimation Techniques 224
   1.2. Expert-Based Effort Estimation 227
   1.3. Algorithmic Techniques 229
   1.4. AI Techniques 235
2. How to Measure a Technique's Prediction Accuracy? 248
3. Which Effort Estimation Technique to Use? 253

4. Web Effort Estimation Literature Survey 254
   4.1. Previous Studies 254
   4.2. Discussion 265
5. Conclusions 266
   References 267

1. Introduction

The Web is used as a delivery platform for numerous types of Web applications, ranging from complex e-commerce solutions with back-end databases using content management systems to online personal static Web pages and blogs [1]. With the great diversity of types of Web applications and technologies employed to develop these applications, there is an ever growing number of Web companies bidding for as many Web projects as they can accommodate. Under such circumstances it is usual that companies, in order to win the bid, estimate unrealistic schedules, leading to applications that are rarely developed on time and within budget.

Note that within the context of this chapter, cost and effort are used interchangeably because effort is taken as the main component of project costs. However, given that project costs also take into account other factors such as contingency and profit [2] we will use most often the word ''effort'' and not ''cost'' throughout.

1.1 An Overview of Effort Estimation Techniques

The reason to estimate effort is to predict the total amount of time it will take one person or a group of people to accomplish a given task/activity/process; this estimate is generally obtained taking into account the characteristics of the new project for which an estimate is needed, and also the characteristics of previous ''similar'' projects for which actual effort is known. The project characteristics employed herein are only those assumed to be relevant in determining effort.

Figure 1 details an effort estimation process.

The input to this process comprises the following:

(1) Data on past finished projects, for which actual effort is known, represented by project characteristics (independent variables) believed to have an effect upon the amount of effort needed to accomplish a task/activity/process.

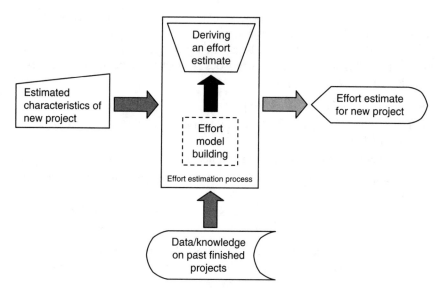

FIG. 1. Steps used to obtain an effort estimate.

(2) Estimated data relating to the new project for which effort is to be estimated. Such data also uses the same project characteristics employed in (1) above.

The estimation process itself includes two subprocesses, detailed below:

(1) Effort model building—This subprocess represents the construction of a tangible representation of the association between project characteristics and effort using data/knowledge from past finished projects for which actual effort is known. Such representation can take several forms, for example, an equation, a binary tree, an acyclic graph. This subprocess is shown using a dashed line because in some instances no concrete model representation exists, as for example, when an effort estimate is obtained from a domain expert based solely on their previous experience.

(2) Deriving an effort estimate—This subprocess represents either the use of the estimated characteristics of a new project as input to a concrete effort model that provides an effort estimate, or the use of the estimated characteristics of a new project as input to [the] subprocess [from step 1 above] that derives an effort estimate using previous knowledge from past projects.

The output to the estimation process outlined in Fig. 1 is an effort estimate (dependent variable) for a new project.

We will use an example to illustrate this process. Suppose a Web company employs the following project characteristics as input to predict the effort necessary to implement a new Web application:

- Estimated number of new Web pages.
- The number of functions/features (e.g., shopping cart) to be offered by the new Web application.
- Total number of developers who will help develop the new Web application.
- Developers' average number of years of experience with the development tools employed.
- The number of different technologies used to develop the new Web application (e.g., relational database, HTML, Javascript, Graphics software).

Of these variables, first two inputs are project characteristics used to estimate the *size* the problem to be solved (Web application). They are called size measures. The other three also represent project characteristics that are associated with effort; however, these three project characteristics do not measure an application's size, and hence are cojointly named "cost drivers."

The task/activity/process to be estimated can be as simple as developing a single function (e.g., creating a Web form with 10 fields) or as complex as developing a large e-commerce application that retrieves data from a Relational database system. Regardless of the application type, in general, the one consistent input (independent variable) believed to have the strongest influence on effort is size (i.e., the total number of Web pages), with cost drivers also playing an influential role.

In most cases, even when effort estimation is mostly based on the developer(s) and project manager(s) current expertise, data and/or knowledge from past finished projects can be used to help estimate effort for new projects yet to start.

Several techniques for effort estimation have been proposed over the past 30 years in software engineering. These fall into three broad categories [3]: expert-based effort estimation, algorithmic models, and artificial intelligence (AI) techniques. Each category is described in the following sections. These techniques are presented herein for two reasons: the first one is to present the reader with a wide view regarding the types of effort estimation techniques used in the literature; the second reason is most of the techniques presented herein have also been used for Web effort estimation, as detailed in Section 4.

1.2 Expert-Based Effort Estimation

Expert-based effort estimation is the process whereby effort is estimated by subjective means, often based on previous experience with developing and/or managing similar projects [4]. This is by far the most commonly used technique for Web effort estimation, with the attainment of accurate effort estimates being directly proportional to the competence and experience of the individuals involved (e.g., project manager, developer). Within the context of Web development, our experience suggests that expert-based effort estimates are obtained using one of the following mechanisms:

- An estimate that is based on a detailed effort breakdown that takes into account all of the lowest level parts of an application and the functional tasks necessary to develop this application. Each task attributed with effort estimates, is repeatedly combined into higher level estimates until we finally obtain one estimate that is considered as the sum of all lower level estimate parts. This type of estimation is called bottom-up [5]. Each estimate can be an educated guess or based on sound previous experience with similar projects.
- An estimate representing an overall process to be used to develop an application, as well as knowledge about the application to be developed, that is, the product. A total estimate is suggested and used to calculate estimates for the component parts (tasks), relative portions of the whole. This type of estimation is called top-down [5].

Estimates can be suggested by a project manager, or by a group of people mixing project manager(s) and developers, usually by means of one or more brainstorming sessions.

A survey of 32 Web companies in New Zealand conducted in 2004 [6], showed that 32% prepared effort estimates during the requirements gathering phase, 62% prepared estimates during their design phase, while 6% did not have to provide any effort estimates to their customers since these were happy to pay for the development costs without the need for a quote.

Of the 32 companies surveyed, 38% did not refine their effort estimate, and 62% did refine their estimates but not often. Therefore, these results suggest that for the majority of the companies we surveyed, the initial effort estimate was used as their "final" estimate, and work was adjusted to fit this initial quote. These results corroborated those published by Jørgensen and Sjøberg [7].

Sometimes Web companies gather only effort data for past Web projects believing it to be sufficient to help obtain accurate estimates for new projects. However, without even gathering data on the project characteristics (factors) that influence effort within the context of a specific company, effort data alone is unlikely to be sufficient to warrant the attainment of accurate effort estimates for new projects.

The drawbacks of expert-based estimation are as follows:

(i) It is very difficult to quantify and to clearly determine the factors that have been used to derive an estimate, making it difficult to apply the same reasoning to other projects (repeatability);
(ii) When a company finally builds up its expertise with developing Web applications using a given set of technologies, other technologies may appear and be rapidly adopted (mostly due to hype), thus leaving behind valuable knowledge that had been accumulated in the past.
(iii) Obtaining an effort estimate based on experience with past similar projects can be misleading when projects vary in their characteristics. For example, knowing that a Web application $W1$ containing 20 new static HTML Web pages and 20 new images, with a development time of 80 person hours, does not mean that a very similar application ($W2$) will also consume 80 person hours. Let us assume that $W1$ was developed by a single person and that $W2$ will be developed by two people. Two people may need additional time to communicate, and may also have different experiences with using HTML, which may affect total effort. In addition, another application eight times the size of $W1$ is unlikely to take exactly eight times longer to complete. This suggests that experts should be fully aware of most, if not all, project characteristics that impact effort in order to make an informed decision; however, it is our experience that often this is not the case.
(iv) Developers and project managers are known for providing optimistic effort estimates for tasks that they have to carry out themselves [8]. Optimistic estimates lead to underestimated effort with the direct consequence of projects being over budget and over time.

To cope with underestimation, it is suggested that experts provide three different estimates [9]: an optimistic estimate, o; a realistic estimate, r; and a pessimistic estimate, p. Based on a beta distribution, the estimated effort E is then calculated as:

$$E = (o + 4r + p)/6 \qquad (1)$$

This measure is likely to be better than a simple average of o and p; however, caution is still necessary.

Although there are problems related to using expert-based estimations, some studies have reported that when used in combination with other less subjective techniques (e.g., algorithmic models, Bayesian networks) expert-based effort estimation can be an effective estimating tool [10–12].

Expert-based effort estimation is a process that has not been objectively detailed, where only tacit knowledge is employed; however, it can still be represented in terms of the diagram presented in Fig. 1, where the order of steps that take place to obtain an expert-based effort estimate are as follows:

(a) An expert/group of developers implicitly look(s) at the estimated characteristics of the new project for which effort needs to be predicted.
(b) Based on the data obtained in (a) they remember or retrieve data/knowledge on past finished projects for which actual effort is known.
(c) Based on the data from (a) and (b) they subjectively estimate effort for the new project.

Therefore, data characterizing the new project for which effort is to be estimated is needed in order to retrieve, from memory and/or a database, data/knowledge on finished similar projects. Once this data/knowledge is retrieved, effort can be estimated.

It is important to stress that within a context where estimates are obtained via expert-based opinion, deriving a good effort estimate is much more likely to occur when the previous knowledge/data about completed projects relates to projects that are very similar to the one having its effort estimated. Here, we use the principle "similar problems have similar solutions." Note that for this assumption to be correct we also need to guarantee that the productivity of the team working on the new project is similar to the team productivity for the past similar projects.

The problems related to expert-based effort estimation aforementioned led to the proposal of other techniques for effort estimation. Such techniques are presented in the following sections.

1.3 Algorithmic Techniques

Algorithmic techniques are to date the most popular techniques described in the Web and software effort estimation literature. Such techniques attempt to build tangible models that represent the relationship between effort and one or more project characteristics via the use of algorithmic models. Such models assume that application size is the main contributor to effort thus in any algorithmic model the central project characteristic used is usually taken to be some notion of application size (e.g., the number of lines of source code, function points, number of Web pages, number of new images). The relationship between size and effort is often translated into an equation. An example of such type of equation is given as Equation 2, where a and b are constants, S is the estimated size of an application, and E is the estimated effort required to develop an application of size S.

$$E = aS^b \qquad (2)$$

In Equation 2, when $b < 1$ we have economies of scale, that is, larger projects use comparatively less effort than smaller projects. The opposite situation ($b > 1$) gives diseconomies of scale, that is, larger projects use comparatively more effort than smaller projects. When b is either $>$ or < 1, the relationship between S and E is nonlinear. Conversely, when $b = 1$ the relationship is linear.

However, as previously discussed, size alone is most unlikely to be the only contributor to effort. Other project characteristics (*cost drivers*), such as developer's programming experience, tools used to implement an application, maximum/average team size, are also believed to influence the amount of effort required to develop an application. Therefore, an algorithmic model should include as input not only application size but also the cost drivers believed to influence effort, such that estimated effort can be obtained by also taking into account these cost drivers (see Equation 3).

$$E = aS^b \text{CostDrivers} \qquad (3)$$

Different proposals have been made in an attempt to define the exact form such algorithmic model should take. The most popular are presented below.

COCOMO

One of the first algorithmic models to be proposed in the literature was the Constructive COst MOdel (COCOMO) [13]. COCOMO aimed to be a generic algorithmic model that could be applied by any organization to estimate effort at three different stages in a software project's development's life cycle: early on in the development life cycle, when requirements have not yet been fully specified (Basic COCOMO); once detailed requirements have been specified (Intermediate COCOMO); and when the application's design has been finalized (Advanced COCOMO). Each stage corresponds to a different model, and all three models take the same form (see Equation 4):

$$\text{EstimatedEffort} = a \ \text{EstSizeNewProj}^b \text{EAF} \qquad (4)$$

where:

- *EstimatedEffort* is the estimated effort, measured in person months, to develop an application;
- *EstSizeNewProj* is the size of an application measured in thousands of delivered source instructions (KDSI);
- a and b are constants which are determined by the class of project to be developed. The three possible classes are:
  - *Organic:* The *organic* class incorporates small, noncomplicated software projects, developed by teams that have a great amount of experience with similar projects, and where software requirements are not strict.

AN OVERVIEW OF WEB EFFORT ESTIMATION 231

- o *Semidetached:* The *semidetached* class incorporates software projects that are half-way between "small to easy" and "large to complex." Development teams show a mix of experiences, and requirements also present a mix of strict and slightly vague requirements.
- o *Embedded*: The *embedded* class incorporates projects that must be developed within a context where there are rigid hardware, software, and operational restrictions.
- *EAF* is an effort adjustment factor, calculated from cost drivers (e.g., developers, experience, tools).

The COCOMO model makes it clear that size is the main component of an effort estimate. Constants a and b, and the adjustment factor *EAF* all vary depending on the model used, and in the following ways:

The Basic COCOMO uses an value *EAF* of 1; a and b differ depending on a project's class (see Table I).

The Intermediate COCOMO calculates *EAF* based on 15 cost drivers, grouped into four categories: product, computer, personnel, and project (see Table II). Each cost driver is rated on a 6-point ordinal scale ranging from "very low importance" to "extra high importance." Each scale rating determines an effort multiplier, and the product of all 15 effort multipliers is taken as the *EAF*.

The Advanced COCOMO uses the same 15 cost drivers as the Intermediate COCOMO; however, they are all weighted according to each phase of the development lifecycle, that is, each cost driver is broken down by development's phase (see example in Table III). This model, therefore, enables the same cost driver to be rated differently depending on development's phase. In addition, it views a software application as a composition of modules and subsystems, to which the Intermediate COCOMO model is applied to.

The four development phases used in the Advanced COCOMO model are requirements planning and product design (RPD), detailed design (DD), coding

TABLE I
PARAMETER VALUES FOR BASIC AND INTERMEDIATE COCOMO

	Class	a	b
Basic	Organic		
	Semidetached		
	Embedded		
Intermediate	Organic	3.2	1.05
	Semidetached	3.0	1.12
	Embedded	2.8	1.20

TABLE II
COST DRIVERS USED IN THE INTERMEDIATE AND ADVANCED COCOMO

	Cost driver
Personnel	Analyst capability
	Applications experience
	Programmer capability
	Virtual machine experience
	Language experience
Project	Modern programming practices
	Software tools
	Development schedule
Product	Required software reliability
	Database size
	Product complexity
Computer	Execution time constraint
	Main storage constraint
	Virtual machine volatility
	Computer turnaround time

TABLE III
EXAMPLE OF RATING IN THE ADVANCED COCOMO

Cost driver	Rating	RPD	DD	CUT	IT
ACAP (analyst CAPability)	Very low	1.8	1.35	1.35	1.5
	Low	0.85	0.85	0.85	1.2
	Nominal	1	1	1	1
	High	0.75	0.9	0.9	0.85
	Very high	0.55	0.75	0.75	0.7

and unit test (CUT), and integration and test (IT). An overall project estimate is obtained by aggregating the estimates obtained for each of the subsystems, which themselves were obtained by combining estimates calculated for each module.

The original COCOMO model was radically improved 15 years later, and renamed as COCOMO II, and incorporates changes that have occurred in software development environments and practices over the previous 15 years [14]. COCOMO II is not detailed in this chapter; however, interested readers are referred to Refs. [14,15].

The COCOMO model is an example of a *general purpose* algorithmic model, where it is assumed it is not compulsory for ratings and parameters to be adjusted (calibrated) to specific companies in order for the model to be used effectively.

We have presented this model herein because it is the best known general-purpose effort estimation algorithmic model proposed to date. However, the original COCOMO model proposed in 1981 was created focusing at contractor developed mostly military projects on remote batch processing mainframe computer systems. This means that the environment assumptions used as basis for the cost drivers proposed in that model are likely not to be applicable to Web development projects.

Despite the existence of *general purpose* models, such as COCOMO, the effort estimation literature has numerous examples of *specialized* algorithmic models that were built using applied regression analysis techniques on datasets of past completed projects (e.g., [4]). Specialized and regression-based algorithmic models are most suitable to local circumstances, such as "in-house" analysis, as they are derived from past data that often represent projects from the company itself. Regression analysis, used to generate regression-based algorithmic models, provides a procedure for determining the "best" straight-line fit (see Fig. 2) to a set of project data that represents the relationship between effort (response or dependent variable) and size and cost drivers (predictor or independent variables) [4].

Using real data on Web projects, Fig. 2 shows an example of a regression line that describes the relationship between log(*Effort*) and log(*totalWebPages*). It should be

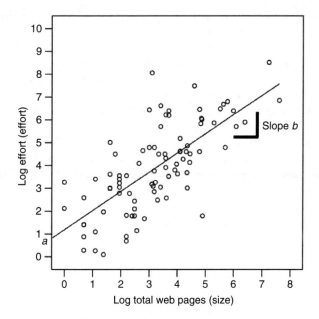

FIG. 2. Example of a regression line.

noted that the original variables *Effort* and *totalWebPages* have been transformed using the natural logarithmic scale to comply more closely with the assumptions of the regression analysis techniques. Details on these assumptions and how to identify variables that need transformation described in more depth in Ref. [1] and Chapter 5. Further details on regression analysis techniques are provided in Ref. [1] and Chapter 10.

The equation represented by the regression line in Fig. 2 is as follows:

$$\log(\text{Effort}) = a + b \log(\text{totalWebPages}) \qquad (5)$$

where a is the point in which the regression line intercepts the Y-axis, known simply as the *intercept* and b represents the slope of the regression line, that is, its inclination.

Equation 5 shows a linear relationship between log(Effort) and log(totalWeb-Pages). However, since the original variables have been transformed before the regression technique was employed, this equation needs to be transformed back such that it uses the original variables. This is done by taking the long of both sides in order to provide a linear fit to the data. The resultant equation is:

$$\text{Effort} = a \ \text{totalWebPages}^b \qquad (6)$$

This equation presents the same relationship as COCOMO's Effort Equation (see Equation 2).

Regarding the regression analysis itself, two of the most widely used techniques in the software and Web effort estimation literature are multiple regression (MR) and stepwise regression (SWR). The difference between these two techniques is that MR obtains a regression line using all the independent variables at the same time, whereas SWR is a technique that examines different combinations of independent variables, looking for the best grouping to explain the greatest amount of variation in effort. Both use least squares regression, where the regression line selected is the one that reflects the minimum values of the sum of the squared errors. Errors are calculated as the difference between actual and estimated effort and are known as "residuals" [4].

In terms of the diagram presented in Fig. 1, an algorithmic model uses constant scalar values based on past project data; however, for anyone wishing to use this model, the steps to use are as follows:

(a) Data on past finished projects, for which actual effort is known, is used to build an effort estimation model using multivariate regression analysis.
(b) The estimated characteristics of the new project for which effort needs to be predicted are entered as input to the Effort Equation built in (a).
(c) An effort estimate for the new Web application is obtained from (b)

How often step (a) is to be carried out varies from one company to another. If a *general-purpose* algorithmic model is in use, step (a) may have taken place only once, given that it is likely that such models are be used by companies without recalibration of values for the constants. Within the context of a *specialized* algorithmic model, step (a) is used whenever it is necessary to recalibrate the model. This can occur after several new projects are finished and incorporated to the company's database of data on past finished projects. However, a company may also decide to recalibrate a model after every new project is finished, or to use the initial model for a longer time period. If the development team remains unchanged (and assumes that the team does not have an excessive learning curve for each new project) and new projects are similar to past projects there is no pressing need to recalibrate an algorithmic model too often.

1.4 AI Techniques

AI techniques have also been used in the field of effort estimation as a complement to, or as an alternative to, the two previous categories. Examples include fuzzy logic [16], classification and regression trees (CART) [17], neural networks [18], case-based reasoning (CBR) [3], and Bayesian networks (BN) [19–23]. Within the context of this chapter, we will cover in detail CBR, CART, and BNs models as these are currently the most popular machine learning techniques employed for Web effort estimation. A useful summary of numerous machine learning techniques can also be found in Refs. [24, 25].

1.4.1 Case-Based Reasoning

CBR uses the assumption that *"similar problems provide similar solutions"* [18]. It provides effort estimates by comparing the characteristics of the current project for which effort is to be estimated, against a library of historical information from completed projects with known actual effort (case base).

Using CBR involves [26]:

(i) Characterizing a new project p, for which an estimate is required, with variables (features) common to those completed projects stored in the case base. In terms of Web and software effort estimation, features represent size measures and cost drivers that have a bearing on effort. This means that if a Web company has stored data on past projects where the, for example, data represents the features *effort*, *size*, *development team size*, and *tools used*, then the data used as input to obtaining an effort estimate will also need to include these same features.

(ii) Use of this characterization as a basis for finding similar (analogous) completed projects, for which effort is known. This process can be achieved by measuring the "distance" between two projects at a time (project p and one finished project), based on the features' values, for all features (k) characterizing these projects. Each finished project is compared to project p, and the finished project presenting the shortest distance overall is the "most similar project" to project p. Although numerous techniques can be used to measure similarity, nearest neighbor algorithms using the unweighted Euclidean distance measure have been the most widely used to date in Web engineering.

(iii) Generation of an effort estimate for project p based on the effort of those completed projects that are similar to p. The number of similar projects taken into account to obtain an effort estimate will depend on the size of the case base. For small case bases (e.g., up to 90 cases), typical values use the most similar finished project, or the two or three most similar finished projects (1, 2, and 3 closest analogues) [26]. For larger case bases no conclusions have been reached regarding the best number of similar projects to use. The calculation of estimated effort is obtained using the same effort value as the closest neighbor, or the mean effort for two or more closest neighbors. This is the common choice in Web engineering.

With reference to Fig. 1, the sequence of steps used with CBR is as follows:

(a) The estimated size and cost drivers relating to a new project p are used as input to retrieve similar projects from the case base, for which actual effort is known.
(b) Using the data from (a) a suitable CBR tool retrieves similar projects to project p, and ranks these similar projects in ascending order of similarity, that is, from "most similar" to "least similar."
(c) A suitable CBR tool provides an effort estimate for project p.

The sequence just described is similar to that employed with expert-based effort prediction, that is, the characteristics of a new project must be known in order to retrieve finished similar projects. Once similar projects are retrieved, effort can then be estimated.

When using CBR there are six parameters that need to be considered, which are as follows [27]:

1.4.1.1 Feature Subset Selection.
Feature subset selection involves determining the optimum subset of features that yield the most accurate estimations. Some existing CBR tools, for example, ANGEL [3] optionally offer this functionality using a brute-force algorithm, searching for all possible feature

subsets. Other CBR tools (e.g., CBR-Works from tec:inno) have no such functionality, and therefore to obtain estimated effort, we must use all of the known features of a new project to retrieve the most similar finished projects.

1.4.1.2 Case Similarity.
Case similarity relates to measuring the level of similarity between different cases. Several case similarity measures have been proposed to date in the literature, where the three most popular measures used in both the Web and software engineering literature have been the unweighted Euclidean distance, the weighted Euclidean distance, and the maximum distance [26–28]. However, there are also other similarity measures available, and presented in Ref. [26]. The three case similarity measures aforementioned are described below.

Unweighted Euclidean distance: The unweighted Euclidean distance measures the Euclidean (straight-line) distance d between two cases, where each case has n features. This measure has a geometrical meaning as the shortest distance between two points in an n-dimensional Euclidean space [26]. The equation used to calculate the distance between two cases x and y is the following:

$$d(x,y) = \sqrt{|x_0 - y_0|^2 + |x_1 - y_1|^2 + \cdots + |x_{n-1} - y_{n-1}|^2 + |x_n - y_n|^2} \quad (7)$$

where x_0 to x_n represent features 0 to n of case x; y_0 to y_n represent features 0 to n of case y.

Given the following example 1:

Project	newWebPages	newImages
1 (new)	100	20
2	300	15
3	560	45

The unweighted Euclidean distance between the new project 1 and finished project 2 would be calculated using the following equation:

$$d = \sqrt{|100 - 300|^2 + |20 - 15|^2} = 200.0625 \quad (8)$$

The unweighted Euclidean distance between the new project 1 and finished project 3 would be calculated using the following equation:

$$d = \sqrt{|100 - 560|^2 + |20 - 45|^2} = 460.6788 \quad (9)$$

Using the weighted Euclidean distance, the distance between projects 1 and 2 is smaller than the distance between projects 1 and 3, thus project 2 is more similar to project 1 than project 3.

Weighted Euclidean distance: The difference between the unweighted and the weighted Euclidean distance is that the latter enables the allocation of weights to features, thus quantifying their relative importance. Herein the equation used to calculate the distance between two cases x and y is the following:

$$d(x,y) = \sqrt{w_0|x_0-y_0|^2 + w_1|x_1-y_1|^2 + \cdots + w_{n-1}|x_{n-1}-y_{n-1}|^2 + w_n|x_n-y_n|^2} \quad (10)$$

where x_0 to x_n represent features 0 to n of case x; y_0 to y_n represent features 0 to n of case y; w_0 to w_n are the weights for features 0 to n.

Based on Example 1 above, if we were now to attribute weight = 5 to feature newImages, and weight = 1 to feature newWebPages, the distances would be as follows:

Distance between the new project 1 and finished project 2:

$$d = \sqrt{1|100-300|^2 + 5|20-15|^2} = 200.3123 \quad (11)$$

The unweighted Euclidean distance between the new project 1 and finished project 3 would be calculated using the following equation:

$$d = \sqrt{1|100-560|^2 + 5|20-45|^2} = 463.3843 \quad (12)$$

The pattern observed herein would be similar to that previously found: the distance between projects 1 and 2 is smaller than the distance between projects 1 and 3, making project 2 more similar to project 1 than project 3.

Maximum distance: The maximum distance computes the highest feature similarity, that is, the one to define the closest analogy. For two points (x_0, y_0) and (x_1, y_1), the maximum measure d is equivalent to the equation:

$$d = \sqrt{\max((x_0-y_0)^2, (x_1-y_1)^2)} \quad (13)$$

This effectively reduces the similarity measure down to a single feature, where the feature that is chosen may differ for each retrieval episode. So, for a given "new" project P_{new}, the closest project in the case base will be the one that has at least one feature with the most similar value to the same feature in project P_{new}.

1.4.1.3 Scaling.

Scaling (also known as standardization) represents the transformation of a feature's values according to a defined rule, such that all features present values within the same range; as a consequence all features have the same degree of influence upon the results [26]. A common method of scaling is to assign "zero" to the minimum observed value and "one" to the maximum observed value [29]. This is the strategy used by ANGEL and CBR-Works, that is, for each feature, its original values are normalized (between 0 and 1) by CBR tools to guarantee that they all influence the results in a similar fashion.

1.4.1.4 Number of Analogies.

The number of analogies refers to the number of most similar analogues (cases) that will be used to generate an effort estimate. With small sets of data, it is reasonable to consider only a small number of most the similar analogues [26]. Several studies in Web and software engineering have used only the closest analogue (CA) ($k=1$) to obtain an estimated effort for a new project [30, 31], while others have also used the two closest and the three closest analogues [5, 26, 28, 32–36].

1.4.1.5 Analogy Adaptation.

Once the most similar analogues have been selected the next step is to identify how to adapt an effort estimate for project P_{new}. Choices of adaptation techniques presented in the literature vary from the nearest neighbor [30, 33], the mean of the CAs [3], the median of the CAs [26], the inverse distance weighted mean and inverse rank weighted mean [29], to illustrate just a few. The adaptations used to date for Web engineering are the nearest neighbor, mean of the CAs [5, 28], and the inverse rank weighted mean [31, 34, 35, 37].

Each adaptation is explained below:

Nearest neighbour: For the estimated effort P_{new}, this type of adaptation uses the same effort of its CA.

Mean effort: For the estimated effort P_{new}, this type of adaptation uses the average of its closest k analogues, when $k > 1$. This is a typical measure of central tendency, often used in the Web and software engineering literature. It treats all analogs as being equally important towards the outcome—the estimated effort.

Median effort: For the estimated effort P_{new}, this type of adaptation uses the median of the closest k analogs, when $k > 2$. This is also a measure of central tendency, and has been used in the literature when the number of selected closest projects is > 2 [26].

Inverse rank weighted mean: This type of adaptation allows higher ranked analogues to have more influence over the outcome than lower ones. For example, if we use three analog, then the CA would have weight $= 3$, the second closest analogue (SC) would have weight $= 2$, and the third closest analogue (LA) would have weight $= 1$. The estimated effort would then be calculated as:

$$\text{Inverse Rank Weighed Mean} = \frac{3CA + 2SC + LA}{6} \qquad (14)$$

1.4.1.6 Adaptation Rules.

Adaptation rules represent the adaptation of the estimated effort, according to a given criterion, such that it reflects the characteristics of the target project (new project) more closely. For example, in the context of effort prediction, the estimated effort to develop an application *a* would be adapted such that it would also take into consideration the size value of application *a*. The adaptation rule that has been employed to date in Web engineering is based on the linear size adjustment to the estimated effort [34,35], obtained as follows:

- Once the most similar analogue in the case base has been retrieved, its effort value is adjusted and used as the effort estimate for the target project (new project).
- A linear extrapolation is performed along the dimension of a single measure, which is a size measure strongly correlated with effort. The linear size adjustment is calculated using the equation presented below.

$$\text{Effort}_{\text{newProject}} = \frac{\text{Effort}_{\text{finishedProject}}}{\text{Size}_{\text{finishedProject}}} \text{Size}_{\text{newProject}} \qquad (15)$$

Given the following example:

Project	totalWebPages (size)	totalEffort (effort)
Target (new)	100 (estimated value)	20 (estimated and adapted value)
Closest analogue	450 (actual value)	80 (actual value)

The estimated effort for the target project will be calculated as:

$$\text{Effort}_{\text{newProject}} = \frac{80}{450} 100 = 17.777 \qquad (16)$$

When we use more than one size measure as feature, the equation changes to:

$$E_{\text{est.}P} = \frac{1}{q} \left(\sum_{q=1}^{q=x} \frac{E_{\text{act}} S_{\text{est.}q}}{S_{\text{act.}q}|_{>0}} \right) \qquad (17)$$

where

q is the number of size measures used as features.
$E_{est.P}$ is the Total Effort estimated for the new Web project P.
E_{act} is the Total Effort for the CA obtained from the case base.
$S_{est.q}$ is the estimated value for the size measure q, which is obtained from the client.
$S_{act.q}$ is the actual value for the size measure q, for the CA obtained from the case base.

This type of adaptation assumes that all projects present similar productivity; however, this may not be an adequate assumption for numerous Web development companies worldwide.

1.4.2 Classification and Regression Trees

CART [38] are techniques whereby binary trees are constructed using data from finished projects for which effort is known, such that each leaf node represents either a category to which an estimate belongs, or a value for an estimate.

The data used to build a CART model is called a *"learning sample,"* and once a tree has been built it can be used to estimate effort for new projects.

In order to obtain an effort estimate for a new project p one has to traverse the tree from root to leaf by selecting the node values/categories that are the closest match to the estimated characteristics of p.

For example, assume we wish to obtain an effort estimate for a new Web project using as its basis the simple binary tree presented in Fig. 3.

This binary tree was built from data obtained from past completed Web applications, taking into account their existing values of effort and project characteristics (independent variables) believed to be influential upon effort (e.g., total number of new Web pages (*newWebPages*), total number of new Images (*newImages*), and the number of Web developers (*numDevelopers*)).

Assuming that the estimated values for *newWebPages*, *newImages*, and *numDevelopers* for a new Web project are 38, 15, and 3, respectively, we would obtain an estimated effort of 75 person hours after traversing the tree from its root down to leaf "*Effort* = 75."

If we now assume that the estimated values for *newWebPages*, *newImages*, and *numDevelopers* are 26, 34, and 7, respectively, we would obtain an estimated effort of 65 person hours after navigating the tree from its root down to leaf "Effort = 65."

In this particular example all the variables that characterize a Web project are numerical. Whenever this is the case the binary tree is called a regression tree.

Now let us look at the binary tree shown in Fig. 4. It uses the same variable names as that shown in Fig. 3; however, these variables are now all categorical, where

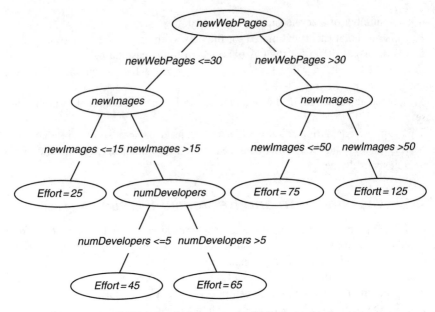

FIG. 3. Example of a binary tree for Web effort estimation.

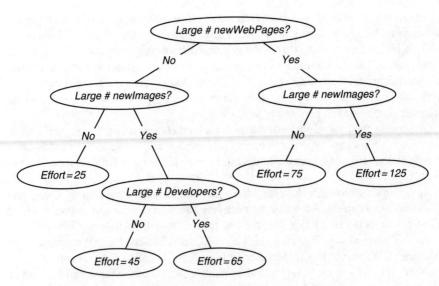

FIG. 4. Example of another binary tree for Web effort estimation.

possible categories (classes) are "Yes" and "No." This binary tree is therefore called a Classification tree.

A CART model constructs a binary tree by recursively partitioning the predictor space (set of all values or categories for the independent variables judged relevant) into subsets where the distribution of values or categories for the dependent variable (e.g., effort) is successively more uniform. The partition (split) of a subset $S1$ is decided on the basis that the data in each of the descendant subsets should be "purer" than the data in $S1$. Thus node "impurity" is directly related to the amount of different values or classes in a node, that is, the greater the mix of classes or values, the higher the node "impurity." A "pure" node means that all the cases (e.g., Web projects) belong to the same class, or have the same value. The partition of subsets continues until a node contains only one class or value. Note that it is not always the case that all the independent variables are used to build a CART model, rather only those variables that contribute effectively to effort are selected by the model. This means that in this instance a CART model can be used not only for effort prediction, but also to obtain insight and understanding of the variables that are relevant to estimate effort. A detailed description of CART models is given in Ref. [1] and Chapter 7.

In terms of the diagram presented in Fig. 1, an effort estimate is obtained according to the following steps:

(a) Data on past finished projects, for which actual effort is known, is used to build an effort estimation model using a binary tree (CART).
(b) The estimated characteristics of the new project for which effort needs to be predicted are used to traverse the tree built in (a).
(c) An effort estimate for the new Web application is obtained from (b).

The sequence of steps described above, corresponds to the same sequence used with the algorithmic techniques. In both situations, data is used to either build an equation that represents an effort model, or to build a binary tree, which is later used to obtain effort estimates for new projects. This sequence of steps contrasts to the different sequence of steps used for expert opinion and CBR, where knowledge about a new project is used to select similar projects.

1.4.3 Bayesian Networks

A BN is a model that supports reasoning with uncertainty due to the way in which it incorporates existing knowledge of a complex domain [39]. This knowledge is characterized using two parts. The first, the *qualitative part*, represents the structure of a BN as depicted by a directed acyclic graph (digraph) (see Fig. 5). The digraph's nodes represent the relevant variables (factors) in the domain being modeled, which

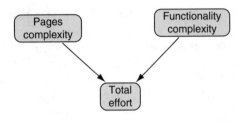

FIG. 5. A small Bayesian network and three CPTs.

can be of different types (e.g., observable or latent, categorical). The digraph's arcs represent the causal relationships between variables, where relationships are quantified probabilistically [39].

The second, the *quantitative part*, associates a conditional probability table (CPT) to each node, its probability distribution. A parent node's CPT describes the relative probability of each state (value) (Fig. 5, nodes "Pages complexity" and "Functionality complexity"); a child node's CPT describes the relative probability of each state conditional on every combination of states of its parents (Fig. 5, node "Total Effort"). So, for example, the relative probability of "Total Effort" being "Low" conditional on "Pages complexity" and "Functionality complexity" being both "Low" is 0.7.

Each row in a CPT represents a conditional probability distribution and therefore its values sum up to 1 [39].

Formally, the posterior distribution of a BN is based on Bayes' rule [39]:

$$p(X|E) = \frac{p(E|X)p(X)}{p(E)} \tag{18}$$

AN OVERVIEW OF WEB EFFORT ESTIMATION 245

where

- $p(X|E)$ is called the *posterior* distribution and represents the probability of X given evidence E;
- $p(X)$ is called the *prior* distribution and represents the probability of X before evidence E is given;
- $p(E|X)$ is called the *likelihood* function and denotes the probability of E assuming X is true.

Once a BN is specified, evidence (e.g., values) can be entered into any node, and probabilities for the remaining nodes are automatically calculated using Bayes' rule [39]. Therefore, BNs can be used for different types of reasoning, such as predictive, diagnostic, and "what-if" analyses to investigate the impact that changes on some nodes have on others [40].

BNs can be built from different sources—data on past finished projects, expert opinion, or a combination of these. In any case, the process used to build a BN is depicted in Fig. 6 and briefly explained below.

This process is called the Knowledge Engineering of Bayesian Networks (KEBN) process, which was initially proposed in Ref. [41] (see Fig. 6). In Fig. 6, arrows represent flows through the different processes, depicted by rectangles. Such processes are executed either by people—the Knowledge Engineer (KE) and the Domain Experts (DEs) (white rectangles), or by automatic algorithms (dark gray rectangles).

The three main steps within the KEBN process are the Structural development, parameter estimation, and model validation. This process iterates over these steps until a complete BN is built and validated. Each of these three steps is detailed below:

Structural development: This step represents the qualitative component of a BN, which results in a graphical structure comprised of, in our case, the factors (nodes, variables) and causal relationships identified as fundamental for effort estimation of Web projects. In addition to identifying variables, their types (e.g., query variable, evidence variable) and causal relationships, this step also comprises the identification of the states (values) that each variable should take, and if they are discrete or continuous. In practice, currently available BN tools require that continuous variables be discretized by converting them into multinomial variables [39]. The BN's structure is refined through an iterative process, and uses the principles of problem solving employed in data modeling and software development. In order to elicit a BN structure, a combination of existing literature in the related knowledge domain, existing data and knowledge from DEs can be used. Throughout this step the KE(s) also evaluate(s) the BN's structure in two stages. The first entails checking whether

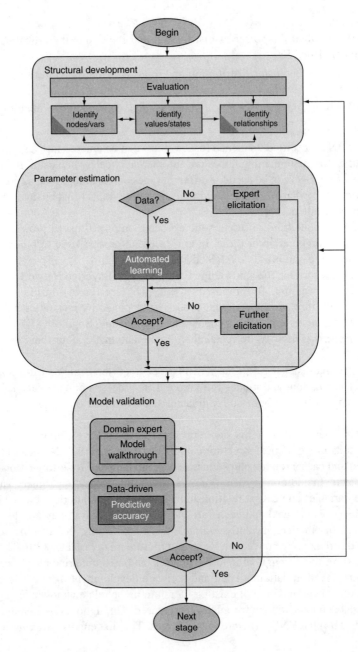

FIG. 6. KEBN, adapted from Woodberry et al. [41].

variables and their values have a clear meaning; all relevant variables have been included; variables are named conveniently; all states are appropriate (exhaustive and exclusive); a check for any states that can be combined. The second stage entails reviewing the BN's graph structure (causal structure) to ensure that any identified d-separation dependencies comply with the types of variables used and causality assumptions. D-separation dependencies are used to identify variables influenced by evidence coming from other variables in the BN [39, 42]. Once the BN structure is assumed to be close to final, KEs may still need to optimize this structure to reduce the number of probabilities that need to be elicited or learnt for the network. If optimization is needed then techniques that change the causal structure (e.g., divorcing) and the use of parametric probability distributions (e.g., noisy-OR gates) are employed [39, 42].

Parameter estimation: This step represents the quantitative component of a BN, where conditional probabilities corresponding to the quantification of the relationships between variables are obtained [39, 42]. Such probabilities can be attained via Expert Elicitation, automatically from data, from existing literature, or using a combination of these. When probabilities are elicited from scratch, or even if they only need to be revisited, this step can be very time consuming.

Model validation: This step validates the BN that results from the two previous steps, and determines whether it is necessary to revisit any of those steps. Two different validation methods are generally used—Model Walkthrough and Predictive Accuracy.

Model Walkthrough represents the use of real case scenarios that are prepared and used by DEs to assess if the predictions provided by a BN correspond to the predictions experts would have chosen based on their own expertise. Success is measured as the frequency with which the BN's predicted value for a target variable (e.g., quality, effort) that has the highest probability corresponds to the experts' own assessment.

Predictive Accuracy uses past data (e.g., past project data), rather than scenarios, to obtain predictions. Data (evidence) is entered on the BN model, and success is measured as the frequency with which the BN's predicted value for a target variable (e.g., quality, effort) that has the highest probability corresponds to the actual past data.

Within the context of Web effort estimation and to some extent software effort estimation, the challenge using Predictive Accuracy is the lack of reliable effort data gathered by Web and Software companies. Most companies, who claim to collect effort data, use manually entered electronic timesheets (or even paper!) which is unreliable when staff rely on their memory and complete their timesheets at the end of the day. Collecting manually entered timesheets every 5 min (assume 1 min/entry) in a bid to improve data accuracy increases data collection cost by as much as

10-fold. The problem here is that "effort accuracy" is inversely related to productivity, that is, the longer one takes filling out timesheets the less time one has to do the real work!

In terms of the diagram presented in Fig. 1, an effort estimate is obtained according to the following steps:

(a) Data on past finished projects/knowledge from DEs/both is/are used to build an effort estimation BN model.
(b) The estimated characteristics of the new project for which effort needs to be predicted are used as evidence, and applied to the BN model built in (a).
(c) An effort estimate for the new Web application is obtained from (b).

The sequence of steps described above, corresponds to the same sequence used with the algorithmic techniques and CART. In all situations, data/knowledge is used to either build an equation, a binary tree, or an acyclic graph that represents an effort model, which is later used to obtain effort estimates for new projects. As previously stated, this sequence of steps contrasts to the different sequence of steps used for expert opinion and CBR, where knowledge about a new project is used to select similar projects.

2. How to Measure a Technique's Prediction Accuracy?

Effort estimation techniques aim to provide accurate effort predictions for new projects. In general, these techniques use data on past finished projects, which are then employed to obtain effort estimates for new projects.

In order to determine how accurate the estimations supplied by a given effort estimation technique are, we need to measure its predictive accuracy, generally using the four-step process described below, and illustrated in Fig. 7.

Step 1: Split the original dataset into two subsets, validation and training. The validation set represents data on finished projects p_n to p_q that will be used to simulate a situation as if these projects were new. Each project p_n to p_q will have its effort estimated using a technique t, and, given that we also know the project's actual effort, we are in a position to compare its actual effort to the estimated effort obtained using t, and therefore ultimately assess how far off the estimate is from the actual.

Step 2: Use the remaining projects (training set) to build an effort estimation model m. There are estimation techniques that do not build an explicit model

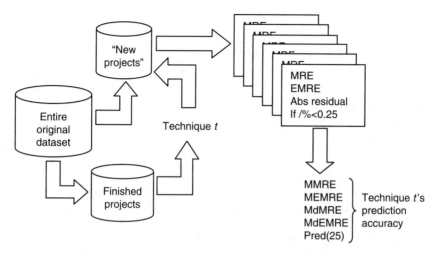

FIG. 7. Overall process used to measure a technique's prediction accuracy.

(e.g., CBR). If this is the case, then the training set becomes a database of past projects to be used by the effort technique t to estimate effort for p_n to p_q.

Step 3: Apply model m to each project p_n to p_q, and obtain estimated effort. Once estimated effort is obtained, accuracy statistics for each of these projects p_n to p_q can also be calculated. If the technique does not build a model, then this step comprises applying this technique t to each project p_n to p_q, to obtain estimated effort, and once estimated effort is obtained, accuracy statistics for each of these projects can be calculated.

Step 4: Once estimated effort and accuracy statistics to p_n to p_q have been attained, aggregated accuracy statistics can be computed, which provide an overall assessment of the predictive accuracy of model m or technique t.

Note that the subset "New projects" (see Fig. 7) comprises data on finished projects, for which actual effort is known. This data is used to simulate a situation where a Web company has a subset of new projects for which they wish to obtain estimated effort. A simulation in this instance allows us to compare effort estimates obtained using technique t, and the actual known effort used to develop these projects. This comparison will indicate how close estimated effort is from actual effort, and consequently, how accurate a technique t is.

Next we describe the most widely used prediction accuracy statistics, both in Software and Web engineering studies, and how they are employed to measure a technique's predictive accuracy.

Project's prediction accuracy statistics

To date the three most commonly used project effort prediction accuracy statistics have been the following:

- The magnitude of relative error (MRE) [43].
- Pred(*l*). A measure that assesses if the relative error associated with a project is not greater than 0.25 [44]. The value for *l* is generally set at 25% (or 0.25).
- Absolute residual, which is the absolute difference between actual and estimated effort [45].

MRE is defined as:

$$\text{MRE} = \frac{|e - \hat{e}|}{e} \quad (19)$$

where e is the actual effort of a project in the validation set, and $\hat{e}$ is the estimated effort that was obtained using technique t.

Prediction statistics for each of the projects in the validation set are calculated, and later used to obtain accuracy statistics for an estimation technique. Thus, projects' prediction accuracy statistics are the basis for obtaining a technique's prediction accuracy statistics.

Technique's prediction accuracy statistics

The three measures of a technique's prediction accuracy that are widely used are the following:

- The mean magnitude of relative error (MMRE) [43]
- The median magnitude of relative error (MdMRE) [46]
- The prediction at level n (Pred(n)) [43]

MMRE, MdMRE, and Pred(*l*) are used so often in effort estimation studies that for some they are considered *de facto* standard evaluation criteria to measure the predictive accuracy (power) of effort estimation techniques [47]. Suggestions have been made that a good prediction model or technique should have MMRE and MdMRE no greater than 0.25, and that Pred(25) should not be smaller than 75% [43].

MMRE and MdMRE are the mean (arithmetic average) and median MRE for a set of projects, respectively.

MMRE is calculated as:

$$\text{MMRE} = \frac{1}{n}\sum_{i=1}^{i=n} \frac{|e_i - \hat{e}_i|}{e_i} \quad (20)$$

Since the mean is calculated by taking into account the value of every estimated and actual effort from the validation set employed, the result may give a distorted assessment of a technique's predictive power when there are extreme MREs.

Therefore, we also recommended using the MdMRE, which is the median MRE. Unlike the mean, the median always represents the middle value v, given a distribution of values, and guarantees that there is the same number of values below v as above v.

The third accuracy measure often used is the prediction at level l, also known as Pred(l). This measure computes, given a validation set, the percentage of effort estimates that are within $l\%$ of their actual values. The choice of l often used is $l=25$.

In addition to MMRE, MdMRE, and Pred(25), it is also suggested that boxplots of absolute residuals should be employed to assess a model's, or technique's prediction accuracy.

Boxplots (see Fig. 8) use the median, represented by the horizontal line in the middle of the box, as the central value for the distribution. The box's height is the interquartile range, three and contains 50% of the values. The vertical lines (whiskers), up or down from the edges, contain observations which are less than 1.5 times interquartile range. Outliers are taken as values greater than 1.5 times the height of the box. Values greater than 3 times a box's height are called extreme outliers [45].

When upper and lower tails are approximately equal and the median is in the center of the box, the distribution is symmetric. If the distribution is not symmetric the relative lengths of the tails and the position of the median in the box indicate the nature of the skewness.

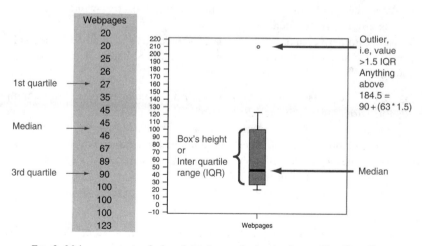

FIG. 8. Main components of a boxplot and example showing how to identify outliers.

If the distribution of data is positively skewed (skewed right distribution), it means that there is a long right tail containing a few high values widely spread apart, and a small left tail where values are compactly crowded together. This is the distribution presented in Fig. 8. Conversely, if the distribution of data is negatively skewed (skewed left distribution), it means that there is a long left tail containing a few low values widely spread apart, and a small right tail where values are compactly crowded together.

The length of the box relative to the length of the tails gives an indication of the shape of the distribution. Thus, a boxplot with a small box and long tails represents a very peaked distribution, meaning that 50% of the data points are all cluttered around the median. A boxplot with a long box represents a flatter distribution, where data points are scattered apart [45].

Cross-validation

Figure 7 showed that whenever we wish to measure a technique's prediction accuracy we should split the complete original dataset into two subsets, one to use as the set of finished projects, and another to simulate as "new projects." This is done because if we use the same set of projects to build a model and then use part of these same projects as "new" projects, our results will be biased in favor of the effort estimation technique.

The splitting of a dataset into training and validation sets is also known as cross-validation. An n-fold cross-validation means the original dataset is divided into n subsets of training and validation sets. Thus, a 10-fold cross-validation represents the splitting of the complete original dataset into 10 different training and validation sets. When the validation set has only one project the cross-validation is called "leave-one-out" cross-validation.

Thus, to summarize, in order to calculate the predictive accuracy of a given effort estimation technique t, based on a given dataset of finished projects d, we do the following:

(1) Divide the dataset d into a training set tr and a validation set v. It is common for a training set to include 66% of the projects in the original complete dataset, thus leaving the remaining 34% to be part of the validation set.

(2) Using tr, produce an effort estimation model m (if applicable). Even when a technique does not build a model, it still uses tr in order to obtain effort estimates for the projects in the validation set.

(3) Using m, or t, to predict the effort for each of the projects in v, simulating "new" projects for which effort is unknown.

Once done, we will have for each project in v, an *estimated* effort $\hat{e}$, calculated using the model m, or technique t, and also the *actual* effort e that the project actually used. We are now able to calculate MRE and absolute residual for each project in the validation set v. The final step, once we have obtained the predictive power for each project, is to aggregate these values to obtain MMRE, MdMRE, and Pred(25) for v, which is taken to be the same for m, or t. We can also use boxplots of absolute residuals to help understand the distribution of residuals values.

In principle, calculated MMREs and MdMREs with values up to 0.25, and Pred (25) at 75% or above, indicate good prediction models [43]. However, each situation needs to be interpreted within its own context, rather than to take a dogmatic approach.

3. Which Effort Estimation Technique to Use?

This chapter has introduced numerous techniques for obtaining effort estimates for a new project, where all have been used in practice, each with a varying degree of success. Therefore, the question that is often asked is: Which of the techniques abovementioned does provide the most accurate prediction for Web effort estimation?

To date, the answer to this question has been simply "it depends."

Algorithmic, CART and BN models have some advantages over CBR and expert opinion, such as:

- They allow users to see how a model derives its conclusions, an important factor for verification as well as theory building and understanding of the process being modeled [24, 25].
- They often need to be specialized relative to the local environment in which they are used. This means that the effort estimations obtained take full advantage of using models that have been calibrated to local circumstances.

Despite these advantages, no convergence for which effort estimation technique has the best predictive accuracy has yet been reached, even though comparative studies have been carried out for nearly 20 years (e.g., [3, 22, 24–27, 29, 30, 32, 33, 36, 37, 48–53]).

One justification is that these studies often use datasets with differing number of characteristics (e.g., number of outliers, amount of collinearity, number of variables, and number of projects) and different comparative designs. Note, an outlier is a value which is far from the others, and collinearity represents the existence of a linear relationship between two or more independent variables.

Shepperd and Kadoda [3] presented evidence showing there is a relationship between the success of a particular technique and factors such as training set size (size of the subset used to derive a model), nature of the "effort estimation" function (e.g., continuous or discontinuous), and characteristics of the dataset. They concluded that the "best" prediction technique that can work on any type of dataset may be impossible to obtain. Note, a continuous function is one in which "small changes in the input produce small changes in the output" (http://e.wikipedia.org/wiki/Continuous_function). If small changes in the input "can produce a broken jump in the changes of the output, the function is said to be discontinuous (or to have a discontinuity)" (http://e.wikipedia.org/wiki/Continuous_function).

4. Web Effort Estimation Literature Survey

This section presents a survey of Web effort estimation models proposed in the literature. Each work is described and finally summarized in Tables IV and V.

4.1 Previous Studies

First Study: Measurement and effort prediction for web applications [28].

Mendes et al. [28] investigated the use of CBR, linear regression, and SWR techniques to estimate development effort for Web applications developed by experienced and inexperienced students. The CBR estimations were generated using a freeware tool—ANGEL—developed at the University of Bournemouth, UK. The most similar Web projects were retrieved using the unweighted Euclidean distance using the "leave one out" cross-validation process. Estimated effort was generated using either the CA or the mean of two or three analogues. The two datasets (HEL and LEL) used, had collected data on Web applications developed by second-year Computer Science students from the University of Southampton, UK, and had 29 and 41 projects, respectively. HEL represented data from students with high experience in Web development, whereas LEL had data from students inexperienced in Web development.

The size measures collected were Page Count (total number of HTML pages created from scratch), Reused Page Count (total number of reused HTML pages), Connectivity (total number of links in the application), Compactness [54] (measured using an ordinal scale from 1 to 5 indicating the level of interconnectedness of an application. A value of 1 represents no connections and a value of 5 represents a totally connected application), Stratum [54], (measured using an ordinal scale from 1 to 5 indicating how "linear" the application is: 1 represents no sequential navigation and 5 represents a completely sequential navigation), and structure (represents the topology of the application's backbone, being either sequential,

TABLE IV
SUMMARY OF LITERATURE IN WEB EFFORT ESTIMATION

Study	Type (case study, experiment, survey)	# datasets (#datapoints)	Subjects (students, professionals)	Size measures	Prediction techniques	Best technique(s)	Measure(s) prediction accuracy
1st	Case study	2 (29 and 41)	2nd year Computer Science students	Page Count, reused Page Count, connectivity, compactness, stratum, structure	Case-based reasoning, linear regression, stepwise regression	Case-based reasoning for high experience group	MMRE
2nd	Case study	1 (46)	Professionals	Web objects	WEBMO (parameters generated using linear regression)	-	Pred(n)
3rd	Case study	1 (37)	Honours and postgraduate Computer Science students	Length size, reusability, complexity, size	Linear regression, stepwise regression	Linear regression	MMRE
4th	Case study	1 (37)	Honours and postgraduate Computer Science students	Structure metrics, complexity metrics, Reuse metrics, size metrics	Generalised linear model	-	Goodness of fit
5th	Case study	1 (25)	Honours and postgraduate Computer Science students	Requirements and Design measures, application measures	Case-based reasoning		MMRE, MdMRE, Pred(25), boxplots of residuals
6th	Case study	1 (37)	Honours and postgraduate Computer Science students	Page count, Media count, Program count, Reused media count, Reused program count, Connectivity density, total page complexity	Case-based reasoning, linear regression, stepwise regression, classification and regression trees	Linear/stepwise regression or case-based reasoning (depends on the measure of accuracy employed)	MMRE, MdMRE, Pred(25), boxplots of residuals

(continued)

TABLE IV (Continued)

Study	Type (case study, experiment, survey)	# datasets (#datapoints)	Subjects (students, professionals)	Size measures	Prediction techniques	Best technique(s)	Measure(s) prediction accuracy
7th	Case study	1 (12)	Professionals	Web objects	COBRA, expert opinion, linear regression	COBRA	MMRE, Pred(25), boxplots of residuals
8th	Case study	2 (37 and 25)	Honours and post-graduate CS students	Page count, media count, program count, reused media count (only one dataset) reused program count (only one dataset), connectivity density, total page complexity	Case-based reasoning	-	MMRE, Pred(25), boxplots of absolute residuals
9th	Experiment	1 (30)	Computer Science students	Information model measures, navigation model measures, presentation model measures	Linear regression	-	-
10th	Unknown	Unknown	Unknown	Functional, navigational structures, publishing and multimedia sizing measures	An exponential model named Metrics Model for Web applications (MMWA)	N/A	Unknown
11th	Case study	1 (15)	Professionals	Web pages, new Web pages, multimedia elements, new multimedia elements, client side Scripts and applications, server side scripts and applications, all the elements that are part of the Web objects size measure	Linear regression, stepwise regression, case-based reasoning, classification and regression trees, combination of CART and CBR	-	MMRE, MdMRE, Pred(25), boxplots of residuals, boxplots of z

12th	Case study	1 (150)	Professionals	Total Web pages, new Web pages, total images, new images, features off-the-shelf (Fots), high and low effort Fots-adapted, high and low effort new features, total high and low effort features	Bayesian networks, stepwise regression, mean and median effort, case-based reasoning classification and regression trees	Bayesian networks	MMRE, MdMRE, MEMRE, MdEMRE, Pred (25), boxplots of residuals, boxplots of z
13th	Case study	1 (195)	Professionals	Total Web pages, new web pages, total images, new images, features off-the-shelf (Fots), high and low effort Fots-adapted, high and low effort new features, total high and low effort features	Bayesian networks, stepwise regression, mean and median effort	Stepwise regression	MMRE, MdMRE, MEMRE, MdEMRE, Pred (25), boxplots of residuals, boxplots of z
14th	Case study	1 (195)	Professionals	Total Web pages, new Web pages, total images, new images, features off-the-shelf (Fots), high and low effort Fots-adapted, high and low effort new features, total high and low effort features	Bayesian networks, stepwise regression, mean and median effort	Stepwise regression	MMRE, MdMRE, MEMRE, MdEMRE, Pred (25), boxplots of residuals, boxplots of z
15th	Case study	Tacit knowledge	Professionals	Total Web pages, features off-the-shelf (Fots), high and low effort Fots-adapted	Bayesian networks	N/A	Number of times the BN selected the effort category that included actual effort
16th	Case study	1 (195)	Professionals	Total Web pages, new Web pages, total images, new images, features off-the-shelf (Fots), high and low effort Fots-adapted, high and low effort new features, total high and low effort features	Support vector regression, Bayesian networks, manual stepwise regression, case-based reasoning, classification and regression trees, mean effort, median effort	SRV (1st hold-out); MSWR (2nd hold-out)	MMRE, MdMRE, MEMRE, MdEMRE, Pred (25), boxplots of residuals

TABLE V
TYPES OF WEB APPLICATIONS USED IN WEB EFFORT ESTIMATION STUDIES

Study	Type of Web application: Web hypermedia, Web software application, or Web application
1st	Web hypermedia applications
2nd	Not documented
3rd	Web hypermedia applications
4th	Web hypermedia applications
5th	Web hypermedia applications
6th	Web hypermedia applications
7th	Web software applications
8th	Web software applications
9th	Web hypermedia applications
10th	Not documented
11th	Web software applications
12th	Web software applications
13th	Web software applications
14th	Web software applications
15th	Web software applications
16th	Web software applications

hierarchical, or network). Prediction accuracy was measured using the MMRE [43] and the MdMRE [43]. Results for the HEL group were statistically significantly better than those for the LEL group. In addition, CBR showed the best results overall.

Second Study: Web development: Estimating quick-to-market software [12]

Reifer [12] proposed a Web effort estimation model—WEBMO—which is an extension of the COCOMO II model. The WEBMO model has nine cost drivers and a fixed effort power law, instead of seven cost drivers and variable effort power law as used in the COCOMO II model. Size is measured in WebObjects, which is a single aggregated size measure calculated by applying Halstead's formula for volume.

The elements of the WebObjects measure are:

- Number of building blocks (Active X, DCOM, OLE, etc.),
- Number of components off-the-shelf (COTS) (includes any wrapper code),
- Number of multimedia files, except graphics files (text, video, sound, etc.),
- Number of object or application points [55] or others proposed (e.g., number of server data tables, number of client data tables),
- Number of xml, sgml, html and query language lines (number of lines including links to data attributes),

- Number of Web components (applets, agents, etc.),
- Number of graphics files (templates, images, pictures, etc.),
- Number of scripts (visual language, audio, motion, etc.) and any other measures that companies find suitable.

Reifer allegedly used data on 46 finished industrial Web projects and obtained predictions that are reported as being "repeatable and robust." However, no information is given regarding the data collection and no summary statistics for the data are presented.

Third Study: Web metrics—Estimating design and authoring effort [5]

Mendes et al. [5] investigated the prediction accuracy of top-down and bottom-up Web effort estimation models, generated using Linear and SWR models. They employed one dataset with data from 37 Web applications developed by Honours and postgraduate Computer Science students from the University of Auckland, New Zealand. Gathered measures were organized into five categories:

- Length size measures.
- Reusability measures.
- Complexity size measures.
- Effort.
- Confounding factors (factors that, if not controlled, could influence the validity of the evaluation).

In addition, measures were also associated with one of the following entities: Web application, Web page, Media, and Program. Effort prediction models were generated for each entity and prediction accuracy was measured using the MMRE measure. Results showed that the best predictions were obtained for the entity Program, based on nonreused program measures (code length and code comment length).

Fourth Study: Measurement, prediction and risk analysis for Web applications [56]

Fewster and Mendes [56] investigated the use of a generalized linear model (GLM) for Web effort estimation and risk management. GLMs provide a flexible regression framework for predictive modeling of effort since they permit nonlinear relationships between the response and predictor variables and in addition allow for a wide range of choices for the distribution of the response variable (e.g., effort).

Fewster and Mendes [56] employed the same dataset used in Mendes et al. [5], however, they reduced the number of size measures targeting only those measures related to the entity type Web application. The measures used were organized into five categories:

- Effort measures.
- Structure measures.

- Complexity measures.
- Reuse measures.
- Size measures.

Finally, they did not use defactor measures of prediction accuracy (e.g., MMRE, MdMRE) to assess the accuracy of their proposed model; instead, they used the Model fit produced for the model as an accuracy measure. However, it should be noted that a model with a good fit to the data is not the same as a good prediction model.

Fifth Study: The application of CBR to early Web project cost estimation [37]

Mendes et al. [37] focused on the harvesting of size measures at different points in a Web application's development life cycle, to estimate development effort, and their comparison based on several prediction accuracy indicators. The rationale for using different measures harvested at different points was as follows:

Most work on Web effort estimation propose models based on late product size measures, such as number of HTML pages, number of images, etc. However, for the successful management of software/Web projects, estimates are necessary throughout the whole development life cycle. Preliminary (early) effort estimates in particular are essential when bidding for a contract or when determining a project's feasibility in terms of cost-benefit analysis.

Mendes et al.'s aim was to investigate if there were any differences in accuracy between the effort predictors gathered at different points during the development life cycle. In addition, they also checked if these differences were statistically dissimilar. Their effort estimates were generated using the CBR technique, where different combinations of parameters were used: Similarity measure; Scaling; Number of closest analogues; Analogy adaptation; and Feature Subset Selection. Their study was based on data from 25 Web applications developed by pairs of postgraduate Computer Science students from the University of Auckland, New Zealand. The measures of prediction accuracy employed were the MMRE, MdMRE, Prediction at 25% (Pred(25)), and Boxplots of residuals. Contrary to the expected, their results showed that late measures presented similar estimation accuracy to early measures.

Sixth Study: A comparison of development effort estimation techniques for Web hypermedia applications [31]

Mendes et al. [31] present an in-depth comparison of Web effort estimation models, where they:

(i) Compare the prediction accuracy of three CBR techniques to estimate the effort to develop Web applications.
(ii) Compare the prediction accuracy of the best CBR technique verses three commonly used prediction models, specifically, multiple linear regression, SWR, and regression trees.

Mendes et al. employed one dataset of 37 Web applications developed by honors and postgraduate Computer Science students from the University of Auckland, New Zealand. The measures used in their study were as follows:

- Page count (number of html or shtml files used in the Web application).
- Media count (number of media files used in the Web application).
- Program count (number of JavaScript files and Java applets used in the Web application).
- Reused media count (number of reused/modified media files).
- Reused program count (number of reused/modified programs).
- Connectivity density (total number of internal links divided by page count).
- Total page complexity (average number of different types of media per Web page).
- Total effort (effort in person hours to design and author a Web application).

Note that Subjects did not use external links to other Web hypermedia applications, that is, all the links pointed to pages within the original application only. Regarding the use of CBR, they employed several parameters, as follows:

- Three similarity measures (unweighted Euclidean, weighted Euclidean, and Maximum).
- Three choices for the number of analogies (1, 2, and 3).
- Three choices for the analogy adaptation (mean, inverse rank weighted mean, and median).
- Two alternatives regarding the standardization of the attributes ("Yes" for standardized and "No" for not standardized).

Prediction accuracy was measured using MMRE, MdMRE, Pred(25), and box-plots of residuals. Their results showed that different measures of prediction accuracy gave different results. MMRE and MdMRE showed better prediction accuracy for MR models whereas boxplots showed better accuracy for CBR.

Seventh Study: Cost estimation for web applications [36]

Ruhe et al.'s [36] employed the COBRA^{TM} (cost estimation benchmarking and risk analysis) method to investigate if this method was adequate for estimating effort of Web applications. They used real project data on 12 Web projects, all developed by a small Web company in Sydney, Australia. COBRA is a registered trademark of the Fraunhofer Institute for Experimental Software Engineering (IESE), Germany, and is a method that aims to develop an effort estimation model that is to be built from on a company-specific dataset. It uses expert opinion and data on past projects to estimate development effort and risks for a new project. The size measure employed was

WebObjects [12], measured for each one of the 12 finished Web applications used in this study. The prediction accuracy obtained using COBRA™ was compared to that obtained using expert opinion and linear regression, all measured using MMRE and Pred(25). They found that COBRA provided the most accurate results.

Eighth Study: Do adaptation rules improve web cost estimation? [34]

Mendes et al. [34] compared several methods of CBR-based effort estimation, investigating the use of adaptation rules as a contributing factor for better estimation accuracy. They used two datasets, where the difference between these datasets was a level of "messiness" each had. "Messiness" was evaluated by the number of outliers and the amount of collinearity [3] that each dataset presented. The dataset considered less "messy" than the other, presented a continuous "cost" function, which also translated out into a strong linear relationship between size and effort. However, the "messiest" dataset presented a discontinuous "cost" function, where there was no linear or log–linear relationship between size and effort.

Both datasets represented data on Web applications developed by Computer Science students from the University of Auckland, New Zealand and two types of adaptation were used; one with weights and another without weights [34]. None of the adaptation rules gave better predictions for the "messier" dataset; however, for the less "messy" dataset one type of adaptation rule (no weights) gave good prediction accuracy, measured using MMRE, Pred(25), and Boxplots of absolute residuals.

Ninth Study: Estimating the design effort of web applications [57]

Baresi et al. [57] investigated the relationship between a number of size measures, obtained from design artifacts that were created according to the W2000 methodology, and the total effort needed to design Web applications. Their size measures were organized into categories which are presented in more detail in Table V. The categories employed were information model, navigation model, and presentation model. Their analysis identified some measures that appear to be related to the total design effort. In addition, they also carried out a finer grained analysis, studying which of the used measures had an impact on the design effort when using W2000. Their dataset was comprised 30 Web applications developed by Computer Science students from Politecnico di Milano, Italy.

Tenth Study: MMWA: A software sizing model for web applications [58]

Mangia and Paiano [58] investigated an effort estimation model and size measures for Web projects based on a specific Web development method, namely the W2000. Their effort estimation model was named metrics model for Web applications (MMWA). This model comprised four complementary submodels, each related to a specific aspect of a Web application, as follows:

- Functional sizing model—aimed to size the functionality of a Web application.
- Navigational structures sizing model—aimed to size a Web application taking into account the complexity of an application's navigational structure.

- Publishing sizing model—aimed to estimate the necessary effort to design, implement, and maintain the application's content.
- Multimedia sizing model—aimed to estimate the effort needed to create the multimedia components of an application.

The chapter does not detail any empirical validation of this model.

Eleventh Study: Effort estimation modelling techniques: A case study for Web applications [59]

Costagliola et al. [59] investigated the use of different size measures and effort estimation techniques for Web effort estimation. They compare these size measures and estimation techniques using data from 15 industrial Web projects developed by a single company. The two sets of size measures employed were: length measures (e.g., number of pages, number of different media, number of client, and server side scripts) and functional measures (e.g., external input, external output, external queries). The estimation techniques employed were linear regression, SWR, regression trees, CBR, and a combination of regression trees and CBR.

Twelfth Study: Predicting web development effort using a BN/the use of a BN for web effort estimation/a comparison of techniques for Web effort estimation [20–22]

These three studies were based on the same dataset and effort models; hence being considered together herein.

The Web effort model used in those three studies was built using a BN. Data on 150 Web projects from the Tukutuku dataset [6] were used for structure and probability learning, and later the BN's structure was also validated by a domain expert who was an experienced Web project manager. This BN model had its prediction accuracy compared with the accuracy of estimates obtained using SWR, mean- and median-based effort models, CBR, and a CART.

Thirteenth Study: The use of BNs for Web effort estimation: Further investigation [23]

Mendes [23] investigated further the use of BN for Web effort estimation when building the BN model using a cross-company dataset. Four BNs were built; two automatically using the Hugin tool with two training sets; two using a structure elicited by a domain expert, with parameters obtained from automatically fitting the network to the same training sets used in the automated elicitation (hybrid models). The accuracy of all four models was measured using two validation sets, and point estimates. As a benchmark, the BN-based predictions were also compared to predictions obtained using manual stepwise regression (MSWR), and CBR.

The BN model generated using Hugin presented similar accuracy to CBR and mean effort-based predictions; however, overall MSWR presented the best results.

Fourteenth Study: BN models for web effort prediction: A comparative study [53]

Mendes and Mosley [53] compared, using a cross-company dataset, several BN models for Web effort estimation. Eight BNs were built; four automatically using Hugin and PowerSoft tools with two training sets, each containing data on 130 Web projects from the Tukutuku database; four using a causal graph elicited by a domain expert, with parameters obtained by automatically fitting the graph to the same training sets used in the automated elicitation (hybrid models). The accuracy of all eight models was measured using two validation sets, each containing data on 65 projects, and point estimates. As a benchmark, the BN-based estimates were also compared to estimates obtained using MSWR, CBR, mean- and median-based effort models.

MSWR presented significantly better predictions than any of the BN models built herein, and in addition was the only technique to provide significantly superior predictions to a median-based effort model. Two BN models—BNAuHu and BNHyHu, presented similar to, or significantly better accuracy than the mean-based effort model, and similar accuracy to the median-based effort model; however, both showed significantly worse accuracy than MSWR. The other two BN models showed worse accuracy than at least mean-based predictions.

Fifteenth Study: Building an expert-based Web effort estimation model using BNs [11]

Mendes et al. [11] described a case study where BNs were used to construct an expert-based Web effort model. They built a single-company BN model solely elicited from expert knowledge, where the domain expert was an experienced Web project manager from a small Web company in Auckland, New Zealand. This model was validated using data from eight past finished Web projects.

This BN model has, to date, been successfully used to estimate effort for several new Web projects, providing effort estimates superior to those based solely on expert opinion, thus suggesting that, at least for the Web Company that participated in this case study, the use of a model that allows the representation of uncertainty, inherent in effort estimation, can outperform expert-based estimates.

Sixteenth Study: Applying support vector regression for Web effort estimation using a cross-company dataset [49]

Corazza et al. [49] investigated the use of a machine learning technique called support vector regression (SVR) for Web effort estimation, using as data the Tukutuku database. To gain a deeper insight on the SVR method, they used: (i) four kernels for SVR, namely linear, polynomial, Gaussian, and sigmoid; (ii) two variables' preprocessing strategies (normalization and logarithmic); (iii) two dependent variables (effort and inverse effort). As a result, SVR was applied using six different configurations for each kernel.

A hold-out approach was adopted to evaluate the prediction accuracy for all the configurations, using two training sets, each containing data on 130 projects randomly selected, and two test sets, each containing the remaining 65 projects. As a benchmark, the SVR-based predictions were also compared to predictions obtained using MSWR, CBR, and BNs.

4.2 Discussion

Table IV summarizes the studies that were presented in the survey literature, and provides us with means to identify a number of trends such as:

- The prediction technique used the most was linear regression.
- The measures of prediction accuracy employed the most were MMRE and Pred (25). However, we can also observe a recent change in trends toward the use of a wider range of prediction accuracy measures, in particular the MMRE relative to the estimate (MEMRE) and the median magnitude of relative error relative to the estimate (MdEMRE) [45]. This is seen as a positive change given that a wider range of measures provides the means for a more detailed interpretation of the results.
- The sizes of the datasets employed varied from 12 to 195. The smaller datasets contained data from a single company, or a single set of students; whereas the larger datasets (in particular with size = 150 or size = 195) contained data volunteered by numerous Web companies.
- Numerous studies employed data from the Tukutuku dataset [6], given this is, as far as we know, the only dataset to date that records data on Web projects volunteered by Web companies worldwide.
- The initial trend toward using project data from student-based Web applications has changed toward the use of project data from real industrial Web projects. This is in our view a positive change given that data on real industrial projects is more adequate, as far as the external validity of the results is concerned, than employing student-based data.
- Size measures differed throughout studies, which indicate the lack of standards to sizing Web applications.

Using the survey literature previously described, we also investigated the type of Web applications that were used in those studies. Our rationale was that we believed earlier studies would have more static Web applications (Web hypermedia applications), and that more recent studies would show a use of more dynamic Web applications (Web software applications, or Web applications).

The classification of Web applications into Web hypermedia applications, Web software applications, and Web applications was proposed by Christodoulou et al. [60]. They define a Web hypermedia application as a nonconventional application characterized by the authoring of information using nodes (chunks of information), links (relations between nodes), anchors, access structures (for navigation), and its delivery over the Web. Technologies commonly used for developing such applications are HTML, JavaScript, and multimedia. In addition, typical developers are writers, artists, and organizations that wish to publish information on the Web and/or CD-ROMs without the need to use programing languages such as Java.

Conversely, Web software applications represent conventional software applications that depend on the Web or use the Web's infrastructure for execution. Typical applications include legacy information systems such as databases, booking systems, knowledge bases, etc. Many e-commerce applications fall into this category. Typically they employ technology such as COTS, components such as DCOM, OLE, ActiveX, XML, PHP, dynamic HTML, databases, and development solutions such as J2EE. Developers are young programers fresh from a Computer Science or Software Engineering degree, managed by more senior staff.

Table V lists the type of application used in each of the studies described in the literature review, and shows that, out of the 16 studies described in this section, 6 (37.5%) have used datasets of Web hypermedia applications, followed by another 8 studies (50%) that used datasets of Web software applications. Therefore, as we expected, studies show a trend toward the use of Web software applications.

5. Conclusions

Effort estimation enables companies to know beforehand and before implementing an application, the amount of effort required to develop this application on time and within budget. To estimate effort, it is generally necessary to have knowledge of previous similar projects that have already been developed by the company, and also to understand the project variables that may affect effort prediction.

These variables represent an application's size (e.g., *number of New web pages and Images, the number of functions/features* (e.g., shopping cart) *to be offered by the new Web application*) and also include other factors that may contribute to effort (e.g., *total number of developers who will help develop the new Web application, developers' average number of years of experience with the development tools employed, main programming language used*).

The mechanisms used to obtain an effort estimate are generally classified as:

Expert-based estimation: Expert-based effort estimation represents the process of estimating effort by subjective means, and is often based on previous experience with developing and/or managing similar projects. This is by far the mostly used technique for Web effort estimation.

Algorithmic-based estimation: Algorithmic-based effort estimation attempts to build models (equations) that precisely represent the relationship between effort and one or more project characteristics via the use of algorithmic models (statistical methods that are used to build equations). These techniques have been, to date, the most popular techniques used in the Web and software effort estimation literature.

Estimation using AI techniques: Finally, AI techniques also aim to obtain effort estimates although not necessarily using a model, such as the ones created with algorithmic-based techniques. AI techniques include fuzzy logic [16], regression trees [17], neural networks [18], CBR [3], BNs [20], and SVR [49].

This chapter also presented a survey of previous work in Web effort estimation, and the main findings were as follows:

- The most widely used prediction technique is linear regression.
- The measures of prediction accuracy employed the most were MMRE and Pred (25).
- The datasets used in the studies varied in size, ranging from 12 to 195 projects.
- Size measures differed throughout studies, which indicate the lack of standards to sizing Web applications.
- Out of the 16 papers, 6 (37.5) have used datasets of Web hypermedia applications, and another 8 (50%) have used datasets of Web software applications.

REFERENCES

[1] E. Mendes, Cost Estimation Techniques for Web Projects, IGI Global Publishers, Hershey, Pennsylvania, 2007, 424pp.
[2] B.A. Kitchenham, L.M. Pickard, S. Linkman, P. Jones, Modelling software bidding risks, IEEE Trans. Soft. Eng. 29 (6) (2003) 542–554.
[3] M.J. Shepperd, G. Kadoda, Using simulation to evaluate prediction techniques, in: Proceedings of the IEEE 7th International Software Metrics Symposium, London, UK, 2001, pp. 349–358.
[4] C. Schofield, An empirical investigation into software estimation by analogy, Unpublished Doctoral Dissertation, Department of Computing, Bournemouth University, 1998.
[5] E. Mendes, N. Mosley, S. Counsell, Web measures—Estimating design and authoring effort, IEEE Multimed. [Special Issue on Web Engineering] 8 (1) (2001) 50–57.
[6] E. Mendes, N. Mosley, S. Counsell, Investigating web size metrics for early web cost estimation, J. Sys. Soft. 77 (2) (2005) 157–172.
[7] M. Jørgensen, D. Sjøberg, Impact of effort estimates on software project work, Inf. Soft. Technol. 43 (2001) 939–948.
[8] T. DeMarco, Controlling Software Projects: Management, Measurement and Estimation, Yourdon Press, New York, 1982.
[9] H.V. Vliet, Software Engineering: Principles and Practice, second ed., Wiley, New York, 2000.
[10] R. Gray, S.G. MacDonell, M.J. Shepperd, Factors systematically associated with errors in subjective estimates of software development effort: The stability of expert judgement, in: Proceedings of the 6th IEEE Metrics Symposium, 1999, pp. 216–226.

[11] E. Mendes, C. Pollino, N. Mosley, Building an expert-based Web effort estimation model using Bayesian networks, in: Proceedings of the EASE Conference, 2009, pp. 1–10.
[12] D.J. Reifer, Web development: Estimating quick-to-market software, IEEE Soft. 17 (6) (2000) 57–64.
[13] B. Boehm, Software Engineering Economics, Prentice-Hall, Englewood Cliffs, 1981.
[14] B. Boehm, COCOMO II, 2000, http://sunset.usc.edu/research/COCOMOII/Docs/modelman.pdf, (accessed January 2006, from The University of Southern California).
[15] B. Bohem, C. Abts, A. Brown, S. Chulani, B. Clark, E. Horowitz, R. Madachy, D. Reifer, B. Steece, Software Cost Estimation with Cocomo II, Pearson Publishers, Upper Saddle River, NJ, 2000.
[16] S. Kumar, B.A. Krishna, P.S. Satsangi, Fuzzy systems and neural networks in software engineering project management, J. Appl. Intell. 4 (1994) 31–52.
[17] L. Schroeder, D. Sjoquist, P. Stephan, Understanding Regression Analysis: An Introductory Guide, No. 57, Sage, Newbury Park, 1986.
[18] M.J. Shepperd, C. Schofield, B. Kitchenham, Effort estimation using analogy, in: Proceedings of ICSE-18, Berlin, 1996, pp. 170–178.
[19] N. Fenton, W. Marsh, M. Neil, P. Cates, S. Forey, M. Tailor, Making resource decisions for software projects, in: Proceedings of the ICSE'04, 2004, pp. 397–406.
[20] E. Mendes, E. Mendes, Predicting Web development effort using a Bayesian network, in: Proceedings of EASE'07, 2007a, pp. 83–93.
[21] E. Mendes, The use of a Bayesian network for Web effort estimation, in: Proceedings of International Conference on Web Engineering, LNCS 4607, 2007b, pp. 90–104.
[22] E. Mendes, A comparison of techniques for Web effort estimation, in: Proceedings of the ACM/IEEE International Symposium on Empirical Software Engineering, 2007c, pp. 334–343.
[23] E. Mendes, The use of Bayesian networks for Web effort estimation: Further investigation, in: Proceedings of ICWE'08, 2008, pp. 2-3–216.
[24] A. Gray, S. MacDonell, Applications of fuzzy logic to software metric models for development effort estimation, in: Proceedings of IEEE Annual Meeting of the North American Fuzzy Information Processing Society—NAFIPS, Syracuse, NY, USA, 1997, pp. 394–399.
[25] A.R. Gray, S.G. MacDonell, A comparison of model building techniques to develop predictive equations for software metrics, Inf. Soft. Technol. 39 (1997) 425–437.
[26] L. Angelis, I. Stamelos, A simulation tool for efficient analogy based cost estimation, Empir. Soft. Eng. 5 (2000) 35–68.
[27] R.W. Selby, A.A. Porter, Learning from examples: Generation and evaluation of decision trees for software resource analysis, IEEE Trans. Soft. Eng. 14 (1998) 1743–1757.
[28] E. Mendes, S. Counsell, N. Mosley, Measurement and effort prediction of Web applications, in: Proceedings of 2nd ICSE Workshop on Web Engineering, June, Limerick, Ireland, 2000, pp. 57–74.
[29] G. Kadoda, M. Cartwright, L. Chen, M.J. Shepperd, Experiences using case-based reasoning to predict software project effort, in: Proceedings of the EASE 2000 Conference, Keele, UK, 2000.
[30] L.C. Briand, K. El-Emam, D. Surmann, I. Wieczorek, K.D. Maxwell, An assessment and comparison of common cost estimation modeling techniques, in: Proceedings of ICSE 1999, Los Angeles, USA, 1999, pp. 313–322.
[31] E. Mendes, I. Watson, C. Triggs, N. Mosley, S. Counsell, A comparison of development effort estimation techniques for Web hypermedia applications, in: Proceedings IEEE Metrics Symposium, June, Ottawa, Canada, 2002, pp. 141–151.

AN OVERVIEW OF WEB EFFORT ESTIMATION 269

[32] R. Jeffery, M. Ruhe, I. Wieczorek, A comparative study of two software development cost modelling techniques using multi-organizational and company-specific data, Inf. Soft. Technol. 42 (2000) 1009–1016.
[33] R. Jeffery, M. Ruhe, I. Wieczorek, Using public domain metrics to estimate software development effort, in: Proceedings of the 7th IEEE Metrics Symposium, London, 2001, pp. 16–27.
[34] E. Mendes, N. Mosley, S. Counsell, Do adaptation rules improve Web cost estimation? in: Proceedings of the ACM Hypertext conference 2003, Nottingham, UK, 2003a, pp. 173–183.
[35] E. Mendes, N. Mosley, S. Counsell, A replicated assessment of the use of adaptation rules to improve Web cost estimation, in: Proceedings of the ACM and IEEE International Symposium on Empirical Software Engineering, Rome, Italy, 2003b, pp. 100–109.
[36] M. Ruhe, R. Jeffery, I. Wieczorek, Cost estimation for Web applications, in: Proceedings of ICSE 2003, Portland, USA, 2003, pp. 285–294.
[37] E. Mendes, N. Mosley, S. Counsell, The application of case-based reasoning to early Web project cost estimation, in: Proceedings of IEEE COMPSAC, 2002, pp. 393–398.
[38] L. Brieman, J. Friedman, R. Olshen, C. Stone, Classification and Regression Trees, Wadsworth, Belmont, 1984.
[39] J. Pearl, Probabilistic Reasoning in Intelligent Systems, Morgan Kaufmann, San Mateo, CA, 1988.
[40] I. Stamelos, L. Angelis, P. Dimou, E. Sakellaris, On the use of Bayesian belief networks for the prediction of software productivity, Info. Soft. Technol. 45 (1) (2003) 51–60.
[41] O. Woodberry, A. Nicholson, K. Korb, C. Pollino, Parameterising Bayesian networks, in: Proceedings of the Australian Conference on Artificial Intelligence, 2004, pp. 1101–1107.
[42] F.V. Jensen, An Introduction to Bayesian Networks, UCL Press, London, 1996.
[43] S. Conte, H. Dunsmore, V. Shen, Software Engineering Metrics and Models, Benjamin/Cummings, Menlo Park, 1986.
[44] M.J. Shepperd, C. Schofield, Estimating software project effort using analogies, IEEE Trans. Soft. Eng. 23 (11) (1997) 736–743.
[45] B.A. Kitchenham, L.M. Pickard, S.G. MacDonell, M.J. Shepperd, What accuracy statistics really measure, IEE Proc. Soft. Eng. 148 (3) (2001) 81–85.
[46] I. Myrtveit, E. Stensrud, A Controlled Experiment to Assess the Benefits of Estimating with Analogy and Regression Models, IEEE Trans. Software Eng. 25 (4) (1999) 510–525.
[47] E. Stensrud, T. Foss, B. Kitchenham, I. Myrtveit, An Empirical Validation of the Relationship Between the Magnitude of Relative Error and Project Size, in: IEEE METRICS 2002, Ottawa Canada, 2002, pp. 3–12.
[48] L.C. Briand, T. Langley, I. Wieczorek, A replicated assessment and comparison of common software cost modeling techniques, in: Proceedings of ICSE 2000, Limerick, Ireland, 2000, pp. 377–386.
[49] A. Corazza, S. Di Martino, F. Ferrucci, C. Gravino, E. Mendes, Applying support vector regression for Web effort estimation using a cross-company dataset, in: Proceedings of the ACM/IEEE Symposium on Empirical Software Measurement and Metrics, accepted for publication.
[50] G.R. Finnie, G.E. Wittig, J.-M. Desharnais, A comparison of software effort estimation techniques: Using function points with neural networks, case-based reasoning and regression models, J. Syst. and Soft. 39 (1997) 281–289.
[51] R.T. Hughes, An Empirical investigation into the estimation of software development effort, Unpublished Doctoral Dissertation, Department of Computing, University of Brighton, 1997.
[52] C.F. Kemerer, An empirical validation of software cost estimation models, Commun. ACM 30 (5) (1987) 416–429.

[53] E. Mendes, N. Mosley, Bayesian network models for web effort prediction: A comparative study, Trans. Soft. Eng. 34 (6) (2008) 723–737.
[54] R. Botafogo, A.E. Rivlin, B. Shneiderman, Structural analysis of hypertexts: Identifying hierarchies and useful measures, ACM Trans. Inf. Syst. 10 (2) (1992) 143–179.
[55] A.J.C. Cowderoy, Measures of size and complexity for web-site content, in: Proceedings of the 11th ESCOM Conference, Munich, Germany, 2000, pp. 423–431.
[56] R.M. Fewster, E. Mendes, Measurement, prediction and risk analysis for Web applications, in: Proceedings of the IEEE METRICS Symposium, 2001, pp. 338–348.
[57] L. Baresi, S. Morasca, P. Paolini, Estimating the design effort of Web applications, in: Proceedings Ninth International Software Measures Symposium, September 3–5, 2003, pp. 62–72.
[58] L. Mangia, R. Paiano, MMWA, A software sizing model for Web applications, in: Proceedings of the Fourth International Conference on Web Information Systems Engineering, 2003, pp. 53–63.
[59] G. Costagliola, S. Di Martino, F. Ferrucci, C. Gravino, G. Tortora, G. Vitiello, Effort estimation modeling techniques: a case study for web applications, in: ICWE 2006, Palo Alto, California, 2006, pp. 9–12.
[60] S.P. Christodoulou, P.A. Zafiris, T.S. Papatheodorou, WWW2000, The developer's view and a practitioner's approach to Web engineering, in: Proceedings of the 2nd ICSE Workshop Web Engineering, 2000, pp. 75–92.

Communication Media Selection for Remote Interaction of *Ad Hoc* Groups

FABIO CALEFATO AND FILIPPO LANUBILE

Dipartimento di Informatica, Collaborative Development Group, University of Bari, Italy

Abstract

Nowadays work is becoming predominantly distributed, bringing significant challenges to effective communication of geographically dispersed groups. In fact, multisite work presents considerable loss of opportunities for rich interaction and a very substantial reduction in frequency of both formal and informal communication between coworkers. While communicating face-to-face (F2F) by speech is easy for individuals, conducting a long-running, productive conversation through the digital medium is difficult, especially as the group size increases. The difficulty of computer-mediated communication (CMC) and collaboration stands in stark contrast to our natural ability to easily communicate and collaborate with one another in the physical world. As such, there is a need to further our understanding of the effectiveness of the many available synchronous and asynchronous communication media (e.g., e-mail, videoconferencing, or specialized collaboration tools) to support activities of distributed teams. However, not only media properties (e.g., synchronicity) affect the performance of groups collaborating from a distance but also the characteristics of groups (e.g., size, history) and tasks (e.g., idea generation, decision making) play a key role. In this chapter, we first present a survey on the group-, task-, and media-related theories that are relevant for the selection of the most appropriate synchronous communication media to better support distributed *ad hoc* groups, that is, short-term groups with neither a history of previous collaborations nor expectation of future ones. Then, we consistently combine all the reviewed theories to create two general models that, respectively, can help researchers to manage the context of experiments on remote group collaboration, and distributed groups themselves to evaluate, compare, and select the most appropriate fits between the task at a hand and the media available.

1. Introduction 272
2. Task-Classification Frameworks 277
   2.1. Task Circumplex 278
   2.2. Complex Tasks Typology 280
3. Group Research 281
   3.1. Teams with No Past and Future: *Ad Hoc* Groups 281
   3.2. Challenges and Needs in Supporting Remote *Ad Hoc* Groups 284
4. CMC Theories 285
   4.1. Social Presence Theory 287
   4.2. Media Richness Theory 288
   4.3. Common Ground Theory 290
   4.4. Media Synchronicity Theory 293
   4.5. Cognitive-Based View 297
5. Development of a Comprehensive Theoretical Framework 300
   5.1. Managing the Context: The Effects of Task, Media, and Group Factors . . . 300
   5.2. Matching Task and Media Characteristics 302
   5.3. Time-Interaction-Performance Theory 302
   5.4. Task/Technology Fit Theory 303
   5.5. Matching Group and Media Characteristics 304
   5.6. Development of a Comprehensive Theoretical Framework 305
6. Conclusions 308
   Acknowledgments 309
   References 309

1. Introduction

Nowadays, no one works completely independently. Almost everyone is part of at least one group, typically several groups at any point in time. Figure 1 shows a typical cooperative work framework [1]. Groups of two or more participants (P) communicate together, share information, generate and organize ideas, build consensus, make decisions, and so on. Being engaged in some common work, participants interact with tools and products (i.e., artifacts of work, A). The main purpose of communication is to establish a common *understanding* of the work shared between participants. The development of the understanding happens both indirectly and directly. The arrows that link participants to the artifacts denote *indirect communication*. It happens

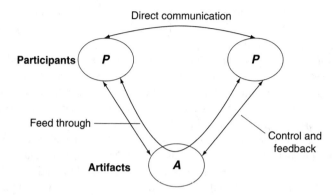

Fig. 1. Cooperative work framework: Communication as the basis of collaboration (adapted from Ref. [1]).

through the manipulation of shared tools and work objects (e.g., a document, a piece of code). *Feedback* represents the information gained by the participant who directly controls a shared artifact. Further, the changes applied to an artifact convey information also to the other participants (*feed through*). *Direct communication* is denoted by the arrow between the participants and happens by speech or over communication media, such as telephone, fax, and e-mail.

While communicating face-to-face (F2F) by speech is easy for individuals, conducting a long-running, productive conversation through the digital medium is difficult, especially as the group size increases. The difficulty of computer-mediated communication (CMC) and collaboration stands in stark contrast to our natural ability to easily communicate and collaborate with one another in the physical world [2]. As such, there is a need to further our understanding of the effectiveness of the many available synchronous and asynchronous communication media (e.g., e-mail, videoconferencing, or specialized collaboration tools) to support activities in distributed teams. However, not only media properties, such as synchronicity, affect the performance of groups collaborating from a distance. In fact, also *who does what* matters, that is, the characteristics of groups (e.g., size, history) and tasks (e.g., idea generation, decision making) play an important role.

Mainly because of economic factors, nowadays work is becoming predominantly distributed, bringing significant challenges to effective communication of geographically dispersed groups. In fact, multisite work presents considerable loss of opportunities for rich interaction and a very substantial reduction in frequency of both formal and informal communication between coworkers [3]. Following the trend to business globalization, also software development has increasingly become distributed, with little or no possibility for developers to meet. Among the software development activities, requirements engineering is one of the most communication-intensive and then, its effectiveness is greatly constrained by the geographical distance between stakeholders [4, 5].

The definition of requirements is a highly collaborative, interactive, and interdisciplinary process, involving heterogeneous teams of stakeholders [6, 7]. It provides another example of a dynamic collaboration that can be accomplished by a virtual, *ad hoc* group, where some members (e.g., representatives from the customer organization) join the developer group when they can add a value (e.g., to take part in the elicitation of the requirements, in a prototype demo session), and disengage at the end of the task. These groups create temporary networks of independent companies and collaborate as virtual organizations, using information technology to share skills and costs. Thus, such teams are *ad hoc* in the sense that they tend to be highly dynamic in creation, participation, and release, other than typically being geographically dispersed and cross-organizational. Other common scenarios of *ad hoc* group collaborations are provided by the partner consortium formed by representatives from different organizations in various sectors (e.g., academic institutions, industry), who have to coauthor a funding proposal for applying to the Framework Programme of the European Commission. Also in the field of software development, several processes, such as document inspections and reviews in general, can be carried out by *ad hoc* groups. The first contribution of this chapter is the proposal of a new definition of *ad hoc* group, which builds on the previous ones given in the existing literature on group research and is compliant with the emerging scenario of short-term distributed collaborations.

Due to their temporary nature, *ad hoc* teams need tools with infrastructure and administration costs kept at minimum. Instead, multipoint audio–video communication poses significant practical barriers to deployment (e.g., expense, infrastructure, support). As such, short-term groups like *ad hoc* teams often fall back to textual communication only. However, rich media theories on CMC, namely *Social Presence* [8], *Media Richness* [9–11], and *Common Ground* [12], have hypothesized group effectiveness to decrease when media other than F2F are used to accomplish equivocal tasks that require relational cues to be exchanged. They have reported about the inadequacy of text-based communication, as compared to rich media, like F2F and video. Lean media, such as e-mail and instant messaging, lack the ability of conveying nonverbal cues that contributes to the level of social presence (e.g., gaze, tone of voice, facial expressions), which in turns fosters individuals' motivation and mutual understanding. Nevertheless, these theories have also been criticized for considering the task to execute as an atomic activity. In addition, both Social Presence and Media Richness theories have generally been supported when tested on traditional media, such as F2F communication and telephone, whereas inconsistent empirical findings have resulted when tested on e-mail and video.

These inconsistencies have encouraged a reconsideration of the descriptive and predictive general validity of such theories. Thus, more recent theories have asserted that the effectiveness of CMC depends also on factors other than media richness, such as the degree of synchronicity, task typology, and group temporal scope. *Media Synchronicity* theory [13–15] and *Cognitive-Based View* [16] have started to investigate on media effects, looking at the underlying communication processes that happen in group tasks. On the one hand, Cognitive-Based View represents a sort of "Copernican revolution," which capsizes the existing perspective of CMC theories, looking at communication as a cognitive process: Not only must the sender's comfort with the communication medium be taken into account, but also the motivation of receivers and, above all, their ability to process the message properly. Furthermore, Cognitive-Based View argued that the use of rich media high in social presence should be used to assure attention for small amounts of information, whereas the use of lean media low in social presence causes a decreased motivation, but increases the ability to process large amounts of information during longer periods of time. On the other hand, Media Synchronicity theory distinguishes between the interplay of two different communication processes (the conveyance of additional information, and the convergence to shared views), which vary with the degree of synchronicity of the medium. Furthermore, since a task is not actually atomic, but rather constituted of several subactivities, Media Synchronicity theory suggests that the synchronicity level of media should be aligned with the degree of conveyance or convergence of each subactivity.

The concept of alignment between task and media characteristics is the very basis of the theories of *Time–Interaction–Performance* [17] and *Task/Technology Fit* [18, 19]. The frameworks proposed by these theories evaluate the appropriateness of task-medium matches, considering tasks no more as somewhat atomic activities, like in Media Richness and Social Presence theories, but rather as complex sets of subactivities and subprocesses, each having different characteristics. Likewise, also group and media characteristics have to be aligned for opportune collaborations to take place. The theories of Common Ground and *Channel Expansion* [20–22] argue that groups without a history of previous collaborations, like *ad hoc* groups, do not share any experience and thus, have not established a level of common ground (i.e., shared understanding) sufficient for communicating effectively over lean media. Conversely, members of long-term groups are expected to communicate more effectively over impoverished media, using their shared experiences to compensate for the media leanness.

Drawing upon these theories, we argue that, by understanding the paradoxical effects of rich media high in social presence, groups may be better able to select and use the most appropriate media to accomplish their goals. Hence, the second contribution of this chapter is presenting a critical review of the very many existing, and often conflicting, theories on CMC, which have been combined in a comprehensive theoretical framework for predicting, evaluating, and comparing the goodness of Task-Technology Fits. The proposed framework also builds on McGrath's Task Circumplex [23], which is the most widely used reference model in group research for task analysis, comparison, and categorization.

Finally, the third and last contribution of this chapter is the definition of a high-level research model, adapted from Ref. [24] (see Fig. 2), which can be used to support empirical studies on distributed group research. The theoretical background outlined later in this chapter will show that providing evidence of group task effectiveness can be overly challenging: The effects of technologies are contingent on many factors that differ from situation to situation, according to the context of a group process—that is, *group composition, task typology*, and *communication medium*. Thus, also the outcome of a group task (e.g., efficiency, effectiveness, product quality) depends upon the interaction between the group process and these varying contextual factors. Therefore, results from empirical study with communication media must be qualified by the context—the group, the task, and the medium—to which they apply. Through the rest of this chapter, we will update such model to include the variables that define the contextual group-, task-, and media-related factors.

To summarize, the three main contributions of this chapter are:

(A) the definition of *ad hoc* groups, which builds on the previous definitions given in the existing literature on group research;
(B) the design of a high-level research model for remote group performance evaluation;
(C) the design of a comprehensive theoretical framework, built upon the Task Circumplex model and the very many existing theories on CMC, which can be used to predict, evaluate, and compare the goodness of Task-Technology Fits.

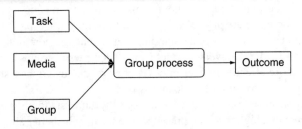

FIG. 2. The general research model adopted to represent the interaction of contextual factors with group process and their effect on the outcome (adapted from Ref. [24]).

The remainder of this chapter is structured as follows. Sections 2 and 3 deal with task-classification frameworks and *ad hoc* groups research, respectively. Section 4, instead, frames the complex background of CMC by reviewing the most prominent theories on media effect. In Section 5, we merge the contribution of the previous sections, thus creating two general frameworks relevant to group research on distributed collaboration. Finally, we conclude in Section 6.

2. Task-Classification Frameworks

When differences in group performance are studied, differences in group tasks must be taken into account with the due regard as well. A widely accepted, general definition of group task is the one given by Campbell, who defined it as "the behavior requirements for accomplishing stated goals, via some process, using given information" [25]. Such definition acknowledges that task characteristics define not only what is to be accomplished, but also how it is to be done. In fact, because required behaviors vary from task to task, it is argued that they can legitimately be viewed as characteristics of tasks themselves [26].

A number of task-classification schemes have been proposed in the literature, such as Hackman's *Task Framework* [26], Wood's *Model of Task Complexity* [27], and Mennecke's *Model of Task Processing in Groups* [28]. A list of historical task classification schemes can be found in Ref. [19]. The two tasks classification frameworks presented here focus on task complexity, the characteristic of task that has been studied the most because it relates to both process and outcomes of task performance, thus playing a key role in categorizing group tasks [19]. As such, the general research model is updated here to include the complexity of tasks factor as influencing the group interaction and outcome (see Fig. 3).

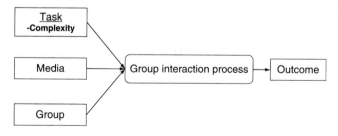

FIG. 3. The research model updated with the task complexity factor.

2.1 Task Circumplex

The most prominent theoretical framework formulated to provide a classification of group tasks is McGrath's Task Circumplex [23]. *Task Circumplex* classification scheme draws upon Hackman's *Task Framework* [26], which defined three types of tasks, namely tasks of idea production, tasks of discussion for group consensus, and tasks of problem solving. In addition, McGrath's classification is based on task as behavior requirements to the extent that each task is characterized not only by its own objective (i.e., *what* the group members are supposed to do to accomplish it), but also by its processes (i.e., *how* the task should be carried out). The Task Circumplex, shown in Fig. 4, categorizes all group tasks as belonging to one of four basic task processes, each of which has in turn two subtypes:

(I)   Generate (ideas or plans);
(II) Choose (correct or preferred answers);
(III) Negotiate (conflicting viewpoints or conflicting interests);
(IV) Execute (in competition against other groups or in evaluation against standards of performance).

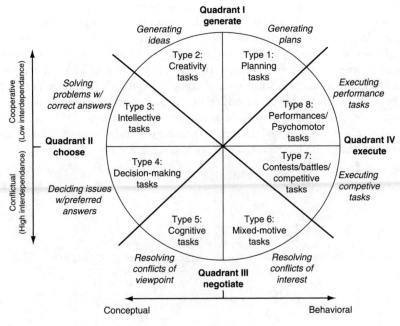

FIG. 4. The Task Circumplex [23].

The four process categories are related to one another and arranged in a circumplex along two dimensions, namely the degree to which processes involve cooperation (i.e., low member interdependence) versus conflict (i.e., high member interdependence), and the degree to which the processes involve conceptual versus behavioral activities. Furthermore, McGrath designed the four process categories to be collectively exhaustive, thus making the Circumplex useful for comparing similarities and differences of tasks used in group research.

As an example, we here use the Task Circumplex to categorize the activities of requirements elicitation (defined as the process of seeking, uncovering, acquiring, and elaborating requirements for computer-based systems [7]) and negotiation (defined as the process of reaching an agreement on requirements by resolving misunderstandings and conflicts due to the conflicting goals and priorities that stakeholders have [6]). According to the framework above, eliciting requirements is mostly a creativity task (Type 2), since it is about generating ideas, with a lower need for problem solving (Type 3) and decision making (Type 4). Conversely, the negotiation of software requirements involves tasks of Types 3–7, namely creativity, intellective, decision making, cognitive, mixed-motive, and competitive tasks [29]. Thus, comparing the two forms of requirements workshops, in Task Circumplex terminology, requirements negotiation is a more complex activity, in that it involves different tasks, both conceptual and behavioral, with medium to high degree of member interdependence. In contrast, elicitation is a simpler activity in that it is mostly a conceptual task of creativity, with low behavioral issues involved and low degree of member interdependence.

The Task Circumplex is not exempt from limitations and criticisms. While it gives a way to compare tasks, it does not provide with an objective means to measure the degree to which tasks in each wedge differ from tasks in both the same category or in different categories [28, 30]. Despite such criticism, the Task Circumplex has been the dominant task-classification scheme in the last two decades. It has been used not only as a task taxonomy, but also as the foundations to develop theories on communication media selection, discussed in Section 4, which encompass the intertwined relationships between tasks and technology, discussed in Section 5.2. Task Circumplex has been adopted by Group Support Systems (GSS) research (see Ref. [31]) for an exhaustive compendium on GSS-related research studies). GSS studies have largely dominated group studies for almost more than two decades, until the end of the 1990s. Christenesen and Fjermstad performed a meta-analysis of 67 GSS studies, conducted until 1997 [32]. They found that more than a half of GSS studies employed creative tasks and that more than one-quarter employed decision-making tasks. Furthermore, most of the laboratory studies reviewed used contrived tasks designed or manipulated for the research purpose. To improve the generalizability of results, Dennis et al. called for the use of tasks as complex as "natural tasks," requiring knowledge already within subjects' knowledge domain [33]. However, since students were and are likely to continue as the most common source of experimental subjects, the usually contrived laboratory tasks were puzzles or games (e.g., lost at sea, the parking problem, the philanthropic foundation task [28]), which required limited or no specialized knowledge to be recalled [34]. These tasks represent a poor surrogate for the complexity of "wicked" natural tasks, and their employment potentially limited the external validity and generalizability of GSS laboratory experiments, and likely accounted for much of the contradictory findings between field and laboratory research [28, 35, 36]. The multifaceted properties and complexity of natural tasks can be achieved by using "realistic tasks," that is, natural tasks replicated in controlled laboratory environments. The flipside of using realistic tasks in place of contrived tasks is the likely higher difficulty in evaluating group interaction processes and task performance. Effectiveness does not have a consistently held definition or interpretation in the group research literature [24]. Satisfaction with both the interaction process and the outcome is an important variable in group research, since it has been acknowledged to be indicative of both individual and group performance [37, 38].

2.2 Complex Tasks Typology

First Campbell [25] and then Zigurs and Buckland [19] identified four fundamental task attributes, from which a *typology of complex tasks* was derived. The first dimension is *outcome multiplicity*, which indicates that a task has more than one desired outcome. All tasks that involve more than one stakeholder with different expectation about the goal provide an example of tasks with outcome multiplicity (e.g., selecting a family home when every family member has different expectations on price, size, position, service proximity, and features alternatives).

The second dimension is *solution scheme multiplicity*, which indicates that more than a solution path exists to accomplish the task and reach the goal. Class scheduling is an example of task with solution scheme multiplicity.

The third dimension is *conflicting interdependence*, which may exist when conflicts are found between alternative task solution schemes or task outcomes. This may also happen when some pieces of information available are conflicting. Examples of tasks with conflicting interdependence are provided by quality versus quantity tasks [19].

The fourth and last dimension is the *solution scheme/outcome uncertainty*, which can be identified in all tasks where there is uncertainty about whether one solution scheme will lead to the desired outcome [19].

Finally, all the possible combination of the four dimension result in 16 distinct tasks categories, which have been narrowed down to 5, as shown in Table I.

TABLE I
AGGREGATED TASK CATEGORIES WITH PRIMARY ATTRIBUTES SHOWN IN BOLD (ADAPTED FROM REF. [19])

Dimension	Simple tasks	Problem tasks	Decision tasks	Judgment tasks	Fuzzy tasks
Outcome multiplicity	**No**	No	**Yes**	No	**Yes**
Solution scheme multiplicity	**No**	**Yes**	No	No	**Yes**
Conflicting interdependence	No	Yes or No	Yes or No	**Yes**	Yes or No
Solution scheme/outcome uncertainty	NA	Low to High	Low to High	Low to High	Low to High

Each of the five task categories is defined in terms of primary attributes, shown in bold in the table. Thus, Simple tasks are primarily characterized by the existence of a single outcome and solution scheme, the opposite of Fuzzy tasks. Problem tasks and Decision task are characterized, respectively, by solution scheme multiplicity and outcome multiplicity. Instead, conflicting interdependence dimension primarily identifies Judgment tasks. Finally, although not applicable in the case of Simple tasks, the dimension of solution scheme/outcome uncertainty does not primarily characterize any of the four remaining categories because it can be present in each of them, ranging from low to high, depending on the nature of the task itself (e.g., when the scope of the task is large or the outcomes hard to measure).

Campbell also observed that other than the four primary complexity attributes, there are also other characteristics that may be associated with task complexity, such as lack of structure, ambiguity, and difficulty. Hence, unlike McGrath's Task Circumplex, the complex tasks typology proposed by Cambpell is not designed to be collectively exhaustive. In addition, as pointed by Zigurs and Buckland, who refined the initial work by Campbell [25], it focuses on the kinds of tasks that are usually found in organizational decision-making groups [19]. As such, the Tasks Typology presented here results less useful than the Task Circumplex for general-purpose group tasks categorization and comparison.

3. Group Research

3.1 Teams with No Past and Future: *Ad Hoc* Groups

Besides task type, another contextual factor that influences group studies is temporal scope, that is, "the extent to which groups have pasts together, and expect to have a future" [17] (p. 149).

Work groups are today increasingly nimble and subject to frequent changes [39]. This underlying idea in *ad hoc* groups is that of a small entity, highly dynamic in creation, participation, and release, formed to accomplish the goal at hand (e.g., solve a specific problem), and then, disband as soon as the collaboration is over. Hence, *ad hoc* teams are also called *goal-oriented* teams [40]. These teams are sometimes associated with strike teams, which are small groups of people with a specialized purpose, such as responding to a critic situation, like a terrorist attack or a natural disaster, in a timely manner. In addition, *ad hoc* groups typically exhibit both loose affiliation and geographical dispersion, that is, they are virtual teams, composed recruiting members from independent departments in different organizations [41]. Virtual organizations of the future will be more and more comprised of flexible, *ad hoc* groups that individuals join when they can add value and disengage when they are no longer needed [42]. Today, a common scenario of *ad hoc* groups collaboration is provided by the partner consortium formed by representatives from different organizations in various sectors (e.g., academic institutions, industry), who have to coauthor a funding proposal for applying to the Framework Programme of the European Commission. Also in the field of software development several processes, such as document inspections and reviews in general, can be carried out by *ad hoc* groups [43]. The scenario of distributed requirements provides another example of a dynamic collaboration that can be accomplished by a virtual, *ad hoc* group, where some members (e.g., representatives from the customer organization) join the developer group, when they can add a value (e.g., to take part in the elicitation of the requirements, in a prototype demo session), and disengage at the end of the task.

The limited group size and temporal scope are the key characteristics of *ad hoc* groups. *Ad hoc* groups do not usually include more than 10 participants. However, every attempt to define the typical size is vain. Even research on small groups reports varying ranges, usually 3–5 participants for small-sized groups, and 6–12 for medium-sized groups [44]. However, in absence of a widely accepted definition of group size, these ranges can be considered reasonable, bearing in mind the research already undertaken. The study of small- and medium-sized groups is important because it has been shown that larger groups do not necessarily produce a proportionally higher number of ideas and thus, there is likely to be an optimal group size, beyond which any further increase in membership does not equate with an increase in contributions [44]. Temporal scope defines group history and future, that is, the shared experience that the group has developed in the past and the expectation of future collaboration, respectively. For *ad hoc* groups, temporal scope corresponds exactly with the time needed to carry out one collaboration. In other words, while traditional groups are conceived as *established*, that is, long-term, standing teams that work together for a long time across several independent projects, *ad hoc* groups are instead teams brought together for a short time to carry out only the collaborative effort in attendance. The meaning of *ad hoc* groups today differs greatly from the earlier definitions provided by researchers over the years. *Ad hoc* groups, also called single-task groups initially, have been studied since the end of the 1950s [45–47] and over the last decades [33, 48–50]. According to the definition given by Mennecke et al. *ad hoc* groups are teams whose "members have no experience working together with other members and little or no expectation that they would work together in the future." In contrast, they defined established groups as "on-going groups, that is, groups where members have a significant history working together as a group and anticipate having a significant future together" [30, 48]. Likewise, Dennis et al. defined *ad hoc* groups as single-task groups whose members have not worked together prior to the study and do not anticipate to continue working together after the study [33]. Although similar to the others, this definition is indicative of how past research considered *ad hoc* groups as single-task, "laboratory groups" of randomly assembled subjects to be studies merely as "experimental, microscopic models" of established groups, seen instead as natural groups [45]. However, Bormann [47], McGrath [23], and Mennecke et al. [30] pointed out the inadequacies associated with using single-tasks groups, in terms of the lower generalizability of results. Nevertheless, single-task groups have almost universally been used in laboratory experimentation, compared to field studies, where established groups are utilized instead.

While previous research has almost exclusively treated *ad hoc* groups as a factor partially accounting for discrepancies between laboratory and field studies, current research cannot continue to neglect the relevance of studying *ad hoc* group *per se*. We cannot continue to refer to established groups as "natural groups," since nowadays *ad hoc* groups are functionally used as well, and no more employed only in laboratory studies. While established groups are still more traditional, they are to be considered as natural as *ad hoc* groups. We suggest to adopt the definitions given by McGrath et al. to distinguish *natural* groups, defined as "groups that exist independently of the researcher's activities" and used in field experiments, from concocted groups, which are instead "brought together only for the purpose of laboratory experiments" [23] (p. 41). Thus, group research studies can employ natural as well as concocted *ad hoc* groups. In addition, compared to concocted established groups, laboratory studies on concocted *ad hoc* groups will suffer from minor problems of results generalizability, since they represent a more adequate experimental model of their natural counterpart. We also suggest the following new definition of *ad hoc* groups.

Definition. An *ad hoc* group is a small- to medium-sized team highly dynamic in creation, participation, and release, whose members have no past experience of working together and little or no expectation of collaborating again in the future. The temporal scope of an *ad hoc* group corresponds exactly to the time needed to carry out the collaboration in attendance.

3.2 Challenges and Needs in Supporting Remote *Ad Hoc* Groups

The definition above voluntarily omits the adjective "distributed," typically used to further characterize an *ad hoc* group, because virtual *ad hoc* teams are more common and of our primary interest, there can be collocated *ad hoc* groups as well.

Our specific interest in supporting collaboration of *ad hoc* groups is twofold. We aim at understanding (1) the key challenges in *ad hoc* group communication processes and (2) the attributes of communication media to use in order to cope effectively with such challenges when *ad hoc* groups are distributed.

Very little is known today about the differences in group dynamics of *ad hoc* groups. In his research study, Tuckman only reported hypotheses on short-term groups development [51] (p. 79). He supposed that "duration of group life would be expected to influence the rate and amount of development." Nevertheless, short-term groups would also be expected to "essentially follow the same course as long-term groups [...] with the requirements that the performing stage be reached quickly," to the detriment of the other phases that are not "as salient as task execution" in task-oriented groups.

The study of short-term groups has been somewhat neglected by group research, especially GSS, since it was only accounted as one of the factors that could explain variance of experimental results. Nevertheless, useful insights have been gained from a review GSS research on the effects of group history and experience, in the comparison between established and *ad hoc* groups. Hall and Williams were among the first to report that conflicts and decision quality in decision-making tasks are moderated by group history [46]. While decision quality resulted positively related to outcome quality in established groups (i.e., the more the conflicts, the higher the decision quality), the relationship resulted reversed for *ad hoc* groups (i.e., the more the conflicts, the lower the decision quality). This result was later confirmed by Dennis et al., who also found that established groups did not communicate more than *ad hoc* groups, which in turn showed a greater equality of members' participation (i.e., no domination as for established groups' communication), but also less openly critic messages (i.e., more inhibited communication) [33]. Mennecke et al. found partial evidence in support of the major quantity of information shared by *ad hoc* groups [48]. Benbasat and Lim performed a meta-analysis of research on the effects of group history and found that decision quality is not significantly affected by group history, which instead was confirmed to negatively affect equality of participation (i.e., the more the past experiences share by a group, the less equal the members' participation) [37]. In addition, with respect to traditional established groups, *ad hoc* groups typically exchange more task-focused, impersonal information, and exhibit less openness and trust [32]. Finally, Alge et al. suggested the need to distinguish between past and future groups for investigating the effects of groups' experience and motivation [49]. *Past* groups are teams nearing to completion of a collaboration, whereas *future* groups, instead, are newly formed teams just starting a collaboration. Past and future groups exhibit different level of motivation. Members of future groups are more likely to be motivated to engage in interactions than members of past teams who feel to be close to the end of the collaboration and thus, tend to exchange a lower amount of information. However, it is unclear how these results relate to *ad hoc* groups. Given our proposed definition, the characteristics of past and future teams blend in the temporary nature of *ad hoc* groups, in the sense that the limited temporal scope makes an *ad hoc* group a newly formed team, also close to the completion.

The technological challenges to be faced in supporting distributed, *ad hoc* groups stem from the limited temporal scope too. Given the rather occasional and temporary nature of *ad hoc* groups' collaboration, the adoption and maintenance costs of complex collaborative platform (groupware) can hardly be justified and sustained. The adoption of such sophisticated collaborative platforms has proved to be problematic even for established groups, in both traditional [52] and virtual organizations [2]. Hence, we argue that *ad hoc* groups, to be effectively nimble, should be supported by communication tools that have a low learning curve, so that dynamic engaging of new members is facilitated, and whose infrastructure and administration costs are minimal, so that dynamic creation is facilitated. This need for supporting dynamism turns out to require the adoption of either commonly available tools, such as instant messaging, e-mail, wikis, issue trackers, or systems that do not require administration and maintenance of any central resource by design [53–55]. In the latter case, P2P collaborative systems can support *ad hoc* groups in the sense that they build overlay networks that sit on top of the Internet, and almost exclusively use resources (e.g., disk storage, bandwidth) already available on the same hosts running the peers (i.e., the edge of the Internet).

To conclude this section, we show in Fig. 5 the research model updated to include the size and temporal scope variables, which characterize the group-related contextual factors.

4. CMC Theories

As geographically dispersed individuals more and more communicate via computer, understanding the effectiveness of the very many available media has become vital. Media are usually classified in the time/space matrix (see Fig. 6), according to

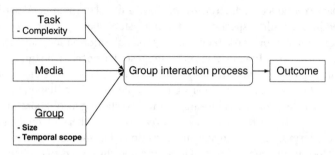

FIG. 5. The research model updated with group-related variables, size, and temporal scope.

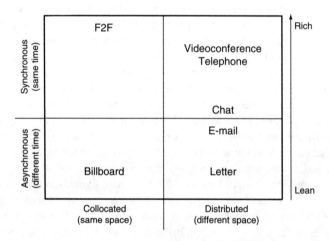

FIG. 6. Rich/lean media ranking in classic time/space matrix (adapted from Ref. [56]).

both the spatial dimension (collocated/distributed, i.e., *where* interaction occurs) and the temporal dimension (synchronous/asynchronous, i.e., *when* the interaction occurs). For instance, F2F communication allows synchronous interaction and requires physical collocation of individuals. Instead, e-mail allows asynchronous interaction and does not require collocation.

Media can also be classified according to another dimension, richness. We can intuitively epitomize *richness* as the ability of media to convey a larger amount of information in different forms. The figure above shows the media along the media richness continuum. F2F is the richest form of communication, since it conveys information via audio and video channels, but also through cues like gesture and posture. Consequently, videoconference is richer than telephone, since the latter lacks video as information channel, whereas e-mail is richer than letter, since electronic mail can also attach multimedia content. Many CMC theories have provided different definitions of media richness, but, despite such differences, the resulting rank of media richness has never changed from the one presented before. Besides, many CMC theories have agreed on the inadequateness of text-based communication for complex, collaborative tasks, suggesting that, as complexity increases, so should the level of richness of the media used.

Despite the negativity of the aforementioned technological and theoretical premises, the last decade has witnessed the success of many open-source projects which are coordinated through the almost-exclusive use of text-based technologies, such as web sites, e-mail, and IM. These technologies, although not novel, have found their own way in supporting collaboration. E-mail is the most used collaborative tool to date, and a place where new collaborations emerge [57]. IM, although initially banned as an application intended only for teenagers, has found a number of uses in the workplace, including opportunistic interactions, and a "signaling" function by which people negotiate their presence and availability [58, 59]. Web sites and their natural evolution, the Wikis, foster collaboration throughout knowledge sharing [60]. Open-source development provides just one of the scenarios where text-based communication is effectively used to perform complex collaborative tasks. As already pointed out, interaction of individuals is deeply influenced not only by media characteristics but also by tasks requirements and group characteristics like history and experience.

In the following, we review the fundamental theories on CMC and media selection. Sections 4.2 and 4.3 discuss the Social Presence Theory and the Media Richness Theory, respectively. The theory of Common Ground is introduced in Section 4.4. Section 4.5 presents the Media Synchronicity Theory. Finally, the Cognitive-Based View is discussed in Section 4.6.

4.1 Social Presence Theory

Social Presence refers to the degree to which one perceives the presence of participants in the communication. Social Presence theory argues that media differ in the ability to convey the psychological perception that other people are physically present, due to the different ability of media to transmit visual and verbal cues (e.g., physical distance, gaze, postures, facial expressions, voice intonation, and so on) [8]. Some mediums (e.g., videoconferencing or telephone) have greater social presence than other mediums (e.g., e-mail), and media higher in social presence are more efficient for relational communication (i.e., building and maintaining interpersonal relationships), as they involve social/personal issues and thoughts.

Social Presence presumes the outcome of an interaction to be determined by the capacity of the selected medium to support the type of communication required. More specifically, Short et al. argue that F2F interaction, thanks to the wider capacity of conveying social presence, is more effective for relational communication than text-based media, such as e-mails, which do not transmit any cue and are then, more effective for task-focused communication.

Finally, Social Presence theory has also been found to be a strong indicator of satisfaction, that is, the higher the sense of social presence conveyed by a medium, the higher the satisfaction perceived by participants when communicating [61].

4.2 Media Richness Theory

One of the most widely applied theories of media selection is Media Richness theory by Daft and Lengel [10, 11]. Media Richness, which builds on the theory of Social Presence, argues that communication media differ in their ability to facilitate understanding. Daft and Lengel have defined information richness as the capacity of information "to change understanding within a time interval" [9]. Thus, in Daft and Lengel's terms, what differentiates richer media from leaner media is the amount of information a medium could convey to change the receiver's understanding within a time interval. This capacity depends on several factors, such as the ability of the medium to transmit multiple cues, immediacy of feedback, and language variety. The perceived sense of social presence of a medium is proportional to the medium richness. As a result, rich media with a wide communication capacity also have a high level of social presence. F2F interaction is the richest media, due to its capability of expressing message context in natural language and conveying at the same time multiple cues via body language and tone of voice, and it is supposed to change understanding of participants in communication in a shorter time interval. The second richest medium is videoconferencing, because, although it still grants the use of natural language, and the access to some visual and verbal cues, it conveys a lower sense of social presence to conversation participants. E-mail, chat/IM, and letters are instead the leanest media because, when adopted, communication exchanged by participants is conveyed on a single channel, that is, text, be it written or typed.

Like Social Presence Theory, also Media Richness theory presumes that the outcome of an interaction is determined by the communication capacity of the selected medium. While Social Presence theory relates performance primarily to the type of interaction required (relational vs. activity focused), Media Richness Theory asserts, instead, that performance depends on the appropriateness of the match between media richness characteristics and information requirements of the task (clarification vs. additional information). Indeed, Media Richness Theory postulates the existence of two complementary forces that act on participants when they process the information exchanged when communicating (see Fig. 7). One force is *uncertainty*, which is defined as the "difference between the amount of information required to perform a task and the amount of information already possessed" [11]. This definition builds on earlier research work about information theory (i.e., as information increases, uncertainty decreases [62]). Uncertainty is reduced obtaining additional data and seeking answers to explicit questions. The other force is *equivocality*, which is the existence of multiple and conflicting interpretations about a situation [11]. As uncertainty is more related to the amount of information available, equivocality is more related on the quality of information available: Equivocality means ambiguity and reflects confusion and lack of common understanding, whereas uncertainty means the absence of sufficient data necessary and reflects the inability to process information properly.

Equivocality is reduced by seeking for clarification, reaching agreement, and deciding what questions to ask. The postulation of the existence of these two complementary forces has also implications on the selection of the most effective medium to use. Media Richness theory posits that rich media are better suited in equivocal communication situations (where there are multiple, even conflicting,

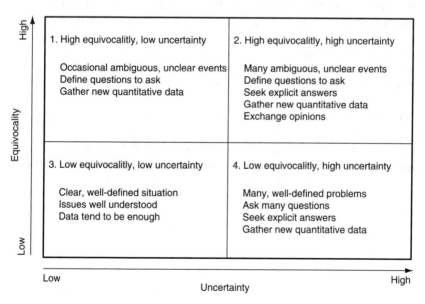

FIG. 7. The uncertainty and equivocality forces that act on individuals during communication (adapted from Ref. [11]).

interpretations for available information), whereas lean media are best suited in uncertain communication situations (where there is a lack of information). Equivocality is often symptomatic of disagreements and, thus, providing sufficient clarifications can reduce it. Rich media interaction (e.g., F2F), is preferred in situations of equivocality, as it allows for rapid feedback and multiple cues, thus facilitating the convergence to a shared interpretation. On the other hand, when messages are not equivocal, lean media are preferred. Thus, uncertainty can be reduced by obtaining sufficient additional information using media like e-mail or written reports. Therefore, in short, Media Richness proposes that task performance will be improved when tasks needs are matched to the medium ability of conveying information.

Finally, we notice that Daft and Lengel have treated equivocality and uncertainty as independent constructs. However, it must be pointed out that a new amount of data can also generate ambiguity, and that equivocal scenarios may need more data to converge as well.

4.3 Common Ground Theory

The Common Ground theory by Clark and Brennan is a fundamental theory in the CMC field [12]. It subsumes all the existing theories of communication in that it describes the basic process of grounding, a process orthogonal to all forms of communication, which encloses the essential goal of communicating: Reaching a common understanding. Indeed, *grounding* is the interactive process by which communicators exchange evidence in order to reach a mutual understanding, updating moment by moment their *common ground*, that is, the amount of shared information already owned.

Communicating is more than simply sending off messages. Speakers must assure themselves that receivers have correctly understood the message. Communication is a collective activity that requires coordinated action of all participants, and grounding is crucial for keeping track of the coordination. Individuals contribute to a conversation repeating two steps, namely *presentation*, that is, the speaker presents an utterance to the receiver(s), and *acceptance*, that is, the receiver(s) accepts (accept) the utterance, giving evidence of correctly understanding what the speaker meant. It takes both phases for a contribution to be complete: Grounding and the communication itself are impaired if the speaker does not get any evidence of acceptance. Evidence can be either positive (the message has been understood, the speaker can go on) or negative (the message is misunderstood and the speaker must repair before proceeding). Such evidence can be provided by different grounding techniques that change with medium. Grounding techniques include, to name but a few, *acknowledgments* (e.g., nodding, saying "yes," or typing "ok"), *spelling* (e.g., spelling one's family name), and *verbatim displays* (e.g., repeating word by word a telephone number). But also speakers can explicitly seek for evidence asking questions (e.g., saying "right?" at the end of an utterance). Questions asked from receivers are usually a form of negative feedback as they represent a request for clarification. However, the positivity or negativity of acknowledgments is not context-free.

Grounding techniques are deeply affected by media characteristics. Since text-based communication does not convey neither visual nor verbal cues (e.g., nodding, face expression, gaze direction are unavailable), it constrains the possible form of evidence that people can seek to acknowledgments (one would never use verbatim displays or spelling in text-based chat). Clark and Brennan go beyond the level of media richness and social presence and present eight properties that act as constraints on the grounding process (see Table II).

Participants in a F2F conversation usually establish common ground on the fly, as they have access to cues like facial expression, gestures, and voice intonation. Instead, when participants communicate over media, the fewer cues they have, the harder to construct it. As a consequence, according to Clark and Brennan's theory, "people who have little common ground benefit significantly from having a video channel" and, conversely, "only people who have previously established a lot of common ground can communicate well over impoverished media" (e.g., e-mail or IM) [2].

From the previous figure we notice that text-based communication lacks key attributes like *copresence* (owned only by F2F communication), *visibility*, and *audibility* that Common Ground Theory claims to be necessary for communicators unknown to each other for developing mutual understanding. *Simultaneity* refers to the ability of the medium to allow for full-duplex communication, that is, individuals can send and receive at once and simultaneously. Simultaneity is strongly related to *synchronicity*, which distinguishes between same time and different time media. However, no medium has all the attributes at the same time. Text-based communication offers two characteristics that even F2F and audio/video communication lack, namely reviewability and revisability. *Reviewability*, also called reprocessability, is the extent to which a message can be reexamined or processed again within the context of the communication event. Text-based media enable the receiver to repeatedly process the message to ensure accurate understanding. *Revisability*, also called rehearseability or editability, is the extent to which media enables the sender to rehearse or fine tune the message before sending. Text-based media enable the sender to carefully edit a message while it is being sent to ensure that the intended meaning is expressed exactly. Erickson and Kellogg [63] have drawn attention to these two powerful characteristics of text-based communication, which make it persistent, traceable, thus enabling the use of search and visualization technologies.

TABLE II
MEDIA CONSTRAINTS AFFECTING GROUNDING (ADAPTED FROM REF. [2])

Medium	Copresence	Visibility	Audibility	Synchronicity	Simultaneity	Sequentiality	Reviewability	Revisability
F2F	•	•	•	•	•	•		
Videoconference		•	•	•	•	•		
Telephone			•	•	•	•		
Chat/IM				•	•	•	•	•
E-mail							•	•
Letter							•	•

When a medium lacks one of these characteristics, it forces people to use alternative grounding techniques. This happens because the costs (i.e., the effort for the speaker, the receiver, or both) of using the different techniques of grounding change. Clark and Brennan count 11 different types of costs. For instance, *delay costs*, that is, the cost of waiting for messages to be completed, are paid by both speakers and receivers. Such costs have to be low in synchronous media, as long pauses would disrupt communication. *Production costs* of messages are paid only by speakers and are much lower in media carrying voice than in those text-based. In contrast, *reception costs* are only paid by receivers. Listening is generally easier than reading. However, reading may be less costly when messages content is particularly complex, to the point that they must be reviewed several times to allow for correct deliberation. Thus, grounding process is also affected by the purpose of communication (i.e., the task). This aspect, however, has not been examined in deep by Common Ground Theory. When individuals communicate, they try to reach understanding minimizing the effort for themselves and the others, paying as few of these costs as possible. This rule is known as the *least collaborative effort* principle.

4.4 Media Synchronicity Theory

Both Social Presence and Media Richness theories presume that the outcome of an interaction is determined by the communication capacity of the selected medium. Media Richness Theory relates performance primarily to type of information required by tasks (clarification vs. additional information), whereas Social Presence theory relates it primarily to the type of interaction (relational vs. activity focused). A number of empirical studies of media use have provided evidence that runs counter to the predictions [15, 22], thus pushing researchers to theorize that media selection is also affected by factors beyond richness.

Social Presence and Media Richness Theories have been refined by Media Synchronicity theory by Dennis and Valacich [13–15]. Social Presence and Media Richness theories are task-centric: A task is the key element to medium selection, but it is considered as a high-level construct—that is, relational or activity focused, equivocal or uncertain. As suggested by McGrath [17], tasks are composed of many subelements, processes, and activities which may need different media. For example, in Daft and Lengel's terms, resolving a task of equivocality would mean developing a shared framework for analyzing the situation, populating the framework with information of a shared meaning, and assessing the results to arrive at a shared conclusion for action. However, each of these steps may have different media needs, such that even tasks of uncertainty may include steps that require rich media [64]. Media Synchronicity theory posits that group communication, regardless of the task (whether equivocal or uncertain, relational or activity focused), is composed of two fundamental communication processes, conveyance and convergence. *Conveyance* is the exchange of information, followed by deliberation on its meaning. It can be divergent, in that not all participants need to focus on the same information at the same time, nor must they agree on its meaning. *Convergence* is the development of shared meaning for information, in that participants must understand each other's views and agree. The constructs of conveyance and convergence are not different from the concepts of uncertainty and equivocality developed by Media Richness theory. However, Daft and Lengel have treated equivocality and uncertainty as independent constructs. Therefore, for resolving equivocality, Media Richness theory emphasizes the need to converge, whereas conveyance is left to tasks of uncertainty. Instead, Media Synchronicity theory argues that conveying information and converging on a shared meaning are equally critical for tasks of equivocality and uncertainty: New amounts of data can also generate ambiguity, and equivocal scenarios may need more data to converge as well. Thus, without adequate conveyance of information, individuals will reach incorrect conclusions, and without adequate convergence, the group cannot move forward.

Social Presence and Media Richness theories assume the existence of the richest medium in absolute, which is F2F communication. According to Dennis and Valacich, ranking media in absolute terms is not practical, though. They argue that media should not be ranked in order of their richness without consideration of context, and that attempting to recommend a single medium based on a high-level task is doomed to failure. Media possess many capabilities, each of which may be more or less important in a given situation. Media Synchronicity theory postulates that media have a set of capabilities, and that performance will be enhanced when such capabilities are aligned with the processes of conveyance and convergence. Thus, in Dennis and Valacich's terms, "the richest medium is that which best provides the set of capabilities needed by the situation," that is, the individuals, the task, and the social context. Table III examines the capabilities of several media.

Symbol variety is the number of ways in which information can be communicated—the "height" of the medium—and subsumes Daft and Lengel's multiplicity of cues and language variety. The importance of symbol variety depends upon the piece of information that needs to be communicated. In general, conveyance should require a greater symbol variety depending upon the task. In contrast, convergence requires understanding others' interpretations, which can usually be communicated using a simpler symbol set. *Parallelism* refers to the number of simultaneous conversations that can exist effectively—the "width" of the medium. In traditional media such as the telephone, only one conversation can effectively use the medium at one time. In contrast, many electronic media can be structured to enable many simultaneous conversations to occur. The importance of parallelism depends upon

TABLE III
CAPABILITIES OF MEDIA (ADAPTED FROM REF. [15])

Medium	Symbol variety	Parallelism	Immediacy of feedback	Rehearseability	Reprocessability
F2F	Low-high	Low	High	Low	Low
Videoconference	Low-high	Low	Medium-high	Low	Low
Telephone	Low	Low	Medium	Low	Low
Letter	Low-medium	High	Low	High	High
E-mail	Low-high	Medium	Low-medium	High	High
Chat	Low-high	High	Low-medium	Medium-high	High

Media are listed as having a range of capabilities because they are configurable (e.g., e-mail may or may not enable the use of tables or graphics).

the number of participants. It is unimportant for small groups. For large groups, however, parallelism is very important to conveyance in enabling all members to participate. Usually, the greater the parallelism, the easier it is to generate divergent information (i.e., conveyance). Conversely, convergence will generally benefit from low parallelism because the focus of the process is on understanding others' viewpoint. As the number of conversations increases, it becomes increasingly difficult for the group to focus on one topic or issue, which may in some circumstances impede the development of mutual understanding (i.e., convergence). *Immediacy of feedback* is the extent to which a medium enables users to give rapid feedback on the communications they receive (i.e., the ability of a medium to support rapid bidirectional communication). It is important in improving understanding because it enables mid-course corrections in message transmission, so that any misleading elements in the message as sent can be quickly corrected. More immediate feedback can have significant benefits in improving the speed and accuracy of communication. Immediacy of feedback and parallelism dimensions define "the level of synchronicity" of media. *Rehearseability* and *reprocessability* match respectively with the attributes of revisability and reviewability defined by Clark and Brennan for the Common Ground Theory. Rehearseability is probably unimportant for simple messages, but becomes more important as the complexity or equivocality of the message increases because increased rehearseability will lead to improved understanding. However, media with high rehearseability tend to have lower feedback. *Reprocessability* enables the receiver to repeatedly process the message to ensure accurate understanding, thus fostering conveyance. Reprocessability becomes more important as the volume, complexity, or equivocality of the message increases. Increased reprocessability will lead to improved understanding, regardless of the information or communication process (conveyance or convergence), although it is often more important to conveyance. Conveyance often produces information requiring deliberation, for which reprocessability is important.

In media selection, one must take into account that most tasks require individuals to both convey information and converge on shared meanings, and media that excel at information conveyance are often not those that excel at convergence. Thus, choosing one single medium for any task may prove less effective than choosing a medium, or set of media, which the group uses at different times in performing the task, depending on the current communication process (convey or converge).

According to Media Synchronicity theory, although the selection of the most appropriate medium (or set of media) depends upon all these five dimensions, the key to effective media usage is matching the synchronicity level to the level of conveyance and convergence required to perform a task. Indeed, Dennis and Valacich posit that media that support high immediacy of feedback and low parallelism encourage the high synchronicity, which is the key to the convergence process. Conversely, media that support low immediacy of feedback and high parallelism provide the low synchronicity, which is the key to the conveyance process. Although the formulation and the constructs names change, the task-media matching suggested by Media Synchronicity theory is the same one suggested by Media Richness Theory. Indeed, high-synchronicity media, with immediate feedback and low parallelism, are exactly F2F, and audio/video conference, that is, the richest media high in social presence that best fit equivocal tasks. High parallelism, instead, is not feasible when audio and video channels are available. Thus, low-synchronicity media with high parallelism are exactly e-mail, chat, and IM, that is, the lean media low in social presence, which best fit uncertain tasks.

Beside synchronicity, there are other factors that influence the effectiveness of media in supporting different groups, even those performing similar tasks. Group history—that is, the extent to which groups have worked together in the past—is a situational factor that can influence effectiveness because it can alter the perception of media richness of time. Established groups are more likely to have established norms (e.g., roles within the group), and well-established processing norms for the task performing. The group will be more likely to move directly to execution with less storming and norming. During performing, group members are able to work separately on their assigned tasks. Thus, performing requires more conveyance than convergence, although some convergence is clearly required. The need for media synchronicity is therefore lower during performing than during forming, storming, and norming. As a group matures they "are likely to become able to carry out all their functions, at least for routine projects, with much less-rich information exchanges" [15]. This means that (1) the communication requirements of groups will likely differ over time, depending upon shared experiences; (2) the perceptions about medium usefulness for a task and the group's ability to perform a task in a given medium change over time. As group members come to know each other better over time, they share common experiences that may be evoked by very simple messages that refer to those shared experiences. Therefore, over time established groups will require less convergence communication processes, or, equivalently, less use of high-synchronicity (high feedback, low parallelism) communication environment. Conversely, newly formed groups (e.g., *ad hoc* groups) will have fewer well-established norms and will likely spend more time in forming, storming, and norming, before moving to performing. This will result in more complex processes requiring more conveyance, and, especially, convergence. Before group members can effectively work together, they often need to have a better understanding of each other, and socially related communication activities that are best developed through media with social presence. Thus, newly formed groups, groups with new members, and groups without accepted norms will require more use of media with high synchronicity (high feedback and low parallelism), and symbols sets with greater social presence.

4.5 Cognitive-Based View

Researchers have long studied the effects of social presence and media richness on media choice, and the effects of media use. However, it is not always the sense of presence that is vital to communication, but also having sufficient information in the appropriate format and the ability to properly process it [65]. Furthermore, the original premise of Daft and Lengel's Media Richness theory was to understand how media effect a change in receivers' understanding. Nevertheless, the influence of media choices on the cognitive processes that underlie communication has been overlooked. Robert and Dennis described a Cognitive-Based View of media choice and media use, based on dual process theories of cognition, which argue that in order for individuals to systematically process messages, they must be motivated to process the message and have the ability to process it [16]. Communication is not only an exchange of information, but also an exchange of attention. Different media have different usage costs to the receivers. Running counter to past research (i.e., the more complex the task, the richer the media to be used), they argued that the use of rich media high in social presence induces increased motivation, but decreases the ability to process information, whereas the use of lean media low in social presence induces decreased motivation but increases the ability to process information (see Fig. 8). Robert and Dennis called the inverse relationship between motivation and attention with the ability to process a *media richness paradox*.

This paradox has profound implications on CMC research, since both Social Presence and Media Richness theories posits that F2F communication, as typical examples of rich/high-social-presence media, is better suited for highly equivocal

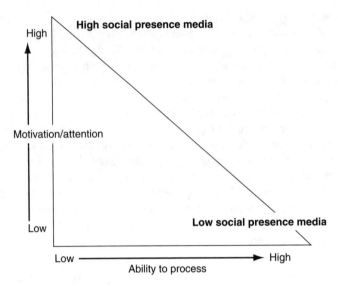

FIG. 8. Cognitive-Based View theory identified the inverse relationship between motivation and attention with the ability to process (adapted from Ref. [16]).

tasks. One of the criticisms often moved against these two theories is that they consider the "perceived" effectiveness of media from a sender's perspective. The cognitive-based model of Robert and Dennis reverses the perspective, analyzing from a receiver's point of view how media affect the change in understanding. In general, the greater the social presence of a medium, the greater the receiver's motivation has to be to participate in the communication process, but also the greater the sender's the ability to monitor attention. Thus, senders will require the use of rich media to ensure that receivers have high levels of attention and are motivated to process the message.

However, the level of social presence provided by media has an inverse relationship with the receiver's ability to process the message. One important media attribute is reviewability (or reprocessability), that is, the ability to allow the receiver to reprocess the information. In general, media with low social presence provide a higher level of reprocessability that allows the receiver to stop and think over important or difficult points. Also, the receiver can repeatedly access extra sources of information, and review the message until it is fully comprehended. In contrast, by social convention, media high in social presence do not allow individuals to elaborate at will, as they are supposed to respond quickly to avoid disrupting the conversation. Rich media high in social presence allow the receiver little ability to access multiple sources of information or reprocess the information. This is a major drawback because individuals have a natural constraint on the amount of information they can accept, process, and recall. Thus, when complex messages are sent over media high in social presence, reducing the amount of time one has to process ends up increasing the information load: A receiver can quickly become overwhelmed with information in a state, commonly referred to as information overload, "in which the amount of information that merits attention exceeds an individual's ability to process it" [66].

Also the number of receivers may impact the relationship between attention, motivation, and ability to process. In large groups or audiences, some receivers may not actively engage in processing the messages and will assume others will do it for them. This is referred to as "free riding." Free riding can go unnoticed because the sender is less able to monitor the behavior. While free riding can occur in either high or low social presence media, it is likely to be worse in low social presence media because monitoring the behavior of the receivers is more difficult than monitoring that of the senders. Past research has shown that members of electronic groups are more likely to ignore information [67].

As a conclusion, the use of rich media high in social presence should be used to assure attention for small amounts of information, whereas the use of lean media low in social presence causes a decreased motivation, but increases the ability to process large amounts of information during longer periods of time. Robert and Dennis argue that different media are needed for complex tasks where information overload may be generated. In such cases, the use of mixed media, or media switching, is motivated by the need to balance attention and motivation required by senders with the ability to process information of receivers. Depending on the task at hand, when senders want to get the attention of the receiver and motivate them for an immediate response, they should use a medium high in social presence. In contrast, when deep thought and deliberation are needed to process the information, the sender should use a medium low in social presence to give the receiver time to objectively elaborate on messages.

However, information overload is not the only risk when groups communicate F2F. The pressure on group members to conform on the view of the group majority has been acknowledged as the most severe dysfunctional aspect in F2F decision making [68]. The studies on group dynamics (e.g., see Ref. [51]) show that in group interactions there is a continuous interplay of task-oriented and relational process, as group members act certain roles while developing and maintaining some personal relationships. Thus, previous research on sociopsychological effects in CMC postulated that the reduction of socioemotional exchange contributes to increase group efficiency in the sense that less-rich communication media allow groups to pay less attention to interpersonal aspects of the interaction, and focus more on task.

Thus, groups interacting using lean media may benefit from using "less social" channels because the restriction imposed on the interpersonal information exchange allows for more-equal participation and greater attention paid to the messages, not to the individuals (i.e., less influenced by high-status member and less susceptible to the pressure of social consensus) [69]. For instance, the effectiveness in generative situations, like requirements elicitation, is less affected by "social noise" in communication. Instead, in problem-solving situations, like requirements negotiations, where social, emotional, and relationship concerns take time and effort away from task resolution, the use of "depersonalized" media may enhance group efficiency by leaving a greater portion of group-work time to task-oriented interaction [70].

To conclude this section, we show in Fig. 9 the research model updated to include the richness and synchronicity level variables, which characterize the media-related contextual factors.

5. Development of a Comprehensive Theoretical Framework

5.1 Managing the Context: The Effects of Task, Media, and Group Factors

The theories discussed in previous sections have framed a complex theoretical background for the selection of communication media for opportune remote group collaboration. Messages communicated to a group on channels that are inappropriate to the context may be misinterpreted by recipients or may be otherwise ineffective with regard to their intended purpose [71, 72]. In group research, context is defined by the group, task, and media factors. In Sections 2–4, we have analyzed each of

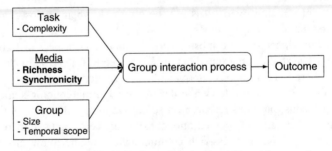

FIG. 9. The research model updated with the media-related variables, richness, and synchronicity.

these situational factors. In addition, the effects of these factors on group process and outcome also depend on their mutual interaction. Figure 10 shows the causal model updated to graphically represent the effects of these interactions. Given a specific group, its interaction process and outcome are heavily affected by the interaction occurring between task and media factors (A). For instance, task-medium mismatches may require communication participants to engage in compensating activities to clarify message content, leading to possible communication inefficiencies [64]. Likewise, given a specific task, group interaction process and outcome are heavily affected by the interaction occurring between group and media factors (B). For instance, group-medium mismatches may cause members of group unknown to each other to misinterpret message content due to the lack of shared experience, leading to possible performance inefficiencies [73].

How to measure group interaction process and its outcome largely depends on the type of task to be accomplished. For instance, if we again consider the definition of software requirements through elicitation and negotiation tasks (which we already compared, applying the Circumplex in Section 2.1), then group interaction can be evaluated through participants' perceptions, measuring the extent to which the process led to open participation of stakeholders, who were able to quickly resolve conflicts and overall, and how much satisfied they are with it. In addition, the outcome quality of both requirements elicitation and negotiation is reflected on a subjective level by the general consensus and satisfaction level attained by stakeholders at the end of the whole process, and, more objectively, by evaluating the quality of the requirements defined (e.g., by identifying defects through requirement documents inspections).

In the remainder of this section, we first discuss the theories for appropriately matching media characteristics with task and group. Then, we finally develop a comprehensive framework for the selection of communication media appropriate for the context, which consistently encompasses all the theories discussed so far.

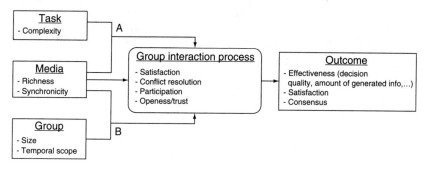

FIG. 10. The intertwined effects of media with task (A) and group (B) factors on group interaction and task outcome.

5.2 Matching Task and Media Characteristics

Although often conflicting, the CMC theories that we reviewed generally agree about the need to consider task characteristics for selecting the most appropriate media. As we already reported in Section 2.1, one of the most acknowledged limitation of McGrath's Task Circumplex is just its limited usefulness for determining technological support for executing groups task when group need to communicate over a medium. Thus, several frameworks have been developed to determine the best-fitting task-technology matches. In this section, we review the two most used frameworks proposed for task/technology fit.

5.3 Time-Interaction-Performance Theory

The Time-Interaction-Performance (TIP) theory, developed by McGrath and Hollingshead [17, 23, 74], has been among the first conceptual frameworks proposed to take into account the interaction of task and technology characteristics, in the evaluation of electronically mediated group interaction. TIP theory builds upon Task Circumplex and Media Richness theories, and hypothesizes that communication that occurs in the four tasks categories of the circumplex can be ordered by complexity and the amount of information required. In other words, the four task categories of the Task Circumplex, ordered by complexity, can be arranged in the same order along the media richness continuum hypothesized by Media Richness Theory (i.e., showing again that the more complex the tasks, the richer the information exchange required). Figure 11 illustrates the task-media fit attempted by the theory, with respect to the communication media.

The best-fitting combinations of information required by tasks and information conveyed by media lie near the main diagonal. Instead, the outer edges that are progressively distant from the diagonal represent less well-fitting to poor-fitting matches. For instance, generating tasks (e.g., brainstorming) may require only the transmission of ideas or plans, hence "less-rich" information. In contrast, tasks requiring groups to negotiate and resolve conflicts may require the transmission not only of facts, but also of affective messages or interpersonal communication, which are best conveyed by rich media. The figure shows that there are two types of poor-fit combinations: (1) when tasks require more information richness than selected media can deliver, groups are expected to suffer from problems of effectiveness and quality, forcing individual to exchange further compensative information; (2) when media provide more information richness than tasks require, groups are expected to suffer from problems of efficiency because media conveys not only facts, but also nonessential communication (e.g., interpersonal and affective messages), which brings distraction. In other words, the theory posits task-media fits are appropriate

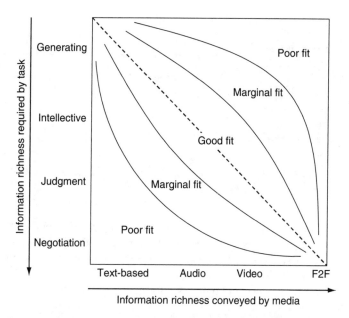

FIG. 11. The task-media fit suggested by the TIP theory (adapted from Ref. [74]).

only when the level of information richness of a medium is adequate to the complexity of the task. Thus, although TIP theory seems to only add to Media Richness theory an objective measure of task complexity, it actually argues that rich media do not always provide the best-fitting combination regardless of the task type.

5.4 Task/Technology Fit Theory

Consistently with what hypothesized by TIP theory, the theory of Task/Technology Fit (TTF), by Goodhue and Thompson [18] and Zigurs and Buckland [19], establishes a correspondence between task requirements and technology. TTF theory posits that, in a scenario of collaboration, the selection of an appropriate technology, which provides features and support "fitting" the task requirements, determines an increase of performance and, to some extent, of technology utilization itself (see Fig. 12).

Hence, TTF theory states that effectiveness of CMC varies on the type of task. For instance, tasks of idea generation that involve divergent thinking and limited member interdependence (e.g., in Task Circumplex, Type 1: planning, and Type 2: brainstorming) do not require information-rich media. On the other hand, more

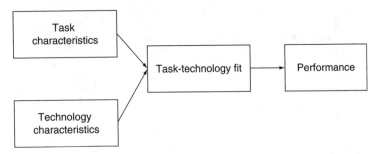

FIG. 12. Matching task and technology characteristics impacts performance and utilization (adapted from Ref. [19]).

intellective tasks (e.g., Type 3/4: problem solving, and Type 5/6: conflict resolution) involve a two-stage process: First, divergent thinking to identify all possible solutions, and secondly, convergent thinking to identify best suited solutions among those identified in the first step. Thus, convergent thinking involves a higher degree of member interdependence and requires information-rich media.

5.5 Matching Group and Media Characteristics

The TTF theory presented above completely neglects the effect of group in recommending the most appropriate matches. Conversely, the theory of Channel Expansion by Carlson and Zmud [20–22] posits that gaining experience with channel use and communication coparticipants[1] increases the perceived richness of that channel and the ability of individuals to communicate more effectively over it. As communication participants acquire these experiences (i.e., have a shared history of collaboration), they enhance their ability to encode/decode richer messages, for instance, referring to shared experiences or using shared jargon [22].

What this theory argues is that the scenario depicted by TTF theory in Fig. 12 describes a group collaboration at time T1, that is, when the group task is performed for the first time by a newly formed group, using a given fit (see Fig. 13). If this group happens to collaborate again for performing the same or a similar task, then the experience acquired on first iteration in collaborating with the same teammates over a medium (called appropriations and adaptations), will be reused in the next

---

[1] Actually, the theory identifies two other forms of relevant experience, namely experience with the messaging topic and the organizational context, for which Carlson and Zmud only found partial support. Besides, these forms of experience are not of interest here.

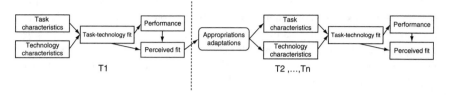

FIG. 13. Experiences influence recurring collaboration for a given task/technology fit.

iterations (T2,...,TN), thus positively influencing the perceived fit and group performance. This is a factor to take into account during empirical validation because group performance and task outcome is often evaluated through subjective data reports. However, it must be also pointed out that Carlson and Zmud found that, over time, the influence of experience (adaptations and appropriations) tends to diminish and eventually stabilizes.

Channel Expansion theory does not address the channel selection process, though. Instead, it is concerned only with the increasing perceived richness level of a given channel and the ability to communicate more effectively over it with time. Nevertheless, the theory can be used as predictive of the effects of temporal scope in matching group and media characteristics. Channel experience is gained through use and thus, it is related to the length of time a channel has been utilized. Likewise, experience with group members is developed through interaction and, thus, it is related to the group history, or the extent to which a group has worked together in the past. Hence, established groups with a shared history of previous collaboration, are expected to be able to communicate effectively also over impoverished media, like e-mail. Conversely, *ad hoc* groups are newly formed and thus do not have any shared experience that can help compensate for the leanness of the medium in use. Consequently, *ad hoc* groups are expected to benefit from the use of rich medium more than established groups. These results are consistent with the theory of Common Ground (see Section 4.3). Group with shared experiences have already established a certain amount of common ground and thus can communicate well even over leaner media.

5.6 Development of a Comprehensive Theoretical Framework

The theoretical frameworks reviewed on media effects, tasks, and group processes have depicted a complex research area. The complexity is reflected by the equivocality of the existing body of knowledge from previous studies conducted to evaluate the (in)effectiveness of computer-mediated group interaction as compared to F2F. The consistent combination of all these group-, task-, and media-related theories resulted in a fully comprehensive framework, which encompasses all the forces, generated from situational factors, which act on the selection process of the most appropriate media for the context. Figure 14 illustrates a graphical representation of our general-purpose framework.

The figure above shows the inversely proportional, main characteristics of rich and lean media. Rich media (e.g., audio and video channels, F2F) are highly synchronous and low parallel, convey a high sense of social copresence of individuals, ensure a higher level of attention and motivation, facilitate mutual understanding (see the top box in the figure). Thus, rich media are more beneficial, especially for groups with no history, whose members are unknown to each other. One risk with rich media is the information overload, due to the multiple channels available at one and the low reprocessability of the information conveyed over them. Conversely, lean media (e.g., e-mail, text chat, IM) are lowly synchronous but highly parallel, convey a low sense of social copresence, motivation, and attention

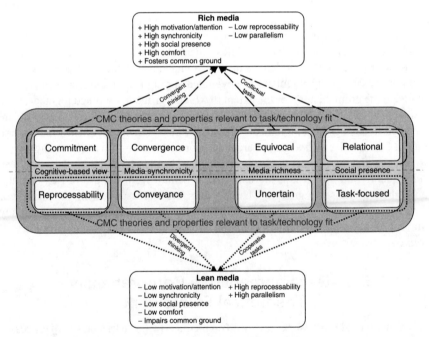

FIG. 14. The comprehensive framework for task/technology fit resulting from the consistent combination of group-, task-, and media-related theories.

(see the bottom box in the figure). Lean media are more effectively used by groups of individuals who share a history of previous collaborations. One advantage of lean media over rich media is the possibility to reprocess the information exchanged, which is otherwise volatile.

The CMC theories reviewed have been divided into task-centric and process-centric theories groups. Task-centric theories (i.e., Social Presence and Media Richness) consider communication as a task to be accomplished by individuals, whereas process-centric theories (i.e., Media Synchronicity and Cognitive-Based View) regard it as a process to be performed by individuals. All these theories, however, define communication through task or process dichotomies. The arrows represent the driving forces that act on the selection process, pushing for the selection of appropriate fits between tasks and synchronous media properties. These forces are not only useful for predicting and evaluating the goodness of TTFs, that is, poor (−), marginal (+/−), and good fits (+). In fact, here we also use the framework to ultimately compare the fits between synchronous text-based communication and distributed requirements workshops.

As an example of application of our framework, we use it to evaluate and suggest the best TTFs for running distributed requirements elicitation and negotiation workshops. According to Task Circumplex classification, negotiating software requirements is a complex, intellective task that involves different subactivities, both conceptual and behavioral, where conflicts have to be resolved to converge readily to one solution among the many identified, thus reaching consensus in a timely manner and enhancing the decision-making process quality. From the point of view of the task-centric theories, a requirements negotiation is a conflictual task characterized by high equivocality and member interdependence, which requires not only task-focused messages, but also social information to be exchanged. From the perspective of communication as a process, resolving ambiguities means that opposing individual views must converge into a single shared view. All these forces consistently drive to the selection of rich media for conducting effective requirements negotiation workshops and, consequently, also show that synchronous text-based communication and requirements negotiation represent a poor TTF. Hence, for instance, videoconferencing negotiation workshops represent a good fit (+), whereas synchronous text-based conferring negotiation workshop is evaluated as a poor fit (−).

According to Task Circumplex classification, elicitation is a creativity task, where new ideas or different solutions to a given problem have to be generated. Idea generation requires a low degree of member interdependence because it involves only divergent thinking. Thus, from the perspective of task-centric theories, elicitation is a cooperative, task-focused activity with limited degree member interdependence and consequently, a little need of communicating social information, which may make participant more susceptible to pressure of social consensus and domination, and take time away from task-oriented interaction. The uncertainty existing in a generative task can only be reduced by conveying additional information. Hence, from the perspective of process-centric theories, the conveyance of information is better supported by lean media, high in parallelism (or low in synchronicity), which foster idea generation by allowing multiple individuals to contribute information at the same time. Thus, all these forces consistently drive to the selection of lean media for conducting effective requirements elicitation workshops. Nevertheless, in the evaluation of Task-Technology Fits, we must also take into account the existing counter forces. In fact, the use of lean mean has a detrimental effect on the level of satisfaction and motivation/attention perceived by participants, which, conversely increases as rich media are used. In addition, compared to established groups, members of *ad hoc* groups are expected to communicate less effectively over impoverished media, since they cannot use any shared experiences to compensate for the media leanness. As a conclusion, the framework evaluates that both lean and rich media (e.g., synchronous text-based and video conferencing) used for running distributed elicitation workshops represent marginal TTFs (+/−).

6. Conclusions

In this chapter, we have reviewed a large body of theories related to group and group tasks, as well as CMC theories. In particular, we reviewed McGrath's Task Circumplex framework, the most widely used model to categorize tasks, and objectively evaluate and compare their complexity in group research. This chapter also contributed to the study of a particular kind of short-term, dynamic groups, namely *ad hoc* groups, for which we have reviewed the existing literature and proposed a new definition (i.e., small- to medium-sized teams, highly dynamic in creation, participation, and release, with no past and future of collaborations, whose temporal scope corresponds exactly to the time needed to carry out the collaboration in attendance). Short-term collaborations represent an emerging scenario and, consequently, a relevant topic to group research.

Besides, we have reviewed the most prominent theories on CMC. We showed that the theoretical background on CMC is rather complex and equivocal. On the one hand, the theories of media richness posit that the more complex the task, the richer the medium to adopt. Namely, Social Presence, Media Richness, and Common Ground have overwhelmingly reported about the inadequateness of text-based communication, as compared to rich media, like F2F or video communication. Such disregard is due to the fact that lean media (e.g., e-mail and instant messaging) lack the ability of conveying nonverbal cues that contributes to the level of social presence (e.g., gaze, tone of voice, facial expressions), which in turns fosters individuals' motivation and mutual understanding. On the other hand, however, sociopsychological and cognitive theories postulate that the depersonalization effect imposed by lean media can be beneficial for reducing both the information overload and the emotional side effects, like domination and social consensus pressure observed with rich media, thus increasing the meeting effectiveness in group communication. Media Synchronicity theory asserts that the effectiveness of CMC depends also on contextual factors other than media richness, such as communication channel synchronicity, task typology, and group temporal scope. Furthermore, Media Richness Paradox argued that the use of rich media high in social presence should be used to assure attention for small amounts of information, whereas the use of lean media low in social presence causes a decreased motivation, but increases the ability to process large amounts of information during longer periods of time. Drawing upon these theories, we have argued that, by understanding the paradoxical effects of rich media high in social presence, groups may be better able to select and use the most appropriate sets of media to accomplish their tasks.

As a result, we have built two general-purpose models, meant to support experiments in the field of distributed group research. The first model is intended to serve as a reference framework to define the context of the empirical study, thus helping to identify the task-, group-, and media-related variables involved. The second model, instead, consistently combines the most prominent theories on CMC and the Task Circumplex to graphically represent a theoretical framework on media effects, for describing, predicting, and comparing the goodness of Task-Technology Fits. These models can serve as references in setting up of experiments on distributed group research, as well as in the discussion of related findings.

Acknowledgments

We are grateful to Daniela Damian for reviewing and helping us to improve the two frameworks proposed here.

References

[1] A. Dix, J. Finley, G. Abowd, R. Beale, Human–Computer Interaction, third ed., Upper Saddle River, NJ: Prentice Hall, 2003.
[2] G.M. Olson, J.S. Olson, Distance matters, Hum. Comput. Interaction 15 (2/3) (2000) 139–178.
[3] J.D. Herbsleb, A. Mockus, T.A. Finholt, R.E. Grinter, An empirical study of global software development: distance and speed, in: Proccedings of the International Conference on Software Engineering (ICSE '01), Toronto, Canada, 2001, pp. 81–90.

[4] D. Damian, D. Zowghi, Requirements engineering challenges in multi-site software development organizations, Requirements Eng. J. 8 (2003) 149–160.
[5] D. Damian, A. Eberlein, M. Shaw, B. Gaines, An exploratory study of facilitation in distributed requirements engineering, Requirements Eng. 8 (1) (2003) 23–41.
[6] B. Nuseibeh, S. Easterbrook, Requirements engineering: A roadmap, in: Proccedings of the International Conference on Software Engineering (ICSE '00), Limerick, Ireland, 2000, pp. 35–46.
[7] D. Zowghi, C. Coulin, Requirements elicitation: A survey of techniques, approaches, and tools, in: A. Aurum, C. Wohlin (Eds.), Engineering and Managing Software Requirements, 2005, pp. 19–46.
[8] J. Short, E. Williams, B. Christie, The Social Psychology of Telecommunications, Wiley, London, 1976.
[9] R.L. Daft, J. Wiginton, Language and organization, Acad. Manage. Rev. 4 (2) (1979) 179–191.
[10] R.L. Daft, R.H. Lengel, Information richness: A new approach to managerial behaviour and organizational design, in: B.M. Staw, L.L. Cummings (Eds.), Research in Organizational Behaviour, vol. 6, CT JAI Press, Greenwich, 1984, pp. 191–233.
[11] R.L. Daft, R.H. Lengel, Organizational information requirements, media richness and structural design, Manage. Sci. 32 (5) (1986) 554–571.
[12] H.H. Clark, S.E. Brennan, Grounding in communication, in: Perspectives on Socially Shared Cognition, American Psychological Association, Washington, DC, 1991, pp. 127–149.
[13] A.R. Dennis, J.S. Valacich, Testing media richness theory in the new media: The effects of cues, feedback, and task equivocality, Inf. Syst. Res. 9 (3) (1998) 156–274.
[14] A.R. Dennis, J.S. Valacich, C. Speier, M.G. Morris, Beyond media richness: An empirical test of media synchronicity theory, in: Proccedings of the 31st Hawaii International Conference on System Sciences (HICSS-31), 1998, pp. 48–57 vol. 1.
[15] A.R. Dennis, J.S. Valacich, Rethinking media richness: Towards a theory of media synchronicity, in: Proccedings of the 32nd Hawaii International Conference on Syst. Sci. (HICSS-32) 1 (1999) 1017–1026.
[16] L.P. Robert, A.R. Dennis, Paradox of richness: A cognitive model of media choice, IEEE Trans. Prof. Commun. 48 (1) (2005) 10–21.
[17] J.E. McGrath, Time, interaction, and performance (TIP): A theory of groups, Small Group Res. 22 (2) (1991) 147–174.
[18] D.L. Goodhue, R.L. Thompson, Task-technology fit and individual performance, MIS Q. 19 (2) (1995) 213–236.
[19] I. Zigurs, B.K. Buckland, A theory of task/technology fit and group support systems effectiveness, MIS Q. 22 (3) (1998) 313–334.
[20] J. Carlson, R. Zmud, Channel expansion theory: A dynamic view of media and information richness perceptions, in: D.P. Moore (Ed.), Academy of Management Best Papers Proceedings, 1994, pp. 280–284.
[21] J. Carlson, PhD ChapterChannel Expansion Theory: A Dynamic View of Media and Information Richness Perceptions, Florida State University, Tallahassee, 1995.
[22] J.R. Carlson, R.W. Zmud, Channel expansion theory and the experiential nature of media richness perceptions, Acad. Manage. J. 42 (2) (1999) 153–170.
[23] J.E. McGrath, Groups: Interaction and Performance, Prentice Hall, Englewood Cliffs, NJ, 1984.
[24] J.F. Nunamaker, A.R. Dennis, J.S. Valacich, D.R. Vogel, J.F. George, Electronic meeting systems to support group work, Commun. ACM 34 (7) (1991) 40–61.
[25] D.J. Campbell, Task complexity: A review and analysis, Academy of Manage. Rev. 12 (1) (1988) 40–52.
[26] J.R. Hackman, Toward understanding the role of tasks in behavioral research, Acta Psychol. 31 (1969) 97–128.

[27] R.E. Wood, Task complexity: Definition of the construct, Organ. Behav. Hum. Decis. Process 37 (1986) 60–82.
[28] B.E. Mennecke, B.C. Wheeler, Tasks matter: Modeling group task processes in experimental CSCW research, in: Proceedings of the 26th Hawaii International Conference on System Sciences (HICSS-26) 4 (1993) 71–80.
[29] D. Damian, Empirical Studies of Computer Support for Distributed Requirements Negotiation, Department of Cs, University of Calgary, 2001 (PhD Chapter).
[30] B.E. Mennecke, J.A. Hoffer, B.E. Wynne, The implications of group development & history for gss theory & practice, Small Group Res. 23 (4) (1992) 524–572.
[31] J. Fjermestad, R. Hiltz, Case and field studies of group support systems: An empirical assessment, in: Proccedings of the International Hawaii Conference on System Science (HICSS-33), Hawaii, 04-January 07, vol. 1, pp. 4–7.
[32] E. Christensen, J. Fjermestad, Challenging group support systems research: The case for strategic decision making, Group Decis. Negotiation 6 (1997) 351–372.
[33] A.R. Dennis, A.C. Easton, G.K. Easton, J.F. George, J.F. Nunamaker, Ad hoc versus established groups in an electronic meeting system environment, in: Proccedings of the 23rd Hawaii International Conference on System Sciences (HICSS-23), vol. 3, 1990, pp. 23–29.
[34] U.S. Murthy, D.S. Kerr, Task/technology fit and the effectiveness of group support systems: Evidence in the context of tasks requiring domain specific knowledge, in: Proccedings of the 33rd Hawaii International Conference on System Sciences (HICSS-33) 2, 2000, pp. 1–10.
[35] A.R. Dennis, J.F. Nunamaker, D.R. Vogel, A comparison of laboratory and field research in the study of electronic meeting systems, J. Manage. Inf. Syst. 7 (2) (1991) 107–135.
[36] A.R. Dennis, R.B. Gallupe, A history of GSS empirical research: Lessons learned and future directions, in: L. Jessup, J. Valacich (Eds.), Group Support Systems: New Perspectives, MacMillan, New York, 1993, 59–77.
[37] I. Benbasat, L. Lim, The effects of group, task, context and technology variables on the usefulness of group support systems: A meta-analysis of experimental studies, Small Group Res. 24 (1993) 430–462.
[38] A.B. Hollingshead, J.E. Mcgrath, K.M. O'Connor, Group task performance and communication technology, Small Group Res. 24 (3) (1993) 307–333.
[39] M. Hauswirth, I. Podnar, S. Decker, On P2P collaboration infrastructures, 14th IEEE International WETICE Workshops on Enabling Technologies: Infrastructure for Collaborative Enterprise (DMC '05), Linkoping, Sweden, June 13–15, 2005, pp. 66–71.
[40] U.M. Borghoff, J.H. Schlichte, Computer-Supported Cooperative Work: Introduction to Distributed Applications, Springer, Heidelberg, 2000.
[41] M. Hefke, L. Stojanovic, An ontology-based approach for competence bundling and composition of Ad-Hoc teams in an organization, in: Proceedings of the 4th International Conference on Knowledge Management (I-KNOW '04), Journal of Universal Computer Science (J.UCS), Graz, Austria, June, 2004, pp. 126–134.
[42] K. Knoll, S. Jarvenpaa, Learning to work in distributed global teams, in: Proccedings of the 28th Hawaii International Conference on System Sciences (HICSS-28), 1995, pp. 92–101, 3–6 Jan. 1995, vol. 4.
[43] F. Lanubile, T. Mallardo, F. Calefato, Tool support for geographically dispersed inspection teams, in: Software Process: Improvement and Practice, vol. 8, No. 4. John Wiley & Sons, Ltd., Hoboken, NJ, Interscience, 2003, pp. 217–231.
[44] R. Davison, Socio-psychological aspects of group processes, 1995, http://www.is.cityu.edu.hk/Research/WorkingPapers/paper/9503.pdf, 21 October 2009.

[45] I. Lorge, D.D. Fox, M. Brenner, A survey of studies contrasting the quality of group performance and individual performance, Psychol. Bull. 6 (5) (1958) 337–357.
[46] J. Hall, M.S. Williams, A comparison of decision making performance in established and *Ad Hoc* groups, J. Pers. Soc. Psychol. 3 (1966) 214–222.
[47] E.G. Bormann, The paradox and promise of small group research, Speech Monogr. 37 (3) (1970) 211–217.
[48] B.E. Mennecke, J.A. Hoffer, J.S. Valacich, An experimental examination of group history and group support system use on information sharing performance and user perceptions, in: Proccedings of the 28th Hawaii International Conference on System Sciences (HICSS-28) 4 (1995) 153–162.
[49] B.J. Alge, C. Wiethoff, H.J. Klein, When does the medium matter? Knowledge-building experiences and opportunities in decision-making teams, Organ. Behav. Hum. Decis. Process 91 (2003) 26–37.
[50] J.P. Birnholtz, T.A. Finholt, D.B. Horn, S.J. Bae, Grounding needs: Achieving common ground via lightweight chat in large, distributed, Ad-Hoc groups, in: Proccedings of the International Conference on Human Factors in Computing Systems (CHI '05), Portland, USA, April 2–7, 2005, pp. 21–30.
[51] B.W. Tuckman, Developmental sequence in small groups (reprint), Group Facilitat. Res. Appl. J. 3 (2001), http://dennislearningcenter.osu.edu/references/GROUP%20DEV%20ARTICLE.doc, 21 October 2009.
[52] W. Orlikowski, Learning from notes: Organizational issues in groupware implementation, in: Proccedings of the International Conference Computer Supported Cooperative Work (CSCW '92), Toronto, Canada, 1992, pp. 362–369.
[53] F. Lanubile, A P2P toolset for requirements elicitation, in: Proccedings of the ICSE International Workshop on Global Software (GSD '03), Portland, Oregon, USA, May 9, 2003.
[54] F. Calefato, F. Lanubile, T. Mallardo, Peer-to-peer remote conferencing, in: Proceedings of ICSE Workshop on Global Software Development (GSD '04), IEE Publishing, Edinburgh, Scotland, UK, 2004, pp. 34–38.
[55] F. Calefato, F. Lanubile, A decentralized conferencing tool for *Ad-hoc* distributed workgroups, in: Proceedings of ASE Workshop on Cooperative Support for Distributed Software Engineering Processes (CSSE '04), Austrian Computer Society, Linz, Austria, 2004, pp. 27–38.
[56] C.A. Ellis, S.J. Gibbs, G. Rein, Groupware: Some issues and experiences, Commun. ACM 34 (1) (1991) 39–58.
[57] W. Geyer, J. Vogel, L.T. Cheng, M. Muller, Supporting activity-centric collaboration through peer-to-peer shared objects, in: Proccedings of the International Conference on Supporting Group Work (Group '03), 2003 Sanibel Island, FL, USA.
[58] M. Handel, J.D. Herbsleb, What is chat doing in the workplace? in: Proccedings of the International Conference on Computer-Supported Cooperative Work (CSCW '02), 2002 New Orleans, LA, USA.
[59] J.D. Herbsleb, D.L. Atkins, D.G. Boyer, M. Handel, T.A. Finholt, Introducing instant messaging and chat into the workplace, in: Proccedings of the International Conference on Computer-Human Interaction (CHI '02), Minneapolis, MN, USA, 2002.
[60] W. Cunningham, B. Leuf, The Wiki Way. Quick Collaboration on the Web, Addison-Wesley, Boston, MA, 2001.
[61] C.N. Gunawardena, F.J. Zittle, Social presence as a predictor of satisfaction within a computer-mediated conferencing environment, Am. J. Distance Educ. 11 (3) (1997) 8–26.
[62] C.E. Shannon, W. Weaver, The Mathematical Theory of Communication, University of Illinois, Urbana, 1949.
[63] T. Erickson, W.A. Kellogg, Social translucence: An approach to designing systems that support social processes, ACM Trans. Comput. Hum. Interaction (TOCHI) 7 (1) (2000) 59–83.

[64] J.E. McGrath, A.B. Hollingshead, Putting the "Group" back in group support systems: Some theoretical issues about dynamic processes in groups with technological enhancements, in: L.M. Jessup, J.S. Valacich (Eds.), Group Support Systems: New Perspectives, Macmillan, New York, 1993, pp. 78–96.
[65] T.B. Sheridan, Musings on telepresence and virtual presence, Presence 1 (1) (1992) 120–126.
[66] U. Schultze, B. Vandenbosch, Information overload in a groupware environment: Now you see it, now you don't, J. Org. Comput. Electron. Commerce 8 (2) (1998) 127–148.
[67] G.M. Phillips, G.M. Santoro, Teaching group discussion via computer-mediated communication, Commun. Educ. 38 (2) (1989) 151–161.
[68] S.R. Hiltz, M. Turrof, The Network Nations: Human Communication via Computer, Addison-Wesley, Reading, MA, 1978.
[69] S. Kiesler, J. Siegel, T.W. McGuire, Social psychological aspects of computer-mediated communication, Am. Psychol. 39 (10) (1984) 1123–1134.
[70] J.B. Walther, Computer-mediated communication: Impersonal, interpersonal, and hyperpersonal interaction, Commun. Res. 23 (1) (1996) 3–43.
[71] L. Trevino, R. Lengel, R. Daft, Media symbolism, media richness, and media choice in organizations, Commun. Res. 14 (1987) 553–574.
[72] L. Trevino, R. Lengel, W. Bodensteiner, E. Gerloff, N. Muir, The richness imperative and cognitive style, Manage. Commun. Q. 4 (1990) 176–197.
[73] J.E. McGrath, Methods for the study of groups, in: R.M. Baecker (Ed.), Groupware and Computer-Supported Cooperative Work, Morgan Kaufmann, San Mateo, 1993, pp. 200–204.
[74] J.E. McGrath, A.B. Hollingshead, Groups interacting with technology: Ideas, evidence, issues and an agenda, Thousand Oaks, CA: Sage Publications, Inc., 1994.

Author Index

A

Abarbanel, H., 88, 133
Abbi, R., 43, 52
Abowd, G., 272–273
Abts, C., 232
Adami, A., 126
Agostini, G., 83, 118, 122, 136–137
Ahrendt, P., 136
Alamouti, S. M., 207
Alge, B. J., 283–285
Al-Khatib, W. G., 113–114, 133, 136–137
Allamanche, E., 92, 117–118, 134, 137
Almasganj, F., 134, 136
Alsteris, L. D., 115, 134, 136
Anderson, C., 155
Anderson, D., 86, 92, 125, 131, 133, 135, 137
Andre-Obrech, F., 134
Andrews, J. G., 158
Angelis, L., 235–237, 239, 245
Arkko, J., 194
Atal, B. S., 126, 136
Atkins, D. L., 287
Atlas, L. E., 86–87, 92, 128, 134–136
Aucouturier, J.-J., 134

B

Bae, S. J., 283
Bai, J., 111, 114, 136–137
Bai, L., 133, 137
Balbinot, R., 134
Bang, S. Y., 134
Baojie, L., 136
Baresi, L., 259
Barone, D., 126
Bartsch, M. A., 121, 135–136
Batlle, E., 118, 135, 138
Bauer, R., 3
Beal, A., 14
Beale, R., 272–273
Becker, R., 134
Behroozmand, R., 134, 136
Benbasat, I., 280, 284–285
Berger, J., 83, 87, 124
Berghel, H., 42
Birnholtz, J. P., 283
Blank, D., 42
Blum, D., 111, 117, 120, 136–137
Bodensteiner, W., 300
Boehm, B., 226, 230, 232, 235–236, 239
Bogert, B., 87, 124
Booth, A. D., 50
Borghoff, U. M., 282
Bormann, C., 193, 203
Bormann, E. G., 283
Botafogo, R., 254
Bourlard, H., 119
Boyer, D. G., 287
Brazil, E., 135
Breebaart, J., 110, 128, 136–137
Breiteneder, C., 111, 134, 136
Brennan, S. E., 274, 290
Brenner, M., 283
Briand, L. C., 239, 250–252, 265
Bridle, J. S., 101, 124, 136
Brieman, L., 241

Broder, A., 44
Brown, A., 232
Brown, M. D., 101, 124, 136
Buckland, B. K., 275, 277, 280–281, 303–304
Burges, C. J. C., 92, 132, 136
Burmeister, C., 193, 203
Burns, E., 4–5
Buscicchio, C. A., 134
Byrd, D., 83

C

Cai, L. H., 110, 114, 118, 122, 134, 136–137
Cai, R., 110, 114, 118, 122, 134, 136–137
Calefato, F., 282, 285
Campbell, D. J., 277, 280–281
Caneel, R., 134
Cano, P., 138
Caponetti, L., 134
Carbonell, J., 56
Carlson, J., 275, 304
Cartwright, L., 239
Casner, S., 181, 195, 204
Castellano, G., 134
Castello, F., 134
Castrucci, M., 213
Cates, P., 235
Cha, J., 170
Champbell, J. P., 113
Chan, C. G., 134, 137
Chang, S. F., 85, 138
Changsheng, X., 136
Chang, S. J., 45
Chatterjee, M., 197
Chau, M., 46–47
Cheng, C. C., 110, 134, 136–137
Cheng, L. T., 287
Cheng, W. H., 111, 134, 136–137
Chen, H., 45–49, 54, 56, 58–59, 63
Chen, J., 133, 137
Chen, K.-C., 158
Chen, M., 239
Chen, T., 111, 118, 127, 136–137

Chen, Y., 46–47
Chia, L., 124
Chien, L.-F., 49
Choi, E. H. C., 125, 134, 136
Choi, J., 170
Choi, S., 134
Chou, W., 122, 128, 136
Cho, Y. C., 134
Cho, Y. D., 120
Christel, M., 47
Christensen, E., 279, 285
Christie, B., 274, 287
Christodoulakis, D., 50
Christodoulou, S. P., 265
Chrzanowski, M. J., 47
Chuang, S.-L., 49
Chuang, Z. J., 110, 120, 134, 137
Chulani, S., 232
Chung, W., 45, 48–49, 54, 56, 58–59, 61, 63
Chu, W. T., 111, 134, 136–137
Clark, B., 232
Clark, H. H., 274, 290
Clausen, M., 92, 121, 135–136
Conte, S., 250, 253, 258
Cook, P., 131, 135–137
Cool, C., 44
Cooper, M., 135
Corazza, A., 264, 267
Corsetti, G., 134
Costa, J. M., 156, 193
Coulin, C., 274, 279
Counsell, S., 227, 239–240, 254, 259, 262–263, 265
Cove, J. F., 44
Cowderoy, A. J. C., 258
Cowling, M., 74
Crawford, T., 83
Cremer, M., 92, 137
Cunningham, P., 131, 134–136
Cunningham, W., 287
Curado, M., 213
Curran, K., 5
Cutts, M., 19, 22, 30, 35

D

Daft, R. L., 274, 288–289, 300
Dai, H. K., 9
Damian, D., 274, 279
David, B., 95
Davison, R., 282
Davis, S., 124, 136
Davy, M., 74, 138
Decker, S., 282
Deering, S., 193, 202
Deerwester, S., 14
Degermark, M., 193, 203
Deller Jr., J. R., 134
De Marca, J. R. B., 158
DeMarco, T., 228
Dennis, A. R., 275–276, 280, 283–284, 293, 295–298
De Silva, L. C., 134
Di Martino, S., 264, 267
Dimitriadis, D., 136
Dimitroff, A., 4, 11, 13
Dimou, P., 245
Divakaran, A., 85, 134, 136, 138
Dix, A., 272–273
Dong, Y., 125, 136
Downie, J. S., 74
Duan, L., 124
Dumais, S., 14
Dunsmore, H., 250, 253, 258

E

Easterbrook, S., 279
Easton, A. C., 280, 283–284
Easton, G. K., 280, 283–284
Eberlein, A., 274
Effelsberg, E., 134
Effelsberg, W., 119
Efthymiou, N., 119, 136
El-Emam, K., 239
Ellis, C. A., 286
El-Maleh, K., 110, 113, 136–137
Ennis, M., 44

Erickson, T., 291
Ermolinskyi, A., 137
Ertel, C., 117, 134, 137
Esmaili, S., 134
Essid, S., 95
Essl, G., 131, 136

F

Fanelli, A. M., 134
Fang, C., 136
Fan, H., 46
Farinas, J., 134
Fastl, H., 83–84, 90, 116–117, 119, 121, 126–128, 137
Fellbaum, C., 50
Fenton, N., 235
Fernström, M., 135
Ferrucci, F., 264, 267
Fewster, R. M., 258–259
Finholt, T. A., 273, 283, 287
Finley, J., 272–273
Firmin, T., 47
Fischer, S., 134
Fitzek, F. H. P., 156–157, 196–197, 204
Fjermestad, J., 279, 285
Fletcher, H., 91
Foo, S. W., 134–135
Foote, J., 129, 135–136
Forey, S., 235
Fox, D. D., 283
Frba, B., 92, 137
Frederick, R., 181, 195
Friedman, J., 241
Fukuda, T., 110, 137
Fukushima, H., 193, 203
Furnas, G., 14

G

Gadde, V. R., 115, 136
Gaines, B., 274
Gajic, B., 86, 92, 136–137
Gallupe, R. B., 280

Ganguly, S., 197
Gay, G, 2
Geevarghese, J., 204
Gehrmann, T., 130, 136
George, J. F., 276, 280, 283–284
Gerloff, E., 300
Geyer, W., 287
Ghulam, M., 110, 137
Gibbs, S. J., 286
Glasberg, B. R., 89–90
Godsill, S. J., 138
Goldstein, J., 56
Goodhue, D. L., 275, 303
Goto, M., 121, 135–136
Granka, L. A., 2
Gravino, C., 264, 267
Gray, A., 235
Gray, R., 228
Greenberg, A., 31
Greenberg, S., 88, 128, 137
Greene, R. J., 134
Greene, S., 47
Grimaldi, M., 131, 134–136
Grinter, R. E., 273
Guaus, E., 118, 135
Guedes Silveira, J., 134
Gu, L., 122, 128, 136
Gunawardena, C. N., 288
Gu, Q. R., 134

H

Hackman, J. R., 277–278
Haitsma, J., 138
Hakenberg, R., 193, 203
Häkkinen, J., 125, 136
Hall, J., 283–284
Handel, M., 287
Handschuh, S., 47
Hanjalic, A., 110, 114, 118, 122, 134, 136–137
Hannu, H., 193, 203
Hansen, J. H. L., 134, 137
Han, Y.-H., 170

Harjula, I., 157, 196–197, 207
Harshman, R., 14
Hauptmann, A., 47
Hauswirth, M., 282
Healy, M., 87, 124
Hearst, M. A., 56
Hefke, M., 282–283
Hegde, M., 115
Hegde, R. M., 115, 136
Heinemann, C., 134, 136
Helmuth, O., 92, 118, 137
Herbsleb, J. D., 273, 287
Hermansky, H., 86, 119, 126, 137
Herre, J., 92, 117–118, 134, 137
Herrera, P., 123, 137
Hess, W., 120
Hiltz, R., 279
Hiltz, S. R., 299
Hinden, R., 193, 202
Hippsley, H., 31
Hirose, K., 134, 136
Hoffer, J. A., 279, 283–284
Hollingshead, A. B., 280, 293, 301, 303
Horikawa, J., 110, 137
Horn, D. B., 283
Horowitz, E., 232
Houston, A., 47
Houtgast, T., 127
Hsu, C. T., 110, 134, 136–137
Hsu, J. Y. J., 111, 134, 136–137
Huang, J. C., 85–86, 118, 138
Huang, R., 134, 137
Huang, Y. C., 137
Huang, Z., 59
Huusko, J., 204
Hu, Y., 133, 137
Hwang, T. H., 92, 125, 134

I

Ikbal, S., 119
Inanoglu, Z., 134
Ingwersen, P., 44
Izmailov, R., 197

J

Jacobson, V., 181, 195
Jana, S., 92, 132, 136
Jang, H., 170
Jansen, B. J., 45
Jarvenpaa, S., 282
Jayant, N. S., 118, 137
Jee, J., 170
Jeffery, R., 261
Jehan, T., 134, 136
Jenor, S. K., 137
Jensen, S., 189
Jeong, J.-H., 110, 137
Jiang, H., 111, 114, 129, 136–137
Jimaa, S., 114, 136–137
Joachims, T., 2
Johnson, D., 194
Johnson, M. T., 88, 133, 137
Jones, G. J. F., 134, 137
Jonsson, L.-E., 193, 203
Jørgensen, M., 227
Juang, B., 74, 134

K

Kabal, P., 110, 113, 136–137
Kadoda, G., 239
Kalker, T., 138
Kankanhalli, M. S., 116, 118, 121, 123, 125, 134–137
Kasten, T., 92, 137
Katz, M. D., 156, 204, 213
Kawaguchi, A., 46
Kazuhiro, K., 37
Kazumi, S., 37
Kedem, B., 110
Kellogg, W. A., 291
Kennedy, D. J., 37
Kerr, D. S., 280
Kerstetter, J., 34
Khan, M. K. S., 113–114, 133, 136–137
Kiernan, B. G., 156, 193
Kiesler, S., 300

Kil, R. M., 110, 136–137
Kim D. S., 110, 136–137
Kim, H., 116, 118, 122–123, 132, 136–137
Kim, J.-W., 110, 137
Kim, M. Y., 120
Kim, S. R., 120
Kimuara, M., 37
Kingsbury, B. E. D., 88, 128, 137
Kinnunen, T., 136
Kirakowski, J., 265
Kitchenham, B. A., 235, 250–252, 265, 267
Klapuri, A., 74
Kleinberg, J., 50
Klein, H. J., 283, 285
Klein, M., 110, 113, 136–137
K∈necht, A., 15
Knoll, K., 282
Kohler, M. A., 134
Kohonen, T., 47, 56
Kokaram, A., 131, 134–136
Kokkinos, I., 132–134, 137
Kokosis, P., 50
Kok, P., 265
Komiya, R., 134
Korb, K., 245–246
Koren, T., 193, 203–204
Krikos, V., 50
Krishna, A. G., 113, 136
Krishna, B. A., 235, 267
Krishnan, S., 114, 116, 118–119, 134–137
Krishnan, V., 36
Kuhlthau, C., 43–44
Kumar, R., 49
Kumar, S., 235, 267
Kümmerer, M., 119, 136
Kuo, C. C. J., 86, 111, 128–129
Kurth, F., 92, 121, 130, 135–136

L

Lally, A., 46–47
Lancini, R., 118
Landauer, T., 14

Landon, M., 47
Lane, C. E., 90
Lanubile, F., 282, 285
Lao, S., 133, 137
Larsen, J., 136
Lee, C.-H., 47, 49
Lee, K. Y., 134
Leem, W. T., 135
Lee, S. Y., 110, 136–137
Lee, Y. W., 45
Le, K., 193, 203
Lengel, R., 300
Lengel, R. H., 274, 288–289
Leroy, G., 46
Leuf, B., 287
Levialdi, S., 45
Lew, M. S., 77
Liang, Q., 50
Lienhart, R., 119
Li, J., 50
Li, K. W., 50
Lim, L., 280, 285
Li, N., 50
Linars, G., 136–137
Lindgren, A. C., 88, 133, 137
Li, Q., 94, 113, 125, 136–137
Li, S. Z., 95, 111, 118, 136–137
Li, T., 94, 113–114, 116, 118, 134–137
Liu, M., 86, 113–114, 116, 124, 134, 136–137
Liu, Z., 85–86, 111, 118, 127, 136–138, 193, 203
Li, Y., 9
Löhken, I., 114, 118–119, 134, 136–137
Loiacono, E., 45
Longari, M., 83, 118, 122, 136–137
Lorge, I., 283
Lu, G., 138
Lu, L., 95, 110–111, 113–114, 118, 122, 129, 134–137
Lupu, E., 134
Lvy, C., 136–137

M

MacDonell, S. G., 228, 235, 250–252, 265
Madanapalli, S., 170
Maddage, N. C., 113–114, 125–126, 136–137
Madsen, L., 181, 190
Malaga, R., 4, 11, 19
Mallardo, T., 282, 285
Mallat, S., 131
Malvar, H., 132
Mandin, J., 170
Mangia, L., 262
Mann, T. M., 47
Mapelli, F., 118
Maragos, P., 132–134, 136–137
Marchionini, G., 43, 45, 47
Marks, R. B., 156, 193
Marshall, B., 49, 54, 56
Marsh, W., 235
Marsico, M. D., 45
Martensson, A., 193, 203
Martufi, G., 213
Maxwell, K. D., 239
McAdams, S., 123, 137
McDonald, D., 47, 49, 54, 56
McGrath, J. E., 275–276, 278, 280–281, 283, 293, 301–303
McGuire, M. W., 300
McKinney, M. F., 110, 128, 136–137
Meddis, R., 120, 137
Mendes, E., 224, 227, 234, 239, 243, 254, 259, 262–265, 267
Meng, A., 136
Mennecke, B. E., 277, 279–280, 283–284
Merkl, D., 88, 90, 119, 131, 135, 137
Mermelstein, P., 124, 136
Mesaros, A., 134
Mierswa, I., 86
Minematsu, N., 134
Misra, H., 119
Mitrovic, D., 111, 134, 136
Miyazaki, A., 193, 203
Mockus, A., 273

Moinuddin, M., 113–114, 136–137
Moore, B. C. J., 90, 111
Moore, C. J., 89
Morasca, S., 262
Mörchen, F., 114, 118–119, 134, 136–137
Moreau, N., 116, 118, 122–123, 132, 136–137
Morgan, N., 86, 88, 126, 137
Morik, K., 86
Morris, M. G., 275, 293
Morris, R. D., 115, 117–118, 135–137
Mosley, N., 224, 227, 234, 239–240, 243, 259, 262–265
Mosley, S., 254
Mowshowitz, A., 46
MuBler, G., 47
Muhamed, R., 158
Muir, N., 300
Muller, M., 287
Müller, M., 92, 121, 130, 135–136
Munson, W. A., 91
Muralishankar, R., 136
Murthy, H. A., 115, 136
Murthy, K. V. M., 115
Murthy, U. S., 280

N

Nagarajan, T., 115, 136
Navratil, J., 134
Neil, M., 235
Neves, P., 213
Newman, E. B., 89, 124
Ng, D., 47
Nicholson, A., 245–246
Nie, Z., 9
Niles, I., 50
Nissilä, T., 157, 196–197
Nitta, T., 110, 137
Nobles, R., 15
Nocera, P., 136–137
Nöcker, M., 119, 136
Noll, A. M., 124
Noll, P., 118, 137

Norton, L., 52
Ntoulas, A., 50
Nunamaker, J. F., 45, 48, 58, 276, 280, 283–284
Nuseibeh, B., 279
Nwe, T. L., 134

O

O'Connor, K. M., 280
Ogihara, M., 94, 113–114, 118, 134–137
Ohye, M., 23
Olshen, R., 241
Olson, G. M., 273, 285, 291–292
Olson, J. S., 273, 285, 291–292
O'Mard, L., 120, 137
Ong, T.-H., 56, 59
O∈Reilly, T., 10
Orlikowski, W., 285
Orwig, R., 47, 49
Owsley, L., 134, 136
Ozmutlu, H. C., 45
Ozmutlu, S., 45

P

Pachet, F., 134
Paiano, R., 262
Paliwal, K. K., 86, 92, 115, 125, 134, 136–137
Palmer, C., 47
Pampalk, E., 88, 90, 119, 131, 135–137
Panagiotakis, C., 110–111, 134, 137
Paolini, P., 262
Papatheodorou, T. S., 265
Park, S. D., 170
Patil, B., 170
Pauws, S., 135
Pearl, J., 243–245, 247
Pease, A., 50
Peeters, G., 94, 115, 117, 122–123, 136–138
Pellegrino, F. C., 134
Pelletier, G., 203

Pentikousis, K., 157–158, 174, 196–197, 204–205, 207, 213
Perkins, C., 194
Pesenti, J., 47
Peters, R. W., 89
Petkovic, D., 114, 137
Petrucci, G., 110, 113, 136–137
Pezzano, R., 118
Pfeiffer, S., 91, 119, 134, 136
Phillips, G. M., 299
Pickard, L. M., 250–252, 265
Pinola, J., 157–158, 196–197, 204–205, 207
Pipino, L. L., 45
Piri, E., 157, 196–197, 204–205, 207, 213
Pitsikalis, V., 133, 137
Pitton, W. J., 86–87, 92, 128, 135–136
Plaisant, C., 47
Platt, J. C., 92, 132, 136
Podnar, I., 282
Pohle, T., 136–137
Pollastri, E., 83, 118, 122, 136–137
Pollino, C., 224, 234, 243, 245–246, 264
Ponceleon, D., 114, 137
Porter, A. A., 235–237, 239
Potamianos, A., 136
Povinelli, R. J., 88, 133, 137
Prasad, R., 158
Proakis, J., 178

Q

Qin, J., 47
Qin, Y., 47
Qi, T., 136

R

Raahemifar, K., 134
Raake, A., 183, 191, 201
Rabiner, L., 74, 112, 134, 136
Radhakrishnan, R., 134, 136
Raghavan, P., 49
Raisinghani, M. S., 4
Rajagopalan, S., 49

Raj, R., 36
Ramakrishnan, A. G., 136
Ramalingam, A., 116, 118–119, 135–137
Rao, G. V. R., 115, 136
Rauber, A., 88, 90, 119, 131, 135, 137
Ravindran, S., 86, 92, 125, 131, 133, 135, 137
Razak, A. A., 134
Reifer, D. J., 224, 258, 262
Rein, G., 286
Reiterer, H., 47
Reynolds, D. A., 134
Rice, R. E., 45
Richard, G., 95
Richens, R. H., 50
Rivlin, A. E., 254
Robert, L. P., 275, 297–298
Rouas, J.-L., 134
Rowse, D., 15
Ruddyand, P., 204
Ruhe, M., 261
Rusu, C., 134

S

Sakellaris, E., 245
Salton, G., 76
Sanderson, M., 47
Sandler, M., 134
Sandlund, K., 203
Santoro, G. M., 299
Saracevic, T., 44
Sarikaya, B., 170
Sato, M., 49
Sato, S., 37, 49
Satsangi, P. S., 235, 267
Saunders, J., 110
Schafer, R., 112, 136
Schatz, B., 47
Scheirer, E., 114, 116, 118, 127, 129–131, 136–137
Schlemmer, K., 86, 92, 125, 131, 133, 135, 137
Schlichte, J. H., 282

AUTHOR INDEX 323

Schofield, C., 227, 233–235, 250, 267
Schroeder, L., 235, 267
Schuffels, C., 47, 49
Schulte, G., 204
Schultze, U., 299
Schulzrinne, H., 181, 195
Seeling, P., 204
Sekiguchi, M., 134
Sengupta, S., 197
Sen, R., 2
Sethares, J. C., 115, 117–118, 135–137
Sethares, W. A., 115, 117–118, 135–137
Sewell, R., 47
Shamma, S., 125
Shannon, B. J., 92, 125, 136
Shannon, C. E., 289
Shao, X., 113–114, 125–126, 136–137
Shaw, M., 274
Shen, V., 250, 253, 258
Shepard, R. N., 120
Shepperd, M. J., 228, 235, 239, 250–252, 265, 267
Sheridan, T. B., 297
Shibata, T., 134
Shneiderman, B., 45, 47, 254
Short, J., 274, 287
Siegel, J., 300
Sikora, T., 116, 118, 122–123, 132, 136–137
Simoes, P., 213
Singer, E., 134
Siohan, O., 125, 136–137
Sitte, R., 74
Sivadas, S., 119
Sjøberg, D., 227
Sjoquist, D., 235, 267
Slaney, M., 83, 87, 114, 116, 118, 124, 127, 129, 136–137
Smaragdis, P., 134, 136
Smith, J., 181, 190
Smits, R., 115
Soerensen, K., 189
Sollaud, A., 200, 204
Soong, F. K., 125, 136–137

Sowa, J. F., 50
Speier, C., 275, 293
Spence, R., 45
Spencer, S., 22
Spink, A., 44–45
Sreenivas, T. V., 113, 136
Srinivasan, H., 116, 118, 123, 134, 136–137
Srinivasan, S., 114, 137
Stamelos, I., 235–237, 239, 245
Stamm, C., 119, 136
Stamou, S., 50
Steeneken, H. J., 127
Stephan, P., 235, 267
Stevens, S. S., 83, 89, 91, 124
Stojanovic, L., 282–283
Stone, C., 241
Strong, D. M., 45
Sukittanon, S., 86–87, 92, 128, 135–136
Sullivan, D., 5
Sung, W.-K., 47
Sun, H., 85, 138
Surmann, D., 239
Sutcliffe, A. G., 44

T

Tailor, M., 235
Terasawa, H., 83, 87, 124
Terhardt, E., 82
Thies, M., 114, 118–119, 134, 136–137
Thompson, B., 204
Thompson, R. L., 275, 303
Tian, Y.-a., 50
Tombros, A., 47
Tomkins, A., 49
Torres-Carrasquillo, P. A., 134
Tourneret, J. Y., 113
Tremain, T., 112
Trevino, L., 300
Triggs, C., 239
Tuckman, B. W., 284, 289, 299
Tukey, J., 87, 124
Turrof, M., 299

Tzanetakis, G., 86, 94, 114, 116, 120, 129, 131, 134–138
Tziritas, G., 110–111, 134, 137

U

Uchihashi, S., 129, 136
Ultsch, A., 114, 118–119, 134, 136–137
Umapathy, K., 114, 136–137

V

Valacich, J. S., 275–276, 280, 283–284, 293, 295–296
Valdes-Perez, R., 47
Valin, J.-M., 188, 191
Vandenbosch, B., 299
Van Meggelen, J., 181, 190
Van Nee, R., 158
Vetro, A., 85, 138
Viikki, O., 125, 136
Vliet, H. V., 228
Vogel, D. R., 276, 280
Vogel, J., 287
Volkmann, J., 89, 124
Vos, K., 189

W

Wactlar, H., 47
Wakefield, G. H., 121, 135–136
Walsh, B. C., 44
Walther, J. B., 300
Wan, C., 86, 113–114, 116, 124, 134, 136–137
Wang, A., 85, 115, 137–138
Wang, G., 59
Wang, H. C., 92, 125, 134
Wang, K., 125, 134
Wang, L., 9
Wang, M., 135
Wang, R. Y., 45
Wang, X., 125, 136
Wang, Y., 85–86, 111, 118, 127, 136–138
Watson, I., 239

Weaver, W., 289
Wegel, R. L., 90
Wen, J., 9
West, M., 203
Wheaton, J., 111, 117, 120, 136–137
Wheeler, B. C., 277, 279–280
Widmer, G., 136–137
Wieczorek, I., 239, 261
Wiethoff, C., 283, 285
Wiginton, J., 274, 288–289
Williams, E., 274, 287
Williams, M. S., 283–284
Wold, T., 111, 117, 120, 136–137
Wong, A., 76
Woodberry, O., 245–246
Wood, R. E., 277
Woods, W. A., 50
Wu, C. H., 110, 120, 134, 137
Wu, J. L., 111, 134, 136–137
Wu, L., 133, 137
Wynne, B. E., 279, 283

X

Xia, F., 170
Xi, S., 136
Xu, B., 111, 114, 136–137
Xu, C., 113–114, 124–126, 134, 136–137
Xu, M., 124

Y

Yang, C. C. 50
Yang, C. S., 76
Yang, H.-C., 47, 49
Yang, X., 125
Yegnanarayana, B., 115, 136
You, K. H., 92, 125, 134
Yusof, M. H. M., 134

Z

Zafiris, P. A., 265
Zeng, D., 46
Zeppelzauer, M., 111, 134, 136

Zhang, H. J., 95, 110–111, 113–114, 118, 122, 129, 134–137
Zhang, J., 4, 11
Zhang, S., 111, 114, 136–137
Zhang, T., 86, 110–111, 128–129, 134, 136–137
Zhang, Y., 59
Zhao, L, 9
Zhou, Y., 47
Zhu, B., 46
Zhu, Y., 121, 135, 137
Zigurs, I., 275, 277, 280–281, 303–304
Zittle, F. J., 288
Zmud, R., 275, 304
Zowghi, D., 274, 279
Zwicker, E., 83–84, 89–90, 116–117, 119, 125–126, 128, 131, 137

Subject Index

A

Access service network (ASN), 168–169, 194
Adaptive time-frequency transform (ATFT), 114
Aggregation_Link, 213–214
Ajeeb portal, 53
Albawaba.com, 53
Anchor directory, 57
Artificial intelligence (AI) techniques
  Bayesian networks (BN)
    conditional probability table (CPT), 244
    knowledge engineering of Bayesian networks (KEBN), 245–247
    Model Walkthrough, 247
    posterior distribution, 244–245
    predictive accuracy, 247–248
    structure, 243–244
  case-based reasoning (CBR)
    adaptation rules, 240–241
    analogy adaptation, 239–240
    case similarity, 237–238
    feature subset selection, 236–237
    number of analogies, 239
    scaling, 238–239
  classification and regression trees (CART)
    binary tree, 241–242
    learning sample data, 241
Audio fingerprinting, 135, 138
Audio spectrum centroid (ASC), 116
Auditory filter bank temporal envelopes, 128
Ayna portal, 53

B

Baidu.com, 51
Bayesian network (BN) model
  conditional probability table (CPT), 244
  knowledge engineering of Bayesian networks (KEBN), 245–247
  literature survey, 263–265
  Model Walkthrough, 247
  posterior distribution, 244–245
  predictive accuracy, 247–248
  structure, 243–244
BIWE.com, 52
Black hat search engine optimization
  cloaking, 22–23
  content generation, 25–26
  doorway pages, 23–24
  Google quality guidelines, 18
  gray hat techniques, 19
  indexing methods, 19
  keyword stuffing, 20–22
  link building
    blog spamming, 26–27
    forum spamming, 28
    guestbook spamming, 26
    HTML injection, 30
    link farms, 29
    paid links, 29–30
    stats page spamming, 28–29
  negative SEO, 30–31
  Webmaster guidelines, 18
Broadband noise, 82–83
Business intelligence explorer (BIE), 48

C

Case-based reasoning (CBR)
  adaptation rules, 240–241
  analogy adaptation, 239–240
  case similarity, 237–238
  feature subset selection, 236–237
  number of analogies, 239
  scaling, 238–239
Cepstral features, 107
  advanced auditory model-based features, 125
  autoregression-based features, 126
  perceptual filter bank-based features, 124–125
  transformation, 87
Chroma energy distribution normalized statistics (CENS), 121
Chromagram, 121
Cognitive-Based View, 275
  dual process theories, 297
  free riding, 299
  lean media, 300
  media richness paradox, 297
  motivation and attention, 297–298
  social presence, 298–299
  sociopsychological effects, 299
Common Ground theory
  acknowledgments, 290–291
  delay costs, 293
  media constraint properties, 291–292
  presentation and acceptance, 290
  production and reception costs, 293
  text-based communication, 291
Communication media selection
  ad hoc groups
    decision quality, 284
    definition, 283–284
    dynamic collaboration, 282
    group-related contextual factors, 285–286
    group size, 282
    natural groups, 283
    past and future groups, 285
    short-term groups, 284
    single-task groups, 283
    temporal scope, 281–282
  comprehensive theoretical framework
    context management, 300–301
    elicitation, 307–308
    equivocality, 305
    graphical representation, 306
    lean media, 307
    matching group and media characteristics, 304–305
    matching task and media characteristics, 302
    rich media, 306
    task- and process-centric theories, 307
    Task/Technology Fit (TTF) theory, 303–304
    time-interaction-performance (TIP) theory, 302–303
  computer-mediated communication (CMC) theories
    Cognitive-Based View, 297–300
    Common Ground theory, 290–293
    F2F communication, 286
    Media Richness theory, 288–290
    Media Synchronicity theory, 293–297
    open-source development, 287
    Social Presence theory, 287–288
    time/space matrix, 285–286
    videoconference, 287
  cooperative work framework, 272–273
  direct communication, 273
  high-level research model, 276
  indirect communication, 272
  multipoint audio–video communication, 274
  software development, 273–274
  task-classification frameworks
    complex tasks typology, 280–281
    group task, 277
    historical task, 277

SUBJECT INDEX

Task Circumplex classification, 278–280
Connectivity service network (CSN), 168–169
Constructive cost model (COCOMO)
    basic model, EAF, 231
    general purpose algorithmic model, 232–233
    intermediate and advanced models, cost drivers, 231–232
Content-based audio retrieval features
    aggregations, 89, 101
    application domains
        audio segmentation, 133–134
        bioacoustic pattern recognition, 134
        emotion detection, 135
        speaker recognition, 134
    architecture
        components, 75–76
        feature database, 75
        feature representations, 79–80
        flute spectrogram, 79
        iterative refinement, 77
        MFCC, 79–80
        query by humming (QBH), 76–77
        query object, 76
        sound amplitude *vs.* time, 77–79
        vector space model, 76–77
    audio attributes
        duration, 82
        loudness, 82, 84
        noise, 82
        pitch, 82–83
        timbre, 83
        tones, 81–82
    automatic speech recognition, 74
    cepstral features, 107
        advanced auditory model-based features, 125
        autoregression-based features, 126
        perceptual filter bank-based features, 124–125
    compact representation, 99–100
    cosine transform, 109
    eigendomain features, 108, 131–132
    environmental sound retrieval, 74
    evaluation, 81
    feature extraction, 73
    filters, 88–89, 100
    literature survey, 135–137
    Mel-filter bank, 101
    modulation frequency features, 107–108
        acoustic frequency features, 128
        hearing sensation, 126
        4-Hz modulation energy, 127–128
        rhythm, 128–131
        short-time Fourier spectrogram, 127
    music information retrieval, 74
    novel taxonomy
        cepstral domain, 98
        frequency domain, 96–97
        method-oriented approach, 95
        modulation frequency domain, 98–99
        music information retrieval, 94
        organizing principles, 94–95
        temporal domain, 96
    numeric challenges, 93–94
    perceptual frequency features, 104–106
        brightness, 116–117
        chroma, 120–121
        harmonicity, 121–123
        loudness, 119
        pitch, 119–120
        tonality, 117–119
    phase space features, 108, 132–133
    physical frequency features, 103–104
        adaptive time-frequency decomposition-based features, 113–114
        autoregression-based features, 112–113
        short-time Fourier transform-based features, 114–115
    properties
        audio representation, 84–85
        domain, 85
        semantic interpretation, 86

Content-based audio retrieval features (Continued)
  temporal scale, 85–86
  underlying model, 86
  psychoacoustic challenges
    auditory masking, 90–91
    frequency selectivity, 90
    human auditory system, 89
    loudness levels, 91
    power law, 91–92
  relevant published surveys, 137–138
  semantic gap, 75
  signal representation, 72
  technical challenges, 92–93
  temporal features, 103
    amplitude-based features, 110–111
    computational complexity, 109
    power-based features, 111–112
    zero crossing-based features, 110
  transformations, 86–88, 100
Correlation domain, 87
Cost estimation benchmarking and risk analysis (COBRA^{TM}) method, 261–262
Cyclic beat spectrum (CBS), 130

D

Daubechies wavelet coefficient histogram (DWCHs), 113–114
Directory Mozilla (DMOZ) Project, 48
Distortion discriminant analysis (DDA), 132
Document categorization techniques, 47

E

Effort adjustment factor (EAF), 230–231
Eigendomain features, 108, 131–132
Extended real-time polling service (ertPS), 209

F

FAST, 45
Fourier transform, 87, 101, 112

Frequency division duplexing (FDD) method, 162

G

G.729.1 codec, 187–188
Generalized linear model (GLM), 259
Group delay function, 115
GTalk, 190

H

Harmonic energy entropy, 123
HelpfulMed portals, 46
Human domain knowledge, 57

I

iAsk.com, 51
Indexical bias, 46
Information quality, 45
Information retrieval system, 46
Information searching, 44
Information seeking strategy, 43–45
INSYDER, 47
Integral loudness, 119
Internet search portals, 46–47
Inverse rank weighted mean, 239–240

J

Johaina search engine, 53

K

Knowledge engineering of Bayesian networks (KEBN), 245–247

L

Least collaborative effort, 293
Linear prediction cepstral coefficients (LPCCs), 126
Linear prediction zero crossing ratio (LP-ZCR), 110

SUBJECT INDEX

Linear predictive coding (LPC), 112–113, 126
Line spectral frequencies (LSF), 113
Link building black hat techniques
　blog spamming, 26–27
　forum spamming, 28
　guestbook spamming, 26
　HTML injection, 30
　link farms, 29
　paid links, 29–30
　stats page spamming, 28–29
Lyapunov exponents, 133

M

Machine translation, 50
MAC protocol data units (MPDUs), 164–165
MAC service data units (MSDUs), 164–165
Manual stepwise regression (MSWR), 263–264
Media Richness theory
　communication capacity, 288
　uncertainty and equivocality force, 289–290
　videoconferencing, 288
Media Synchronicity theory, 275
　conveyance and convergence processes, 294, 296
　forming, storming, norming, and performing, 296–297
　immediacy of feedback, 295
　parallelism, 294–295
　rehearseability and reprocessability, 295
　symbol variety, 294
　task-centric theory, 293
Medium access control (MAC) layer
　ARQ, retransmission method, 165
　connection identifier (CID), 166
　convergence sublayer (CS), 164–165
　data delivery and scheduling services, 166–167
　power consumption, 166
　structure, 164–165
　transmission capacity, 165
　user terminal, 166
MedTextus portals, 46
Mel-filter bank, 101
Mel-frequency cepstral coefficients (MFCCs), 73, 79–80, 101, 124–125
Metasearching, 46–47, 57
Metrics model for Web applications (MMWA), 262–263
Modified group delay function (MGDF), 115
Modulation frequency features, 107–108
　acoustic frequency features, 128
　hearing sensation, 126
　4-Hz modulation energy, 127–128
　rhythm
　　band periodicity, 129
　　beat histogram, 131
　　beat spectrum, 129–130
　　beat tracker, 130–131
　　cyclic beat spectrum (CBS), 130
　　definition, 128
　　patterns, 131
　　pulse metric, 129
　short-time Fourier spectrogram, 127
　transformation, 87–88
MPEG-7
　audio harmonicity, 122
　audio spectrum basis/projection, 132
　audio waveform, 111
　harmonic spectral deviation (HSD), 123
　harmonic spectral spread (HSS), 123
　harmonic spectral variation (HSV), 123
　log attack time, 111–112
　spectral timbral descriptors, 123
　temporal centroid, 111
Multimedia, WiMAX reference scenario
　environmental monitoring, 175–176
　　fire prevention, 177–178
　　volcanic activity, 177
　fixed and mobile, generic scenarios
　　broadband connectivity, 171–172
　　cost-effective solutions, 172
　　NLOS signal propagation, 173

Multimedia, WiMAX reference scenario (Continued)
  relay system, 173–174
  subscriber stations, 173
  ubiquitous connectivity, 173
  Wi-Fi, 172
 telemedicine
  always-on medical assistance scenario, 175–176
  e-health, 174
  remote follow-up, 174–175
Multiresolution entropy, 119
Musical content features, 94

N

NanoPort portals, 47
Narrow-band noise, 82
Noise-robust audio features (NRAF), 125
Non-line-of-sight (NLOS) signal propagation, 173

O

Online Commercial Intention (OCI), 37
On-page black hat techniques, 19–22
Open Directory Project. *See* Directory Mozilla (DMOZ) Project
Openfind.com.tw, 51
Orthogonal frequency division multiple access (OFDMA), 161
Orthogonal frequency division multiplexing (OFDM), 160–161

P

Perceptual frequency features, 104–106
  brightness, 116–117
  chroma, 120–121
  harmonicity, 121–123
  loudness, 119
  pitch, 119–120
  tonality, 117–119
Perceptual linear prediction (PLP), 126

Phase space features, 108, 132–133
Physical frequency features, 103–104
  adaptive time-frequency decomposition-based features, 113–114
  autoregression-based features, 112–113
  short-time Fourier transform-based features, 114–115
Pitch histogram, 120
Pitch synchronous zero crossing peak amplitudes (PS-ZCPA), 110
Postretrieval analysis, 47–48
Psychoacoustics
  auditory masking, 90–91
  frequency selectivity, 90
  human auditory system, 89
  loudness levels, 91
  pitch, 120
  power law, 91–92
Pulse-code modulation (PCM), 185

Q

Query by humming (QBH), 76–77, 135
Query-categorization approach, 49

R

Rate-scale-frequency features, 131–132
Real-time polling service (rtPS), 209
Relative spectral—perceptual linear prediction (RASTA-PLP), 126

S

Search engine optimization (SEO)
  Ask.com, 36
  black hat approach
    blog spamming, 26–27
    cloaking, 22–23
    content generation, 25–26
    doorway pages, 23–24
    forum spamming, 28
    Google quality guidelines, 18
    gray hat techniques, 19

SUBJECT INDEX 333

guestbook spamming, 26
HTML injection, 30
indexing methods, 19
keyword stuffing, 20–22
link farms, 29
negative SEO, 30–31
paid links, 29–30
stats page spamming, 28–29
Webmaster guidelines, 18
history and current statistics, 4–5
indexing
   definition, 9
   social bookmark, 11
   spiders, 9–10
   user-developed content, 10
   Web 2.0 content sites, 10–11
   Wikipedia, 10
integrated Google search results, 36
keyword research
   commercial online intent tool, 9
   commercial viability, 8–9
   consumer purchase model, 8
   Google external keyword tool, 6
   level of competition, 7
   query, 5
   R/S ratio, 7–8
   words/phrases, 5–6
legal and ethical considerations
   copyright issues, 31–33
   PPC ads, 33–34
   SEMPO, 33
   sponsored links, 33–34
   Webmaster guidelines, 33
   Yahoo!, 34
link building
   backlinks, 15–16
   internal linking structure, 16
   NoFollow meta tag, 17
   Page Rank (PR), 16–17
   ranking algorithm, 15
Online Commercial Intention (OCI), 37
on-site optimization
   definition, 11

formatting, 15
latent semantic indexing, 13–14
meta tags, 11–13
updated content, 14–15
paid search engine, 2
paid *vs*. organic search results, 2–3
search engine marketing (SEM), 2
search engine results page (SERP), 5
Traffic Power company, 35
Web pages, 4
white hat SEO methods, 35
Self-organizing feature map (SOM)
   category, 47, 49, 56–57
Semantic network technologies, 50
Short-time energy (STE), 111
Short-time Fourier transform (STFT), 98,
   102, 109, 114–115
SILK codec, 189
Simultaneity, 291
Single_Cam_Link, 213–214
Skype, 189–190
Social Presence theory, 287–288
Speaker recognition, 134
Specific loudness sensation, 119
Spectral centroid (SC), 116
Spectral crest factor, 118
Spectral flux, 114
Speex project, 188–189
Subband energy ratio, 114
Subband spectral flux (SSF), 118
Support vector regression
   (SVR), 264–265

T

Task Circumplex classification, 278–280
   elicitation, 279, 307–308
   Group Support Systems (GSS), 279–280
   group tasks, 278
   natural tasks, 280
   TIP theory, 302
Task/Technology Fit (TTF) theory, 275–276,
   303–304

Temporal features, 103
  amplitude-based features, 110–111
  computational complexity, 109
  power-based features, 111–112
  transformation, 87
  zero crossing-based features, 110
Terra.com, 51–52
Time division duplexing (TDD) method, 162
Time-interaction-performance (TIP) theory, 302–303
Time/space matrix, 285–286
Tukutuku database, 263–264

U

Unsolicited Grant Service (UGS), 208–209
Unweighted Euclidean distance, 237–238

V

Voice over IP (VoIP)
  aggregation
    application- and network-layer, 195–196
    wirelessMAN-OFDM, 196–199
    wirelessMAN-OFDMA, 199–202
  all-IP architectures, 191
  Asterisk voicemail, iPhone application, 181, 183
  codecs, 181, 183, 191–192
  crossbar switch, 178–179
  digital telephony (ITU-T G.711), 185–186
  high-quality voice over packet data networks (ITU-T G729.1), 187–188
  IEEE 802.16 standard, 191, 193
  integrated services digital network (ISDN), 179–180
  multiple-input multiple-output (MIMO) techniques
    Alamouti space–time block coding, 207
    beamforming, 207
    channel state information at the transmitter (CSIT), 206–207
    IEEE 802.16–2009, 206
    multiple antennas, 206
    spatial multiplexing, 207
    TCP goodput, 207–208
  open-source option, voice coding, 188–189
  packet generation, 181, 184
  plain old telephone service (POTS), 178
  private branch exchange (PBX), 181
  public switched telephone network (PSTN), 178
  quality of service (QoS), 208–209
  robust header compression (ROHC)
    context state, packet flow, 203
    performance evaluation, 204
    redundancy, 203
    wirelessMAN-OFDM, 204–205
    wirelessMAN-OFDMA, 205–206
  source code, Asterisk's codec_speex.c file, 181–182
  testbed description, 193–194
  video telephony and teleconferencing (ITU-T G.723.1), 186–187
  voice codecs, 189–190

W

Web directory, 49
Web effort estimation
  algorithmic techniques
    constructive cost model (COCOMO), 230–233
    log(effort) and log(total Web-pages) relationship, 234
    recalibration, 235
    regression line, 233–234
    size and effort relationship, 229–230
  applications, 253–254
  artificial intelligence (AI) techniques
    adaptation rules, 240–241
    analogy adaptation, 239–240
    Bayesian networks (BN), 243–248
    case similarity, 237–238

classification and regression trees
    (CART), 241–243
  feature subset selection, 236–237
  number of analogies, 239
  scaling, 238–239
"cost drivers," 226
derivation, 225
effort model building, 225
expert-based effort estimation
  bottom-up estimation, 227
  data/knowledge retrieval, 229
  drawbacks, 228
  top-down estimation, 227
  Web companies survey, 227
literature survey
  Bayesian network (BN) model,
    263–265
  CBR, 254
  design effort estimation, 262
  measurement, prediction and risk
    analysis, 259–260
  number of trends, 265
  software sizing model, 262–263
  summarization, 254–257, 265
  types, web applications, 254, 258, 266
  Web development, quick-to-market
    software, 258–259
  Web hypermedia, 260–261, 265–266
  Web metrics, 259
  Web project cost estimation, 260–262
prediction accuracy
  cross-validation, 252
  four-step process, 248–249
  magnitude of relative error (MRE), 250
  main components, boxplot, 251–252
  mean magnitude of relative error
    (MMRE), 250–251
  median magnitude of relative error
    (MdMRE), 250–251
  statistics, 249–250
project characteristics, 226
size measures, 226
steps, 224–225

Web pages, 47, 56–57
Web portals, multilingual world
  Arabic-speaking regions, 52–53
  Chinese-speaking regions, 51
  Spanish-speaking regions, 51–52
Web searching and browsing
  Arabic Medical Web Directory
    (AMedDir)
    DMOZ directory, 61
    experiment, 63
    metasearching, 61–62
    screen shots, 63–64
    statistics, 63
  Chinese Business Intelligence Portal
    (CBizPort)
    Chinese phrase lexicon, 59
    encoding conversion, 58
    information sources, 58
    screen shots, 59–60
    summarizer and categorizer, 59
  information-seeking models, 43–44
  limitations, 65
  multilingual perspective, 55
    analysis modules, 56–57
    collection building and metasearching,
      54
    domain analysis, 54
    web directory building, 57
  multinational corporations (MNCs), 65
  regional impacts and information quality,
    45
  Spanish Business Intelligence Portal
    (SBizPort), 59, 61–62
  technologies and approaches
    automated approaches, 49–50
    manual categorization, 48–49
    metasearching, 46–47
    postretrieval analysis, 47–48
    search engines components, 46
    semantic network technologies, 50
Web portals, multilingual world
  Arabic-speaking regions, 52–53
  Chinese-speaking regions, 51

Web searching and browsing (Continued)
  Spanish-speaking regions, 51–52
Web spider/crawler, 46, 54
Weighted Euclidean distance, 238
Windowing process, 88
WirelessMAN-OFDM
  application- and network-layer aggregation, 197–198
  G.723.1 VoIP mean, WiMAX downlink, 197–198
  G.723.1 VoIP mean, WiMAX uplink, 198–199
  64 QAM FEC, 196
  robust header compression (ROHC), 204–205
  VoIP quality, 197
WirelessMAN-OFDMA
  downlink voice sample frame loss rate, 201–202
  G.729.1 VoIP cumulative goodput, 200–201
  IPv6 anycasting, 202
  one-way delays, 202
  16 QAM FEC, 200
  robust header compression (ROHC), 205–206
Wireless metropolitan area networks. *See* Worldwide interoperability for microwave access (WiMAX)
Worldwide interoperability for microwave access (WiMAX)
  audio/video (A/V) material, 155
  commercial off-the-shelf (COTS) equipment, 157
  deployment map, 157–158
  deployment schematic, 217–218
  digital divide, 155
  Greenfield operators, 219
  medium access control (MAC) layer
    ARQ, retransmission method, 165
    connection identifier (CID), 166
    convergence sublayer (CS), 164–165
    data delivery and scheduling services, 166–167
    power consumption, 166
    structure, 164–165
    transmission capacity, 165
    user terminal, 166
  multimedia, reference scenario
    environmental monitoring, 175–178
    fixed and mobile, generic scenarios, 171–174
    telemedicine, 174–176
  network reference model
    air interface, 169
    ASN and CSN, 168–169
    gateway, 169
    Internet engineering task force (IETF), 170
    mobility management, 169
    Network Working Group (NWG), 168–169
    reference points, 169
  physical (PHY) layer
    adaptive antenna systems (AASs), 163
    adaptive beamforming, 163
    hybrid automatic repeat-request (HARQ), 163–164
    IEEE 802.16 standards, 159–160
    intersymbol interference (ISI), 161
    orthogonal frequency division multiplexing (OFDM), 160–161
    subchannelization scheme, 161
    WiMAX Forum, 162
  remote surveillance and IPTV
    equipment configuration, 212, 216
    fire prevention testbed, 211–214
    Lousã Mountain surveillance tower, 212, 216
    network topology, 210
    propagation conditions and measured throughput, 213, 217
    surveillance cameras, 211–212, 215
    terrain profiles, 209–210
    testbed antennas location, 211

voice over IP (VoIP)
   aggregation, wirelessMAN-OFDM, 196–199
   aggregation, wirelessMAN-OFDMA, 199–202
   digital telephony (ITU-T G.711), 185–186
   high-quality voice over packet data networks (ITU-T G729.1), 187–188
   multiple-input multiple-output (MIMO) techniques, 206–208
   open-source option, voice coding, 188–189
   quality of service, 208–209
   robust header compression, 203–206
   testbed description, 193–194
   video telephony and teleconferencing (ITU-T G.723.1), 186–187
   voice codecs, 189–190

Y

Yam.com, 51

Z

Zero crossing peak amplitudes (ZCPA), 110
Zero crossing rate (ZCR), 95, 110

Contents of Volumes in This Series

Volume 42

Nonfunctional Requirements of Real-Time Systems
    TEREZA G. KIRNER AND ALAN M. DAVIS
A Review of Software Inspections
    ADAM PORTER, HARVEY SIY, AND LAWRENCE VOTTA
Advances in Software Reliability Engineering
    JOHN D. MUSA AND WILLA EHRLICH
Network Interconnection and Protocol Conversion
    MING T. LIU
A Universal Model of Legged Locomotion Gaits
    S. T. VENKATARAMAN

Volume 43

Program Slicing
    DAVID W. BINKLEY AND KEITH BRIAN GALLAGHER
Language Features for the Interconnection of Software Components
    RENATE MOTSCHNIG-PITRIK AND ROLAND T. MITTERMEIR
Using Model Checking to Analyze Requirements and Designs
    JOANNE ATLEE, MARSHA CHECHIK, AND JOHN GANNON
Information Technology and Productivity: A Review of the Literature
    ERIK BRYNJOLFSSON AND SHINKYU YANG
The Complexity of Problems
    WILLIAM GASARCH
3-D Computer Vision Using Structured Light: Design, Calibration, and Implementation Issues
    FRED W. DEPIERO AND MOHAN M. TRIVEDI

Volume 44

Managing the Risks in Information Systems and Technology (IT)
    ROBERT N. CHARETTE
Software Cost Estimation: A Review of Models, Process and Practice
    FIONA WALKERDEN AND ROSS JEFFERY
Experimentation in Software Engineering
    SHARI LAWRENCE PFLEEGER
Parallel Computer Construction Outside the United States
    RALPH DUNCAN
Control of Information Distribution and Access
    RALF HAUSER

Asynchronous Transfer Mode: An Engineering Network Standard for High Speed Communications
    RONALD J. VETTER
Communication Complexity
    EYAL KUSHILEVITZ

Volume 45

Control in Multi-threaded Information Systems
    PABLO A. STRAUB AND CARLOS A. HURTADO
Parallelization of DOALL and DOACROSS Loops—a Survey
    A. R. HURSON, JOFORD T. LIM, KRISHNA M. KAVI, AND BEN LEE
Programming Irregular Applications: Runtime Support, Compilation and Tools
    JOEL SALTZ, GAGAN AGRAWAL, CHIALIN CHANG, RAJA DAS, GUY EDJLALI, PAUL HAVLAK, YUAN-SHIN HWANG, BONGKI MOON, RAVI PONNUSAMY, SHAMIK SHARMA, ALAN SUSSMAN, AND MUSTAFA UYSAL
Optimization Via Evolutionary Processes
    SRILATA RAMAN AND L. M. PATNAIK
Software Reliability and Readiness Assessment Based on the Non-homogeneous Poisson Process
    AMRIT L. GOEL AND KUNE-ZANG YANG
Computer-Supported Cooperative Work and Groupware
    JONATHAN GRUDIN AND STEVEN E. POLTROCK
Technology and Schools
    GLEN L. BULL

Volume 46

Software Process Appraisal and Improvement: Models and Standards
    MARK C. PAULK
A Software Process Engineering Framework
    JYRKI KONTIO
Gaining Business Value from IT Investments
    PAMELA SIMMONS
Reliability Measurement, Analysis, and Improvement for Large Software Systems
    JEFF TIAN
Role-Based Access Control
    RAVI SANDHU
Multithreaded Systems
    KRISHNA M. KAVI, BEN LEE, AND ALLI R. HURSON
Coordination Models and Language
    GEORGE A. PAPADOPOULOS AND FARHAD ARBAB
Multidisciplinary Problem Solving Environments for Computational Science
    ELIAS N. HOUSTIS, JOHN R. RICE, AND NAREN RAMAKRISHNAN

Volume 47

Natural Language Processing: A Human–Computer Interaction Perspective
    BILL MANARIS

Cognitive Adaptive Computer Help (COACH): A Case Study
 EDWIN J. SELKER
Cellular Automata Models of Self-replicating Systems
 JAMES A. REGGIA, HUI-HSIEN CHOU, AND JASON D. LOHN
Ultrasound Visualization
 THOMAS R. NELSON
Patterns and System Development
 BRANDON GOLDFEDDER
High Performance Digital Video Servers: Storage and Retrieval of Compressed Scalable Video
 SEUNGYUP PAEK AND SHIH-FU CHANG
Software Acquisition: The Custom/Package and Insource/Outsource Dimensions
 PAUL NELSON, ABRAHAM SEIDMANN, AND WILLIAM RICHMOND

Volume 48

Architectures and Patterns for Developing High-Performance, Real-Time ORB Endsystems
 DOUGLAS C. SCHMIDT, DAVID L. LEVINE, AND CHRIS CLEELAND
Heterogeneous Data Access in a Mobile Environment – Issues and Solutions
 J. B. LIM AND A. R. HURSON
The World Wide Web
 HAL BERGHEL AND DOUGLAS BLANK
Progress in Internet Security
 RANDALL J. ATKINSON AND J. ERIC KLINKER
Digital Libraries: Social Issues and Technological Advances
 HSINCHUN CHEN AND ANDREA L. HOUSTON
Architectures for Mobile Robot Control
 JULIO K. ROSENBLATT AND JAMES A. HENDLER

Volume 49

A Survey of Current Paradigms in Machine Translation
 BONNIE J. DORR, PAMELA W. JORDAN, AND JOHN W. BENOIT
Formality in Specification and Modeling: Developments in Software Engineering Practice
 J. S. FITZGERALD
3-D Visualization of Software Structure
 MATHEW L. STAPLES AND JAMES M. BIEMAN
Using Domain Models for System Testing
 A. VON MAYRHAUSER AND R. MRAZ
Exception-Handling Design Patterns
 WILLIAM G. BAIL
Managing Control Asynchrony on SIMD Machines—a Survey
 NAEL B. ABU-GHAZALEH AND PHILIP A. WILSEY
A Taxonomy of Distributed Real-time Control Systems
 J. R. ACRE, L. P. CLARE, AND S. SASTRY

Volume 50

Index Part I
Subject Index, Volumes 1–49

Volume 51

Index Part II
Author Index
Cumulative list of Titles
Table of Contents, Volumes 1–49

Volume 52

Eras of Business Computing
    ALAN R. HEVNER AND DONALD J. BERNDT
Numerical Weather Prediction
    FERDINAND BAER
Machine Translation
    SERGEI NIRENBURG AND YORICK WILKS
The Games Computers (and People) Play
    JONATHAN SCHAEFFER
From Single Word to Natural Dialogue
    NEILS OLE BENSON AND LAILA DYBKJAER
Embedded Microprocessors: Evolution, Trends and Challenges
    MANFRED SCHLETT

Volume 53

Shared-Memory Multiprocessing: Current State and Future Directions
    PER STEUSTRÖM, ERIK HAGERSTEU, DAVID I. LITA, MARGARET MARTONOSI, AND MADAN VERNGOPAL
Shared Memory and Distributed Shared Memory Systems: A Survey
    KRISHNA KAUI, HYONG-SHIK KIM, BEU LEE, AND A. R. HURSON
Resource-Aware Meta Computing
    JEFFREY K. HOLLINGSWORTH, PETER J. KELCHER, AND KYUNG D. RYU
Knowledge Management
    WILLIAM W. AGRESTI
A Methodology for Evaluating Predictive Metrics
    JASRETT ROSENBERG
An Empirical Review of Software Process Assessments
    KHALED EL EMAM AND DENNIS R. GOLDENSON
State of the Art in Electronic Payment Systems
    N. ASOKAN, P. JANSON, M. STEIVES, AND M. WAIDNES
Defective Software: An Overview of Legal Remedies and Technical Measures Available to Consumers
    COLLEEN KOTYK VOSSLER AND JEFFREY VOAS

Volume 54

An Overview of Components and Component-Based Development
　ALAN W. BROWN
Working with UML: A Software Design Process Based on Inspections for the Unified Modeling Language
　GUILHERME H. TRAVASSOS, FORREST SHULL, AND JEFFREY CARVER
Enterprise JavaBeans and Microsoft Transaction Server: Frameworks for Distributed Enterprise Components
　AVRAHAM LEFF, JOHN PROKOPEK, JAMES T. RAYFIELD, AND IGNACIO SILVA-LEPE
Maintenance Process and Product Evaluation Using Reliability, Risk, and Test Metrics
　NORMAN F. SCHNEIDEWIND
Computer Technology Changes and Purchasing Strategies
　GERALD V. POST
Secure Outsourcing of Scientific Computations
　MIKHAIL J. ATALLAH, K. N. PANTAZOPOULOS, JOHN R. RICE, AND EUGENE SPAFFORD

Volume 55

The Virtual University: A State of the Art
　LINDA HARASIM
The Net, the Web and the Children
　W. NEVILLE HOLMES
Source Selection and Ranking in the WebSemantics Architecture Using Quality of Data Metadata
　GEORGE A. MIHAILA, LOUIQA RASCHID, AND MARIA-ESTER VIDAL
Mining Scientific Data
　NAREN RAMAKRISHNAN AND ANANTH Y. GRAMA
History and Contributions of Theoretical Computer Science
　JOHN E. SAVAGE, ALAN L. SALEM, AND CARL SMITH
Security Policies
　ROSS ANDERSON, FRANK STAJANO, AND JONG-HYEON LEE
Transistors and 1C Design
　YUAN TAUR

Volume 56

Software Evolution and the Staged Model of the Software Lifecycle
　KEITH H. BENNETT, VACLAV T. RAJLICH, AND NORMAN WILDE
Embedded Software
　EDWARD A. LEE
Empirical Studies of Quality Models in Object-Oriented Systems
　LIONEL C. BRIAND AND JÜRGEN WÜST
Software Fault Prevention by Language Choice: Why C Is Not My Favorite Language
　RICHARD J. FATEMAN
Quantum Computing and Communication
　PAUL E. BLACK, D. RICHARD KUHN, AND CARL J. WILLIAMS
Exception Handling
　PETER A. BUHR, ASHIF HARJI, AND W. Y. RUSSELL MOK

Breaking the Robustness Barrier: Recent Progress on the Design of the Robust Multimodal System
    SHARON OVIATT
Using Data Mining to Discover the Preferences of Computer Criminals
    DONALD E. BROWN AND LOUISE F. GUNDERSON

Volume 57

On the Nature and Importance of Archiving in the Digital Age
    HELEN R. TIBBO
Preserving Digital Records and the Life Cycle of Information
    SU-SHING CHEN
Managing Historical XML Data
    SUDARSHAN S. CHAWATHE
Adding Compression to Next-Generation Text Retrieval Systems
    NIVIO ZIVIANI AND EDLENO SILVA DE MOURA
Are Scripting Languages Any Good? A Validation of Perl, Python, Rexx, and Tcl against C, C++, and Java
    LUTZ PRECHELT
Issues and Approaches for Developing Learner-Centered Technology
    CHRIS QUINTANA, JOSEPH KRAJCIK, AND ELLIOT SOLOWAY
Personalizing Interactions with Information Systems
    SAVERIO PERUGINI AND NAREN RAMAKRISHNAN

Volume 58

Software Development Productivity
    KATRINA D. MAXWELL
Transformation-Oriented Programming: A Development Methodology for High Assurance Software
    VICTOR L. WINTER, STEVE ROACH, AND GREG WICKSTROM
Bounded Model Checking
    ARMIN BIERE, ALESSANDRO CIMATTI, EDMUND M. CLARKE, OFER STRICHMAN, AND YUNSHAN ZHU
Advances in GUI Testing
    ATIF M. MEMON
Software Inspections
    MARC ROPER, ALASTAIR DUNSMORE, AND MURRAY WOOD
Software Fault Tolerance Forestalls Crashes: To Err Is Human; To Forgive Is Fault Tolerant
    LAWRENCE BERNSTEIN
Advances in the Provisions of System and Software Security—Thirty Years of Progress
    RAYFORD B. VAUGHN

Volume 59

Collaborative Development Environments
    GRADY BOOCH AND ALAN W. BROWN
Tool Support for Experience-Based Software Development Methodologies
    SCOTT HENNINGER
Why New Software Processes Are Not Adopted
    STAN RIFKIN

Impact Analysis in Software Evolution
 MIKAEL LINDVALL
Coherence Protocols for Bus-Based and Scalable Multiprocessors, Internet, and Wireless Distributed Computing Environments: A Survey
 JOHN SUSTERSIC AND ALI HURSON

Volume 60

Licensing and Certification of Software Professionals
 DONALD J. BAGERT
Cognitive Hacking
 GEORGE CYBENKO, ANNARITA GIANI, AND PAUL THOMPSON
The Digital Detective: An Introduction to Digital Forensics
 WARREN HARRISON
Survivability: Synergizing Security and Reliability
 CRISPIN COWAN
Smart Cards
 KATHERINE M. SHELFER, CHRIS CORUM, J. DREW PROCACCINO, AND JOSEPH DIDIER
Shotgun Sequence Assembly
 MIHAI POP
Advances in Large Vocabulary Continuous Speech Recognition
 GEOFFREY ZWEIG AND MICHAEL PICHENY

Volume 61

Evaluating Software Architectures
 ROSEANNE TESORIERO TVEDT, PATRICIA COSTA, AND MIKAEL LINDVALL
Efficient Architectural Design of High Performance Microprocessors
 LIEVEN EECKHOUT AND KOEN DE BOSSCHERE
Security Issues and Solutions in Distributed Heterogeneous Mobile Database Systems
 A. R. HURSON, J. PLOSKONKA, Y. JIAO, AND H. HARIDAS
Disruptive Technologies and Their Affect on Global Telecommunications
 STAN MCCLELLAN, STEPHEN LOW, AND WAI-TIAN TAN
Ions, Atoms, and Bits: An Architectural Approach to Quantum Computing
 DEAN COPSEY, MARK OSKIN, AND FREDERIC T. CHONG

Volume 62

An Introduction to Agile Methods
 DAVID COHEN, MIKAEL LINDVALL, AND PATRICIA COSTA
The Timeboxing Process Model for Iterative Software Development
 PANKAJ JALOTE, AVEEJEET PALIT, AND PRIYA KURIEN
A Survey of Empirical Results on Program Slicing
 DAVID BINKLEY AND MARK HARMAN
Challenges in Design and Software Infrastructure for Ubiquitous Computing Applications
 GURUDUTH BANAVAR AND ABRAHAM BERNSTEIN

Introduction to MBASE (Model-Based (System) Architecting and Software Engineering)
    DAVID KLAPPHOLZ AND DANIEL PORT
Software Quality Estimation with Case-Based Reasoning
    TAGHI M. KHOSHGOFTAAR AND NAEEM SELIYA
Data Management Technology for Decision Support Systems
    SURAJIT CHAUDHURI, UMESHWAR DAYAL, AND VENKATESH GANTI

Volume 63

Techniques to Improve Performance Beyond Pipelining: Superpipelining, Superscalar, and VLIW
    JEAN-LUC GAUDIOT, JUNG-YUP KANG, AND WON WOO RO
Networks on Chip (NoC): Interconnects of Next Generation Systems on Chip
    THEOCHARIS THEOCHARIDES, GREGORY M. LINK, NARAYANAN VIJAYKRISHNAN, AND MARY JANE IRWIN
Characterizing Resource Allocation Heuristics for Heterogeneous Computing Systems
    SHOUKAT ALI, TRACY D. BRAUN, HOWARD JAY SIEGEL, ANTHONY A. MACIEJEWSKI, NOAH BECK,
    LADISLAU BÖLÖNI, MUTHUCUMARU MAHESWARAN, ALBERT I. REUTHER, JAMES P. ROBERTSON,
    MITCHELL D. THEYS, AND BIN YAO
Power Analysis and Optimization Techniques for Energy Efficient Computer Systems
    WISSAM CHEDID, CHANSU YU, AND BEN LEE
Flexible and Adaptive Services in Pervasive Computing
    BYUNG Y. SUNG, MOHAN KUMAR, AND BEHROOZ SHIRAZI
Search and Retrieval of Compressed Text
    AMAR MUKHERJEE, NAN ZHANG, TAO TAO, RAVI VIJAYA SATYA, AND WEIFENG SUN

Volume 64

Automatic Evaluation of Web Search Services
    ABDUR CHOWDHURY
Web Services
    SANG SHIN
A Protocol Layer Survey of Network Security
    JOHN V. HARRISON AND HAL BERGHEL
E-Service: The Revenue Expansion Path to E-Commerce Profitability
    ROLAND T. RUST, P. K. KANNAN, AND ANUPAMA D. RAMACHANDRAN
Pervasive Computing: A Vision to Realize
    DEBASHIS SAHA
Open Source Software Development: *Structural Tension in the American Experiment*
    COSKUN BAYRAK AND CHAD DAVIS
Disability and Technology: Building Barriers or Creating Opportunities?
    PETER GREGOR, DAVID SLOAN, AND ALAN F. NEWELL

Volume 65

The State of Artificial Intelligence
    ADRIAN A. HOPGOOD
Software Model Checking with SPIN
    GERARD J. HOLZMANN

Early Cognitive Computer Vision
    JAN-MARK GEUSEBROEK
Verification and Validation and Artificial Intelligence
    TIM MENZIES AND CHARLES PECHEUR
Indexing, Learning and Content-Based Retrieval for Special Purpose Image Databases
    MARK J. HUISKES AND ERIC J. PAUWELS
Defect Analysis: Basic Techniques for Management and Learning
    DAVID N. CARD
Function Points
    CHRISTOPHER J. LOKAN
The Role of Mathematics in Computer Science and Software Engineering Education
    PETER B. HENDERSON

Volume 66

Calculating Software Process Improvement's Return on Investment
    RINI VAN SOLINGEN AND DAVID F. RICO
Quality Problem in Software Measurement Data
    PIERRE REBOURS AND TAGHI M. KHOSHGOFTAAR
Requirements Management for Dependable Software Systems
    WILLIAM G. BAIL
Mechanics of Managing Software Risk
    WILLIAM G. BAIL
The PERFECT Approach to Experience-Based Process Evolution
    BRIAN A. NEJMEH AND WILLIAM E. RIDDLE
The Opportunities, Challenges, and Risks of High Performance Computing in Computational Science and Engineering
    DOUGLASS E. POST, RICHARD P. KENDALL, AND ROBERT F. LUCAS

Volume 67

Broadcasting a Means to Disseminate Public Data in a Wireless Environment—Issues and Solutions
    A. R. HURSON, Y. JIAO, AND B. A. SHIRAZI
Programming Models and Synchronization Techniques for Disconnected Business Applications
    AVRAHAM LEFF AND JAMES T. RAYFIELD
Academic Electronic Journals: Past, Present, and Future
    ANAT HOVAV AND PAUL GRAY
Web Testing for Reliability Improvement
    JEFF TIAN AND LI MA
Wireless Insecurities
    MICHAEL STHULTZ, JACOB UECKER, AND HAL BERGHEL
The State of the Art in Digital Forensics
    DARIO FORTE

Volume 68

Exposing Phylogenetic Relationships by Genome Rearrangement
 YING CHIH LIN AND CHUAN YI TANG
Models and Methods in Comparative Genomics
 GUILLAUME BOURQUE AND LOUXIN ZHANG
Translocation Distance: Algorithms and Complexity
 LUSHENG WANG
Computational Grand Challenges in Assembling the Tree of Life: Problems and Solutions
 DAVID A. BADER, USMAN ROSHAN, AND ALEXANDROS STAMATAKIS
Local Structure Comparison of Proteins
 JUN HUAN, JAN PRINS, AND WEI WANG
Peptide Identification via Tandem Mass Spectrometry
 XUE WU, NATHAN EDWARDS, AND CHAU-WEN TSENG

Volume 69

The Architecture of Efficient Multi-Core Processors: A Holistic Approach
 RAKESH KUMAR AND DEAN M. TULLSEN
Designing Computational Clusters for Performance and Power
 KIRK W. CAMERON, RONG GE, AND XIZHOU FENG
Compiler-Assisted Leakage Energy Reduction for Cache Memories
 WEI ZHANG
Mobile Games: Challenges and Opportunities
 PAUL COULTON, WILL BAMFORD, FADI CHEHIMI, REUBEN EDWARDS, PAUL GILBERTSON, AND
  OMER RASHID
Free/Open Source Software Development: Recent Research Results and Methods
 WALT SCACCHI

Volume 70

Designing Networked Handheld Devices to Enhance School Learning
 JEREMY ROSCHELLE, CHARLES PATTON, AND DEBORAH TATAR
Interactive Explanatory and Descriptive Natural-Language Based Dialogue for Intelligent Information Filtering
 JOHN ATKINSON AND ANITA FERREIRA
A Tour of Language Customization Concepts
 COLIN ATKINSON AND THOMAS KÜHNE
Advances in Business Transformation Technologies
 JUHNYOUNG LEE
Phish Phactors: Offensive and Defensive Strategies
 HAL BERGHEL, JAMES CARPINTER, AND JU-YEON JO
Reflections on System Trustworthiness
 PETER G. NEUMANN

CONTENTS OF VOLUMES IN THIS SERIES

Volume 71

Programming Nanotechnology: Learning from Nature
    BOONSERM KAEWKAMNERDPONG, PETER J. BENTLEY, AND NAVNEET BHALLA
Nanobiotechnology: An Engineer's Foray into Biology
    YI ZHAO AND XIN ZHANG
Toward Nanometer-Scale Sensing Systems: Natural and Artificial Noses as Models for Ultra-Small, Ultra-Dense Sensing Systems
    BRIGITTE M. ROLFE
Simulation of Nanoscale Electronic Systems
    UMBERTO RAVAIOLI
Identifying Nanotechnology in Society
    CHARLES TAHAN
The Convergence of Nanotechnology, Policy, and Ethics
    ERIK FISHER

Volume 72

DARPA's HPCS Program: History, Models, Tools, Languages
    JACK DONGARRA, ROBERT GRAYBILL, WILLIAM HARROD, ROBERT LUCAS, EWING LUSK, PIOTR LUSZCZEK, JANICE MCMAHON, ALLAN SNAVELY, JEFFERY VETTER, KATHERINE YELICK, SADAF ALAM, ROY CAMPBELL, LAURA CARRINGTON, TZU-YI CHEN, OMID KHALILI, JEREMY MEREDITH, AND MUSTAFA TIKIR
Productivity in High-Performance Computing
    THOMAS STERLING AND CHIRAG DEKATE
Performance Prediction and Ranking of Supercomputers
    TZU-YI CHEN, OMID KHALILI, ROY L. CAMPBELL, JR., LAURA CARRINGTON, MUSTAFA M. TIKIR, AND ALLAN SNAVELY
Sampled Processor Simulation: A Survey
    LIEVEN EECKHOUT
Distributed Sparse Matrices for Very High Level Languages
    JOHN R. GILBERT, STEVE REINHARDT, AND VIRAL B. SHAH
Bibliographic Snapshots of High-Performance/High-Productivity Computing
    MYRON GINSBERG

Volume 73

History of Computers, Electronic Commerce, and Agile Methods
    DAVID F. RICO, HASAN H. SAYANI, AND RALPH F. FIELD
Testing with Software Designs
    ALIREZA MAHDIAN AND ANNELIESE A. ANDREWS
Balancing Transparency, Efficiency, AND Security in Pervasive Systems
    MARK WENSTROM, ELOISA BENTIVEGNA, AND ALI R. HURSON
Computing with RFID: Drivers, Technology and Implications
    GEORGE ROUSSOS
Medical Robotics and Computer-Integrated Interventional Medicine
    RUSSELL H. TAYLOR AND PETER KAZANZIDES

Volume 74

Data Hiding Tactics for Windows and Unix File Systems
  HAL BERGHEL, DAVID HOELZER, AND MICHAEL STHULTZ
Multimedia and Sensor Security
  ANNA HAĆ
Email Spam Filtering
  ENRIQUE PUERTAS SANZ, JOSÉ MARÍA GÓMEZ HIDALGO, AND JOSÉ CARLOS CORTIZO PÉREZ
The Use of Simulation Techniques for Hybrid Software Cost Estimation and Risk Analysis
  MICHAEL KLÄS, ADAM TRENDOWICZ, AXEL WICKENKAMP, JÜRGEN MÜNCH, NAHOMI KIKUCHI, AND YASUSHI ISHIGAI
An Environment for Conducting Families of Software Engineering Experiments
  LORIN HOCHSTEIN, TAIGA NAKAMURA, FORREST SHULL, NICO ZAZWORKA, VICTOR R. BASILI, AND MARVIN V. ZELKOWITZ
Global Software Development: Origins, Practices, and Directions
  JAMES J. CUSICK, ALPANA PRASAD, AND WILLIAM M. TEPFENHART

Volume 75

The UK HPC Integration Market: Commodity-Based Clusters
  CHRISTINE A. KITCHEN AND MARTYN F. GUEST
Elements of High-Performance Reconfigurable Computing
  TOM VANCOURT AND MARTIN C. HERBORDT
Models and Metrics for Energy-Efficient Computing
  PARTHASARATHY RANGANATHAN, SUZANNE RIVOIRE AND JUSTIN MOORE
The Emerging Landscape of Computer Performance Evaluation
  JOANN M. PAUL, MWAFFAQ OTOOM, MARC SOMERS, SEAN PIEPER AND MICHAEL J. SCHULTE
Advances in Web Testing
  CYNTRICA EATON AND ATIF M. MEMON

Volume 76

Information Sharing and Social Computing: Why, What, and Where?
  ODED NOV
Social Network Sites: Users and Uses
  MIKE THELWALL
Highly Interactive Scalable Online Worlds
  GRAHAM MORGAN
The Future of Social Web Sites: Sharing Data and Trusted Applications with Semantics
  SHEILA KINSELLA, ALEXANDRE PASSANT, JOHN G. BRESLIN, STEFAN DECKER, AND AJIT JAOKAR
Semantic Web Services Architecture with Lightweight Descriptions of Services
  TOMAS VITVAR, JACEK KOPECKY, JANA VISKOVA, ADRIANMOCAN, MICK KERRIGAN, AND DIETER FENSEL
Issues and Approaches for Web 2.0 Client Access to Enterprise Data
  AVRAHAM LEFF AND JAMES T. RAYFIELD
Web Content Filtering
  JOSÉMARÍA GÓMEZ HIDALGO, ENRIQUE PUERTAS SANZ, FRANCISCO CARRERO GARCÍA, AND MANUEL DE BUENAGA RODRÍGUEZ

Volume 77

Photo Fakery and Forensics
    HANY FARID
Advances in Computer Displays
    JASON LEIGH, ANDREW JOHNSON, AND LUC RENAMBOT
Playing with All Senses: Human–Computer Interface Devices for Games
    JÖRN LOVISCACH
A Status Report on the P Versus NP Question
    ERIC ALLENDER
Dynamically Typed Languages
    LAURENCE TRATT
Factors Influencing Software Development Productivity—State-of-the-Art and Industrial Experiences
    ADAM TRENDOWICZ AND JÜRGEN MÜNCH
Evaluating the Modifiability of Software Architectural Designs
    M. OMOLADE SALIU, GÜNTHER RUHE, MIKAEL LINDVALL, AND CHRISTOPHER ACKERMANN
The Common Law and Its Impact on the Internet
    ROBERT AALBERTS, DAVID HAMES, PERCY POON, AND PAUL D. THISTLE